Asymmetric Synthesis

Asymmetric Synthesis

Garry Procter

Department of Chemistry and Applied Chemistry
University of Salford

OXFORD   NEW YORK   TOKYO

OXFORD UNIVERSITY PRESS

1996

Oxford University Press, Walton Street, Oxford OX2 6DP

Oxford New York
Athens Auckland Bangkok Bombay
Calcutta Cape Town Dar es Salaam Delhi
Florence Hong Kong Istanbul Karachi
Kuala Lumpur Madras Madrid Melbourne
Mexico City Nairobi Paris Singapore
Taipei Tokyo Toronto
and associated companies in
Berlin Ibadan

Oxford is a trade mark of Oxford University Press

Published in the United States by
Oxford University Press Inc., New York

A catalogue record for this book is available from the British Library

Library of Congress Cataloging in Publication Data
(Data available)

ISBN 0 19 855726 4 (Hbk)
ISBN 0 19 855725 6 (Pbk)

Typeset by the author

Printed in Great Britain by
Biddles Ltd, Guildford and King's Lynn

Preface

Molecular asymmetry has been a topic of great interest to chemists, and organic chemists in particular, for many years. The consequences of such asymmetry in the structure of molecules pervade many areas of daily life. For example, the interaction between one chiral molecular species and another can be important in such apparently diverse properties as the smell of a fruit and the antibacterial activity of a drug. It is not surprising then that many chemists interested in organic synthesis find the challenge of preparation of single enantiomers of such chiral molecules, asymmetric synthesis, both interesting and rewarding.

Given the importance of molecular asymmetry in Nature and in industry, the topic of asymmetric synthesis is finding its way increasingly into undergraduate chemistry courses. It is my involvement in just such a course which is responsible for this book. It is my hope that this book will be of some use to those involved in the preparation and teaching of such courses, and to the students themselves. In addition, research workers starting out in this area might also find it of some interest. With this in mind, I have tried to keep the 'jargon' to a minimum, and to keep the stereochemical descriptions as simple as possible. I expect that purists will be unhappy at my sacrificing the precision of strict nomenclature and stereochemical descriptors in my attempt to keep the text as simple as possible. To any offended parties, I apologise. No doubt I will be hearing from you.

The book is structured along the lines of reaction types, and important methods for achieving asymmetric synthesis in the particular reaction type are presented in each individual chapter. I have attempted to make each chapter on reaction types stand alone, with the references presented at the end of each chapter. This inevitably leads to some repetition both in the text and references, but I hope that this makes the book easier to use.

Given the timescale involved, it is inevitable that this book cannot be an up-to-the-minute account of the art of asymmetric synthesis, and I am conscious that much progress has been made during the time in which it has been written. I am also conscious of the fact that I have had to choose examples from a literature which is rich with outstanding achievements. I am lucky enough to be able to count some of the leading exponents in asymmetric synthesis amongst my friends, and to them and all other researchers in the area, I apologise for the inevitable omission of many fine examples of the art. Again, no doubt I will be hearing from you!

This book has taken a long time to write, and I am indebted to the patience of all who have been involved with it, especially the editorial staff at Oxford University Press. A good part of the text and references have been read and commented on by a number of students at Salford, and particular thanks go to Adrian Gill. I should be grateful to hear of the errors which will inevitably have crept into this book, in spite of the best efforts of all involved. As I have typed the text, drawn the diagrams, and 'typeset' the book myself, I bear all the blame for such errors. No doubt I shall be hearing about them.

Finally, I could not finish this without thanking my family, especially Mandy, for putting up with the 'unsocial hours' which I have had to keep at times during the writing and preparation of the manuscript.

Salford   Garry Procter
1995

To Mandy, Adam, and Rachel

Contents

1 Introduction

Organic compounds play an important part in modern life, not least in the area of pharmaceuticals, agrochemicals, and other materials which possess useful biological activity. Often such biological activity arises through the interaction of the organic compound with a 'biomolecule' such as an enzyme or a receptor. Such sites of action are constructed from chiral building blocks such as amino acids or carbohydrates, which means that these sites of action are themselves chiral. Being natural chiral compounds, these building blocks are present as single enantiomers, and it follows of course that the resulting biomolecules are single enantiomers. If the organic compound itself is chiral, then one consequence of this is that the two enantiomers are likely to interact differently with a given biomolecule of the type described above. In general, when considering the interaction of two such chiral systems, the stereochemistry of both systems can have a profound effect on the magnitude of the interaction, and therefore on the biological response.

This principle can be illustrated by considering the natural antibiotic vancomycin. Vancomycin is an orally active antibiotic which is widely used in treatment of postoperative diarrhoea, and is a heptapeptide with the absolute stereochemistry shown in Fig. 1.1.[1]

The molecular basis for its mode of action has been shown to involve its binding to cell-wall mucopeptide precursors terminating in the sequence -D-Ala-D-Ala (Fig. 1.2). As might be expected from this mode of action, vancomycin will bind to the model dipeptide PhCO-D-Ala-D-Ala (**1.1**).

Changing the stereochemistry of the terminal alanine residue in the model (**1.2**) causes a decrease in binding energy of 10 kJ mol^{-1}. This means that

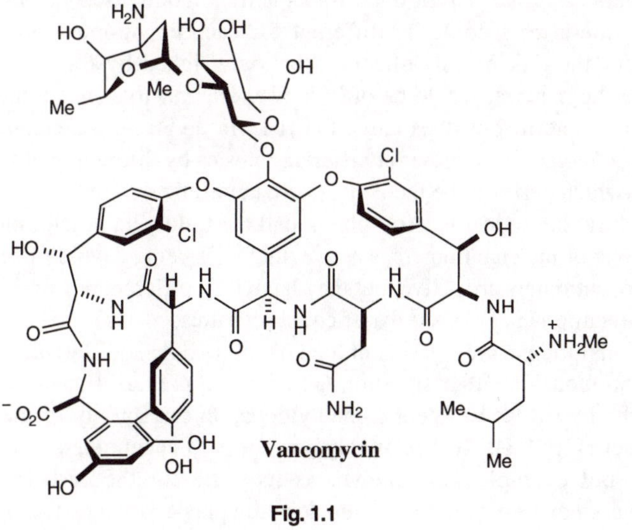

Vancomycin

Fig. 1.1

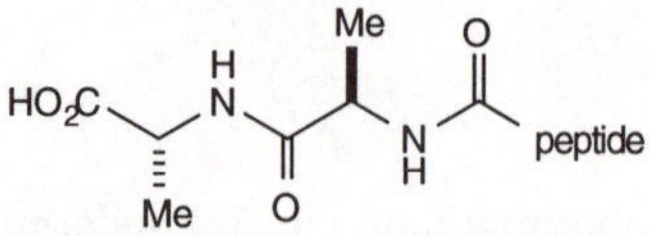

D-Ala-D-Ala terminus

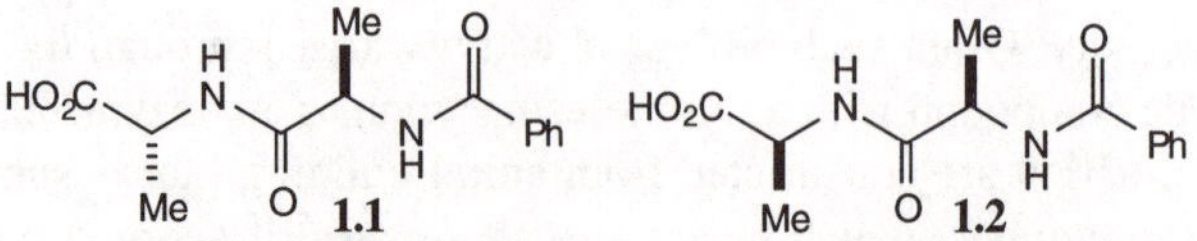

1.1
D-Ala-D-Ala model

1.2
L-Ala-D-Ala model

Fig. 1.2

vancomycin can 'discriminate' between a D- and an L-amino acid at the terminus of the dipeptide as a result of this difference in binding energy. The binding constant for the model with the natural stereochemistry is fifty times greater than that for the model with the unnatural stereochemistry. For the related antibiotic ristocetin, the corresponding ratio of binding constants is between five hundred and four thousand to one.

As stated earlier, if a pharmaceutical, or any biologically active compound, is chiral then the enantiomers are likely to interact differently with the natural biomolecule. The enantiomers will probably possess different levels of biological activity, and could also exhibit quite different types of activity. In effect, the two enantiomers should be viewed as two distinct compounds. It follows that using the racemate of a particular biologically active compound is equivalent to using a mixture of two different compounds.[2]

Usually, one enantiomer is far more active than the other. This being the case it is clearly undesirable to use a racemic biologically active compound. Only one of the enantiomers possesses the desired beneficial activity, but *both* enantiomers carry the risk of undesirable activity (side-effects). Moreover, the possible side-effects could be different for each enantiomer. For the active enantiomer the risk of side-effects is far outweighed by the positive effect, otherwise the material would be of little use. The inactive enantiomer provides little or no benefit, but does carry the risk. In addition, the enantiomers are likely to be metabolized either at different rates or by different pathways, as the enzymes which perform the metabolism are themselves chiral.

For a chiral biologically active compound the following possibilities exist:
(1) only one of the enantiomers is active, the other being devoid of activity;
(2) both enantiomers are active, but they have very different potencies;
(3) both enantiomers have similar or equal activities;
(4) both enantiomers are active, but the type of activity is different.

It is common for either situation (1) or (2) to prevail. For example, in the case of the hypertensive agent α-methyldopa, all the activity resides in the L-enantiomer (Fig. 1.3).[3] It is relatively rare for both enantiomers to have similar potency, but examples are known, as they are for the case in which the enantiomers have different activities. Propoxyphene is interesting in that both

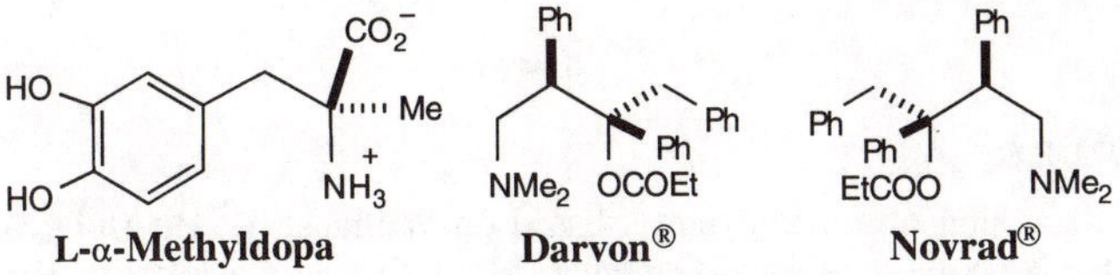

L-α-Methyldopa   **Darvon**®   **Novrad**®

Fig. 1.3

enantiomers have useful but different biological activities. The D-enantiomer is an analgesic, whereas the antipode possesses antitussive properties but no analgesic effect. To reflect the mirror image relationship between these two compounds, they have been given the trade names Darvon® and Novrad® (Fig. 1.3).[4]

The obvious solution to the potential problems related to the use of a racemate is to use only the enantiomer which possesses the desired beneficial biological activity. An equally obvious prerequisite for this general solution is that the pure single enantiomers be available. Two possible methods for achieving this are immediately apparent: resolution of the racemate or an intermediate on the synthetic route, and the use of an enantiomerically pure starting material. Both these are valuable methods, and are currently in use, but both have associated drawbacks as general solutions. Resolution is often expensive as a suitable resolving agent is required and the unwanted enantiomer has to be disposed of. The use of an enantiomerically pure starting material requires that such a compound be readily available, possesses the desired absolute configuration, and that a convenient and practical synthetic route to the desired compound can be developed.

In principle, a general solution to the problem of obtaining enantiomerically pure organic compounds would be to have available an array of synthetic methods which result in the desired transformation *and* control the absolute stereochemistry of chiral centres which are created as a result of the synthetic operation. This is the realm of asymmetric synthesis.

At present, there are relatively few enantioselective synthetic methods which come up to the high standards which would bring them into everyday use and allow for the cost-effective preparation of enantiomerically pure compounds on a reasonable scale.[5] Organic chemists are fortunate in that this area is full of challenges which make scientific demands at the highest level, and that success in this area can have immediate application in the chemical industry. Developing solutions to these challenges involves working at the leading edge of the subject in a highly creative manner. The rest of this volume is concerned with some of the more important methods for asymmetric synthesis and the principles which lie behind them.

References

1. This discussion of vancomycin is based on Williams, D. H., Doig, A. J., Cox, J. P. L., Nicholls, I. A., and Gardener, M. (1990), in *Chirality in Drug Design and Synthesis*, (ed. C. Brown), pp. 101–113, Academic Press, London, and references cited therein.
2. For a discussion of this and related problems, see Ariens, E. J. (1990), in *Chirality in Drug Design and Synthesis*, (ed. C. Brown), pp. 29–43, Academic Press, London.
3. Gillespie, L., Oates, J. A., Crout, J. R., and Sjoerdsma, H. (1962), *Circulation*, **25**, 281.
4. Drayer, D. E. (1986), *Clin. Pharmacol. Ther.*, **40**, 125.
5. For a discussion of the industrial synthesis of single enantiomers, see Sheldon, R. (1990), *Chem. and Ind.*, 212.

2 Principles

In order to achieve asymmetric synthesis, at least one component of the reaction must be chiral and non-racemic. If there is no asymmetric component in the reaction, then transition states which lead to enantiomers will themselves be enantiomeric, equal in energy, and a racemate must be formed. In principle, the use of a chiral, non-racemic substrate, reagent, solvent, or catalyst should lead to asymmetric synthesis. In general terms, *any* feature of the reacting system which would cause the possible transition states for the reaction to be *diastereoisomeric* (where they would normally be *enantiomeric*) could lead to the preferential formation of one diastereoisomer or enantiomer. This follows because transition states which are diastereoisomeric need not be of the same energy and consequently one of the possible products could be formed more rapidly. The various possibilities will be considered in this chapter.

If the substrate itself is chiral and non-racemic, then creation of another chiral centre using this substrate provides the possibility of diastereoisomeric products. If the products themselves are diastereoisomers, then the transition states which lead to them are diastereoisomeric, and a diastereoselective reaction should be expected.

This principle is illustrated in Fig. 2.1, which shows the two possible paths for alkylation of the chiral enolate **2.1**. The use of this type of chiral enolate for asymmetric synthesis will be covered in detail later. If the electrophile attacks from above the plane of the enolate as drawn, the product will be **2.2**, and attack from below will lead to **2.3**, which is diastereoisomeric with **2.2**. In this case the ratio of **2.2**:**2.3** is 99:1, often expressed as a diastereoisomeric excess (Fig. 2.1).[1] This particular reaction is an example of an extremely powerful method for asymmetric synthesis based on the use of 'chiral auxiliary' controlled enolate alkylation.

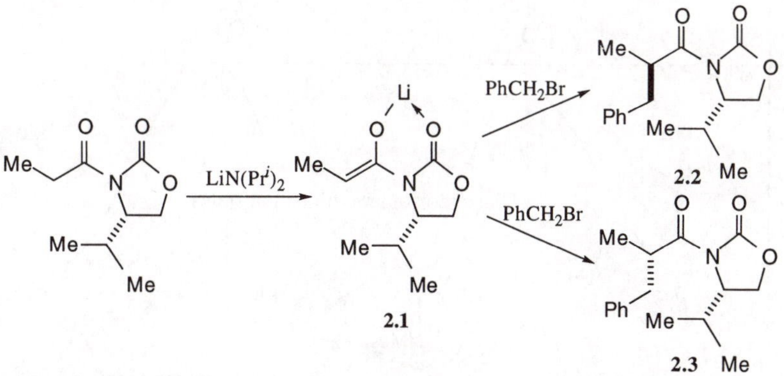

Diastereoisomeric excess (d.e.)= major diastereoisomer (%) – %minor diastereoisomer (%)
= **2.2**(%) – **2.3**(%) = 99 – 1 = 98%

Fig. 2.1

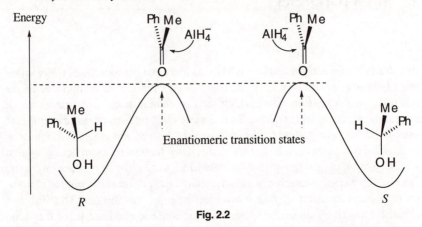

Fig. 2.2

The use of a chiral, non-racemic reagent may be understood by considering the reduction of a prochiral ketone to the corresponding chiral alcohol using either an achiral or a chiral reducing agent. For the achiral reagent, reduction *must* give a racemic mixture since the two transition states are enantiomeric and therefore of equal energy. Clearly this must lead to equal amounts of each enantiomer as the rates of both reactions will be the same (Fig. 2.2).

If a chiral, non-racemic reducing agent is used, because this is involved in the transition states these become diastereoisomeric and need not be of the same energy. The reaction will then produce an excess of the enantiomer which is formed via the lower energy transition state (Fig 2.3).

The example shown in Fig. 2.3 involves the prior reaction of the achiral reducing agent (LiAlH$_4$) with the enantiomerically pure 1,3-aminoalcohol

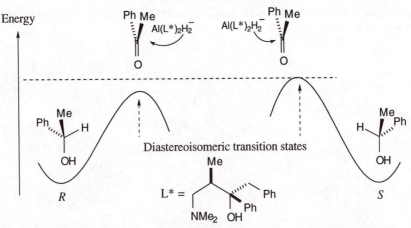

Enantiomeric excess (e.e.) = major enantiomer(%) - minor enantiomer(%)
= R(%) - S(%) = 68%

Fig. 2.3

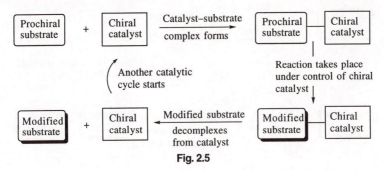

Fig. 2.4

shown (commonly known as 'Darvon alcohol') which produces a chiral, non-racemic complex represented as $Al(L^*)_2H_2$. Using this as the reducing agent results in the transition states which had previously been enantiomeric (Fig. 2.2) becoming diastereoisomeric and therefore of different energies. Transition states *enantiomeric* with those represented in Fig. 2.3 would require the *enantiomer* of the 1,3-aminoalcohol, which is not present as enantiomerically pure 1,3-aminoalcohol was used. In this particular case the enantiomeric excess of the product is 68 per cent, in favour of the (R)-alcohol.[2]

The components of a reaction which takes place in solution will be solvated. Consequently, the use of a solvent which is chiral and non-racemic should lead to asymmetric synthesis as the solvent is likely to be involved in the transition states. As only one enantiomer of the solvent is present, transition states which would be enantiomeric in an achiral solvent become diastereoisomeric and asymmetric synthesis becomes possible. In spite of the attractive nature of this approach to asymmetric synthesis it is currently of little general use, as the level of stereoselectivity induced is often low and unpredictable. Moreover, there are very few enantiomerically pure compounds which are available in sufficient quantity and which possess the properties required to be a useful solvent.

Rather more promising is the use of chiral, non-racemic solvating agents (in a normal solvent) which preferentially solvate a reaction component. Fig. 2.4 shows an example of this type of asymmetric synthesis in which a tin(II) enolate is complexed with a chiral, non-racemic diamine **2.4** before reaction with the aldehyde. The product is obtained in 75 per cent enantiomeric excess.[3]

A chiral, non-racemic catalyst can be used for asymmetric synthesis and the principles of this approach can be understood using arguments similar to those

```
┌─────────────┐     ┌──────────┐   Catalyst–substrate   ┌─────────────┐┌──────────┐
│ Prochiral   │  +  │ Chiral   │  ──────────────────→   │ Prochiral   ││ Chiral   │
│ substrate   │     │ catalyst │    complex forms        │ substrate   ││ catalyst │
└─────────────┘     └──────────┘                         └─────────────┘└──────────┘
                    ⎛  Another catalytic                  Reaction takes place
                    ⎝  cycle starts                       under control of chiral
                                                          catalyst
┌─────────────┐     ┌──────────┐   Modified substrate    ┌─────────────┐┌──────────┐
│ Modified    │  +  │ Chiral   │  ←──────────────────    │ Modified    ││ Chiral   │
│ substrate   │     │ catalyst │    decomplexes          │ substrate   ││ catalyst │
└─────────────┘     └──────────┘    from catalyst        └─────────────┘└──────────┘
```

Fig. 2.5

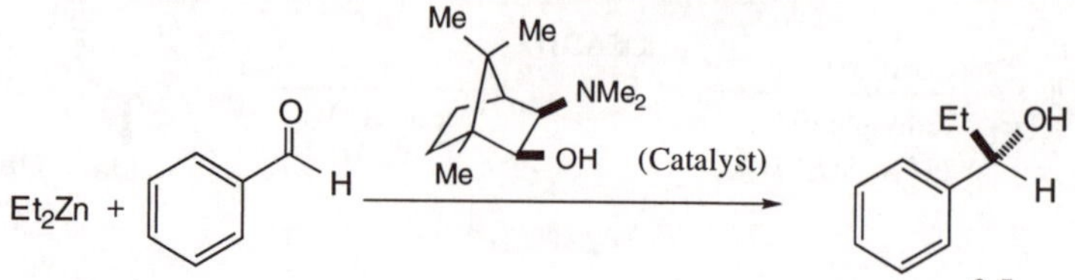

Me Me

NMe₂

Me OH   (Catalyst)

Et₂Zn +

Et OH

H

2.5

Fig. 2.6

above. In this case the stoichiometric reagent is achiral but does not react with the prochiral substrate in the absence of the catalyst. The reaction in which the new chiral centre is created only occurs when the catalyst brings together the reagent and substrate. The catalyst is involved in the transition states and it follows that these will be diastereoisomeric, which should lead to asymmetric synthesis. A schematic representation of such asymmetric catalysis is shown in Fig. 2.5.

This type of asymmetric synthesis is extremely attractive, as a small amount of the chiral, non-racemic catalyst leads to stoichiometric amounts of the desired enantiomerically enriched product. Indeed this type of asymmetric synthesis is widespread; Nature uses chiral, non-racemic catalysts (enzymes) to carry out many enantioselective (and diastereoselective) reactions. In the laboratory and in industry both enzymes and synthetic catalysts are used to achieve asymmetric synthesis. Examples of asymmetric catalysis will be discussed where appropriate, and one such is illustrated in Fig. 2.6. The product alcohol **2.5** is formed in 97 per cent yield and 98 per cent enantiomeric excess.[4]

Another general approach to asymmetric synthesis involves the use of chiral auxiliaries. The overall strategy is shown in Fig. 2.7 and has clear similarities with the asymmetric catalysis cycle shown in Fig. 2.5. In this approach the prochiral substrate is attached to a chiral, non-racemic group, known as the chiral auxiliary, prior to reaction. The two (or more) possible products then become diastereoisomeric and one should be formed in excess. The major diastereoisomer can then be isolated and the chiral auxiliary removed to provide the chiral non-racemic product. The requirements for a chiral auxiliary to be practically useful are listed in Table 2.1, and at present there are relatively few

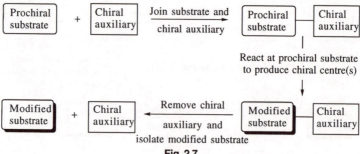

Fig. 2.7

Table 2.1 Requirements for chiral auxiliaries

Enantiomerically pure
Cheap and easy to obtain in quantity
Easy to attach to substrate
Control of stereoselectivity high and predictable
Easy to purify major diastereoisomer
Removal easy without loss of diastereoisomeric or enantiomeric purity
Easily separated from product and recovered

---

chiral auxiliaries which meet all these demands.

The oxazolidinone chiral auxiliaries introduced at the beginning of this chapter (Fig. 2.1) provide an excellent example of what can be achieved using chiral auxiliaries, and will be discussed in detail in Chapters 4, 5, and 6. A simple example which uses the alkylation reaction shown in Fig. 2.1 is given below (Fig. 2.8).

An alternative approach to asymmetric synthesis is that of kinetic resolution, in which a resolution of a racemic substrate is achieved at the same time as an asymmetric reaction. This approach relies on the difference in the rate of reaction of the individual enantiomers of the racemate with an enantiomerically pure reactant, reagent, or catalyst. In an ideal case this rate difference is so large that one enantiomer of the racemate is effectively inert, while the other reacts rapidly. Routine separation of the product and the unreacted enantiomer would then provide both in an enantiomerically pure form. An example of kinetic resolution using the Sharpless asymmetric epoxidation (discussed in Chapter 7) is shown in Fig. 2.9.

The examples considered so far have involved reactions in which only one of the components is chiral. Only the stereoselectivity induced by this component

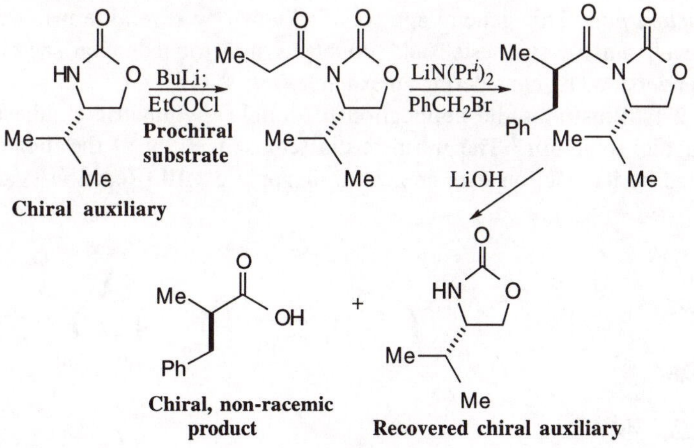

Fig. 2.8

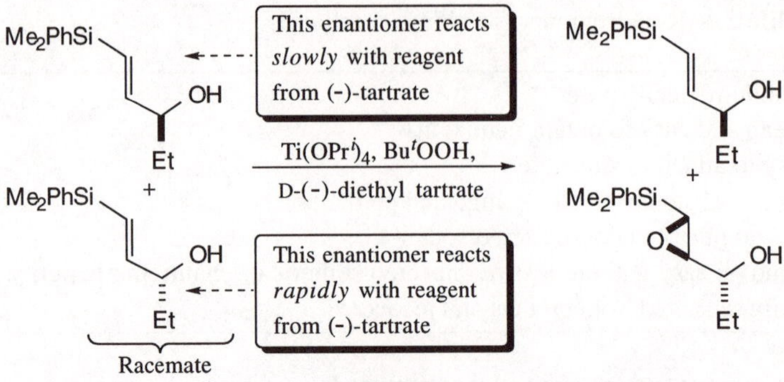

Fig. 2.9

needs to be considered. For a reaction in which two components are chiral then the 'intrinsic stereoselectivity' of each is important. To illustrate the possibilities when two chiral components are used, two extreme cases will be considered.

In the simplest of these, the 'intrinsic' reaction stereoselectivity of one of the components dominates the reaction. In this case the absolute configuration of the new chiral centre(s) is independent of the chirality of the other component. This is often the case in reactions of the oxazolidinone chiral auxiliaries discussed above. The aldol reaction of **2.6** (Fig. 2.10) with the chiral aldehyde **2.7** provides **2.8** with a selectivity of >400:1,[5] in this case the stereoselectivity of the enolate far outweighs the intrinsic diastereoselectivity of the aldehyde.

At the other extreme the two chiral components have similar intrinsic diastereoselectivities. In this case the diastereoselectivities of one pair of chiral components will usually be complementary and will give high stereoselectivity. This combination is often known as the *matched pair*. The other combination will usually result in low stereoselectivity and is often referred to as the *mismatched pair*. This general approach is known by several terms, including 'double asymmetric synthesis'[6] and 'double asymmetric induction', and is more easily understood by considering an example (Fig. 2.11).

Fig. 2.11 illustrates the application of double asymmetric synthesis to a Diels–Alder reaction.[7] The intrinsic diastereoselectivity of the diene **2.9** is estimated by its reaction with an achiral dienophile **2.10** (Reaction 1) and that

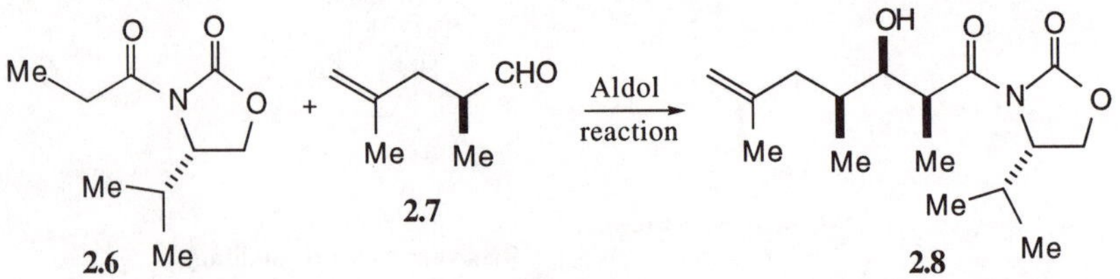

Fig. 2.10

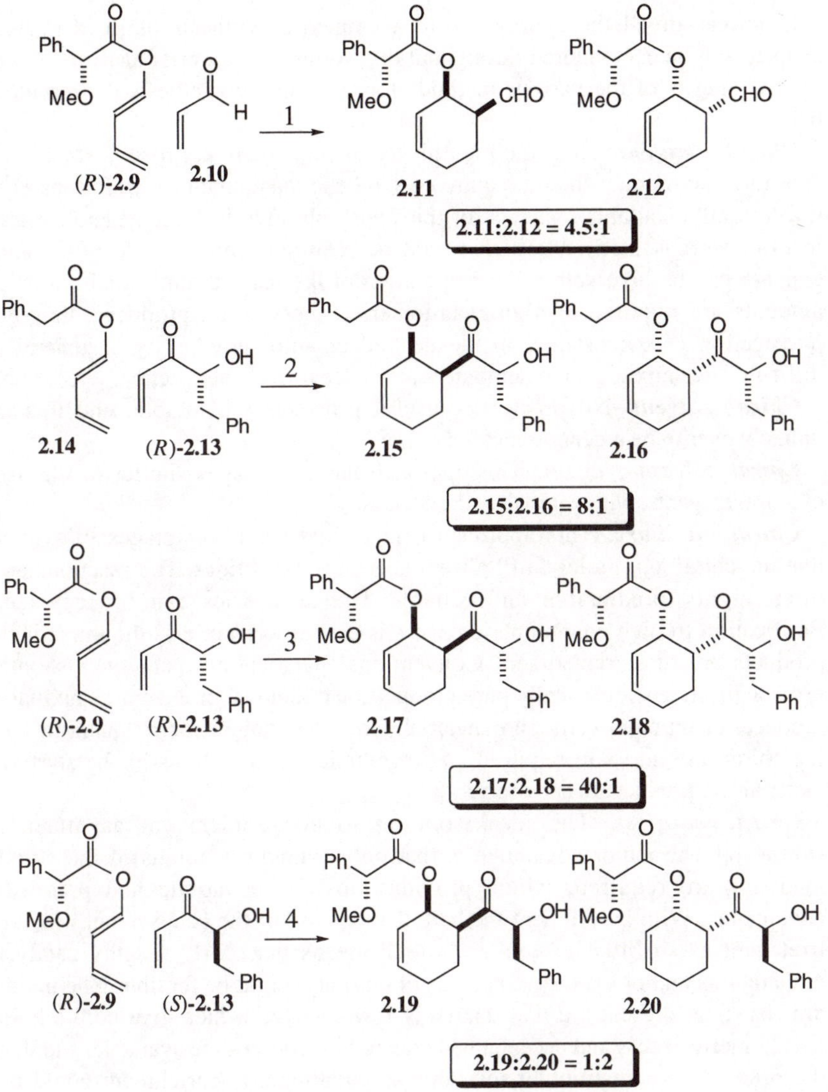

Fig. 2.11

of the dienophile is estimated by reaction with an achiral diene (Reaction 2).

It can be seen that in this case diene (*R*)-**2.9** and dienophile (*R*)-**2.13** both favour the formation of the *same* absolute configuration of the newly formed chiral centres. Reaction of (*R*)-**2.9** and (*R*)-**2.13** (Reaction 3) then leads to a much higher stereoselectivity than either of the previous reactions. This corresponds to reaction of a matched pair. A mismatched pair would correspond to reaction of (*R*)-**2.9** with (*S*)-**2.13** or reaction of (*S*)-**2.9** with (*R*)-**2.13**. The outcome of reaction of the former combination is illustrated in Reaction 4.

Examples of all the approaches to asymmetric synthesis outlined in this chapter will be encountered throughout this volume, and a brief analysis of the relative merits of the various methods for asymmetric synthesis is presented below.

Chiral reagent. In principle this is an excellent approach, since the substrate and product should require no synthetic manipulations. Unfortunately, the currently available reagents for this approach often lack the generality and level of stereoselectivity which would be required. Considerable effort and expense can be involved in the preparation of the reagent, and stoichiometric amounts are required. If high enantiomeric excess is not produced directly, purification of the product to the desired enantiomeric purity is generally difficult since mixtures of enantiomers are isolated from the reaction.

Chiral solvent. No practically useful procedures available, and this is unlikely ever to be a general method.

Chiral solvating agent. This approach has advantages similar to the use of chiral reagents, with equivalent drawbacks.

Chiral auxiliary. This approach offers significant advantages, provided that the chiral auxiliaries fulfil all the necessary conditions. The reactions are often highly predictable and reliable. The auxiliaries can be recycled. Purification to high enantiomeric excess is easy in principle as the immediate products are diastereoisomers. Conventional purification techniques should provide diastereoisomerically pure products, and removal of the chiral auxiliary produces enantiomerically pure material. However, stoichiometric quantities of the chiral auxiliary are required, and synthetic manipulations of the starting material and product are necessary.

Chiral catalyst. The ideal solution to all problems of asymmetric synthesis! The major advantage is that only catalytic amounts of the chiral mediator are required, which provides obvious economic and practical advantages. With an efficient method, the expense of the catalyst can become irrelevant as so little is required, which means that fairly lengthy catalyst preparations and/or expensive sources of chirality might be feasible. The major drawbacks at present are that relatively few catalysts which give both a high enantiomeric excess and accept a wide range of substrates are available, and that the products are enantiomer mixtures so enantiomeric enrichment could be difficult.

In conclusion, in the absence of a general and reliable method for asymmetric catalysis, the method of choice for a particular asymmetric synthesis is likely to be the use of a chiral auxiliary. A major advantage of this method for the practising organic chemist is that the products are diastereoisomers, providing relatively easy purification. An additional advantage is that because the absolute configuration of the chiral auxiliary is known, determination of the absolute configuration of the product is possible by X-ray crystallography. If the product diastereoisomers are easily separated, and the yield is high, then even modest levels of stereoselectivity can be useful. An 80 per cent yield of a 4:1 mixture (only 60 per cent diastereoisomeric excess) could provide a yield of up to 64 per cent of diastereoisomerically pure material.

The rest of this volume concerns selected examples of methods for asymmetric synthesis which are of intrinsic scientific interest and practical utility. Wherever possible the currently accepted reasoning behind the chosen processes will be discussed, although it is inevitable that some of this will be no more than speculation. Emphasis will be laid on the predictability of the methods selected, a factor which is most important for the widespread use of a particular method, and occasionally some of the practical aspects of each will be discussed.

References

1. Evans, D. A., Ennis, M. D., and Mathre, D. J. (1982), *J. Amer. Chem. Soc.*, **104**, 1737.
2. Yamaguchi, S. and Mosher, H. S. (1973), *J. Org. Chem.*, **38**, 1870; Reich, C. J., Sullivan, G. R., and Mosher, H. S. (1973), *Tetrahedron Lett.*, 1505; Grandbois, E. R., Howard, S. I., and Morrison, J. D. (1983), in *Asymmetric Synthesis*, (ed. J. D. Morrison), Vol. 2, pp. 71–90, Academic Press, New York.
3. Iwasawa, N. and Mukaiyama, T. (1982), *Chem. Lett.*, 1441.
4. Kitamura, M., Okada, S., Suga, S., and Noyori, R. (1989), *J. Amer. Chem. Soc.*, **111**, 4028.
5. Evans, D.A. and Bartroli, J. (1982), *Tetrahedron Lett.*, **23**, 807.
6. Masamune, S., Choy, W., Petersen, J.S., and Sita, L.R. (1985) *Angew. Chem. Int. Ed. Engl.*, **24**, 1.
7. Ref. 6, pp. 2–3.

3 Additions to carbonyl compounds

The addition of a nucleophile to a carbonyl group is a fundamental reaction of organic chemistry, and a cornerstone of organic synthesis. As such it is impossible to cover all such reactions which are relevant to asymmetric synthesis in one chapter. Enantioselective reduction of ketones is covered in Chapter 7 along with the reduction of other functional groups, and asymmetric aldol reactions of aldehydes is the subject of Chapter 5. Accordingly, this chapter is concerned mainly with the addition of nucleophiles to ketones and aldehydes in which one or more new chiral centres are formed and which result in non-racemic products.

The most straightforward and commonly encountered example of this type of process is the addition of non-chiral nucleophiles to chiral aldehydes and ketones, and this will be the subject of the first part of this chapter. This in itself constitutes a large area of organic synthesis, and selected examples will be used to illustrate the important general principles and applications.

In the simplest case, addition of an achiral nucleophile to a prochiral aldehyde or ketone could give rise to two diastereoisomers from addition to either the lower or upper face of the carbonyl group to give **3.1** or **3.2** respectively (Fig. 3.1). If the nucleophile is such that two chiral centres are produced in the reaction, one from the nucleophile and one from the aldehyde or ketone, then there are four possible diastereoisomeric products **3.3**, **3.4**, **3.5**, and **3.6** (Fig. 3.1). Products **3.3** and **3.4** arise from addition of the nucleophile on the upper face of the carbonyl group, but differ in that reaction has occurred on either face of the nucleophile. Isomers **3.5** and **3.6** represent attack of the nucleophile on the lower face of the carbonyl group. In this latter case, addition of a prochiral

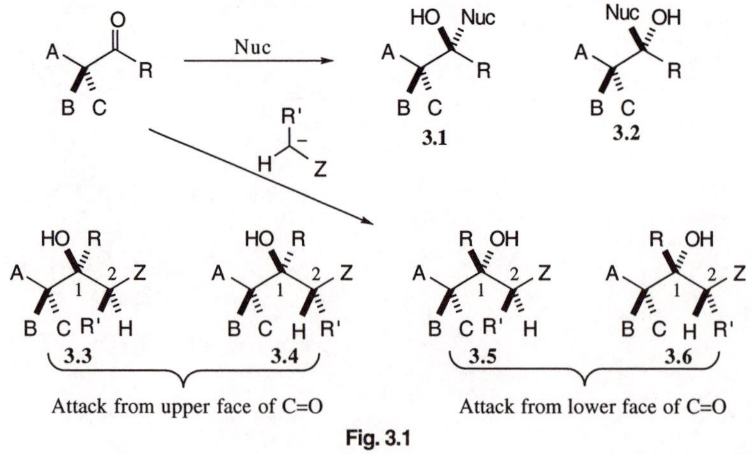

Fig. 3.1

nucleophile to a chiral aldehyde or ketone, the intrinsic selectivity of *both* reacting species needs to be considered, whereas in the former case only the facial selectivity of the carbonyl compound is involved.

The possibility of four stereoisomeric products related in the way outlined for those above arises whenever new chiral centres arise from both reactants, and is particularly important in aldol and Diels–Alder reactions (Chapters 5 and 6).

A great many examples of the addition of achiral nucleophiles to chiral aldehydes and ketones are known, and much effort has gone into developing an understanding of the factors which influence the diastereoselectivity of this reaction.

Cram's rule[1] was developed to rationalize and predict the stereochemical outcome of the addition of organometallic reagents to chiral ketones and aldehydes. The formalization used in the absence of chelation is illustrated below (Fig. 3.2). Several alternative schemes have been proposed, but from theoretical work it is generally thought that the Felkin–Anh model[2] is the most

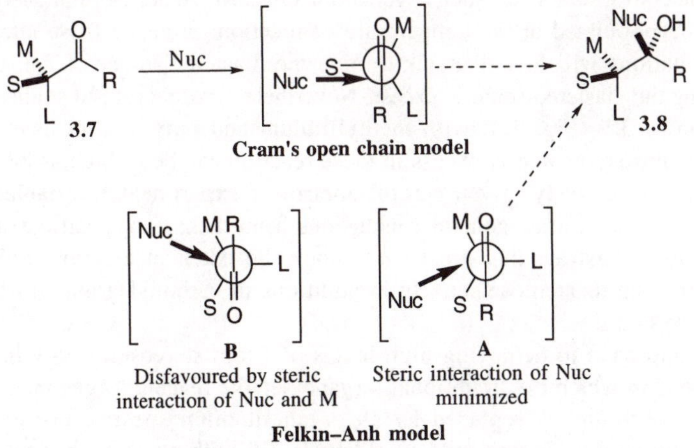

Cram's open chain model

B
Disfavoured by steric interaction of Nuc and M

A
Steric interaction of Nuc minimized

Felkin–Anh model
Fig. 3.2

Table 3.1 Cram addition to **3.7**

R	S	M	L	Nuc	d.e.[a] (%)
H	H	Me	Ph	MeMgI	33
H	H	Et	Ph	MeMgI	43
H	H	Me	Ph	EtMgBr	50
H	H	Me	Ph	PhMgBr	>60
Me	H	Me	Ph	MeMgI	66
Me	H	Me	Ph	EtMgI	75
Me	H	Me	Ph	PhMgI	83

[a]Major diastereoisomer corresponds to **3.8**.

appropriate to account for experimental observations and to provide a reasonable model for the reacting conformation (Fig. 3.2). In both these models the three substituents on the asymmetric carbon adjacent to the carbonyl group are classified in terms of their size, S=small, M=medium, and L=large.

The conformation which undergoes reaction is different in the two models, but the predicted major diastereoisomer of the product is the same (Fig. 3.2). In the original Cram model the conformation was proposed to be as shown, with nucleophilic attack from the less hindered face of the carbonyl group.

In the Felkin–Anh model the most stable conformations for the transition state are proposed to be those in which the C–L bond is orthogonal to the plane of the carbonyl group (A and B in Fig. 3.2). The known preference for the nucleophile to approach along a trajectory which maintains an angle of ~109° with the carbonyl group results in A being favoured over B.

Examples of so-called 'Cram selective' (or 'non-chelation controlled') additions are given in Table 3.1,[3] and as can be seen, the diastereoselectivity can be relatively low. This is not too surprising, since there are often many possible conformations of such acyclic molecules which are close in energy, and therefore populated at the temperature of reaction. Some of these alternative conformations will favour attack to give the diastereoisomer of **3.8**, thereby lowering the diastereoisomeric excess. Nevertheless, recent careful studies on the reaction of **3.9** (Fig. 3.3) with methyllithium and butyllithium have shown that the diastereoisomeric excess in these reaction can be as high as 88 and 79 per cent respectively, given careful control of experimental variables.[4] It is clearly unwise to draw general conclusions from these two specific examples, but it does illustrate that great care should be taken in making predictions regarding the diastereoselectivity in additions to carbonyl groups in flexible acyclic systems.

One approach to achieving high levels of Cram stereoselectivity has been developed in which the 'traditional' organometallic reagent (organomagnesium or organolithium) is replaced by an organotitanium species. For example, trialkoxytitanium reagents add to aldehyde **3.9** with much higher selectivity than the corresponding Grignard reagent (Fig. 3.3).[5]

The models described above apply only when there is no likelihood of coordination of the metal ion of the nucleophile which could lead to chelation.

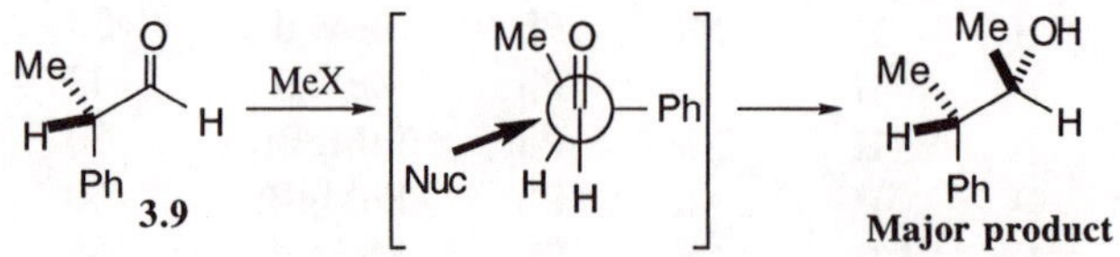

X = MgI, d.e. 33%

X = Ti(OPr^i)$_3$, d.e. 76%

X = Ti(OPh)$_3$, d.e. 86%

Fig. 3.3

When the aldehyde or ketone does contain a group (Ÿ) which includes an atom (usually oxygen) capable of ligating the metal ion of the nucleophile, then chelation can occur (Fig. 3.4), provided that Ÿ is close enough to the carbonyl group. This will often dominate all other effects which determine the conformation of acyclic ketones and aldehydes and result in a conformationally rigid chelate. These conditions should favour high diastereoselectivity and predictability, and this is generally the case. The coordinating group Ÿ is usually OH, OR, or NR$_2$, and some typical examples are provided in Table 3.2.[6] It has been demonstrated experimentally for several systems that chelates are indeed true intermediates in nucleophilic additions of this type.[7]

Table 3.2 shows that several factors can influence the diastereoselectivity of such a reaction, including solvent, counter-ion, and the nature of the ligating group. In the most common case the ligating atom is oxygen in the form of a protected hydroxyl group, and the nature of the protecting group can have a profound influence on the level of diastereoselectivity observed. Highest diastereoisomeric excesses are usually obtained when the protecting group is attached to the oxygen atom via sp^3 carbon. Best results are usually obtained when the hydroxyl is protected as the benzyl, BOM (CH$_2$OCH$_2$Ph), MOM (CH$_2$OMe), or MEM (CH$_2$OCH$_2$CH$_2$OMe) ether. Ketones such as **3.12** and

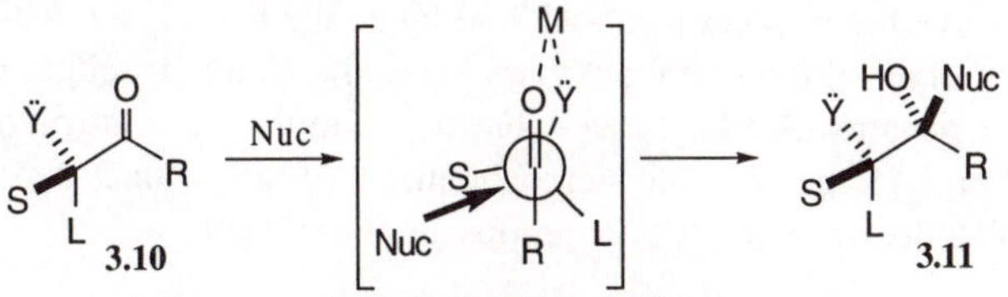

Cram's chelation model
Fig. 3.4

Table 3.2 Chelation controlled addition to **3.9**

R	S	L	Ÿ	Nuc(solvent)	d.e.[a] (%)
Ph	Me	Ph	OH	MeLi(Et$_2$O)	84
Ph	Me	Ph	OH	Me$_2$Mg(Et$_2$O)	66
Ph	Me	Ph	OH	MeMgBr(Et$_2$O)	50
Ph	Me	Ph	OH	MeMgBr(THF)	80
Ph	Me	Ph	OMe	MeLi(Et$_2$O)	34
Ph	Me	Ph	OMe	MeMgBr(Et$_2$O)	34
Ph	Me	Ph	OMe	MeMgBr(THF)	84
Me	Me	Ph	OH	Ph$_2$Mg(Et$_2$O)	74
Me	Me	Ph	OMe	Ph$_2$Mg(THF)	86
Me	H	C$_7$H$_{15}$	OMEM	C$_4$H$_9$MgBr(THF)	100
H	Me	H	OH	PhLi(Et$_2$O)	46

[a]Major diastereoisomer corresponds to **3.11**.

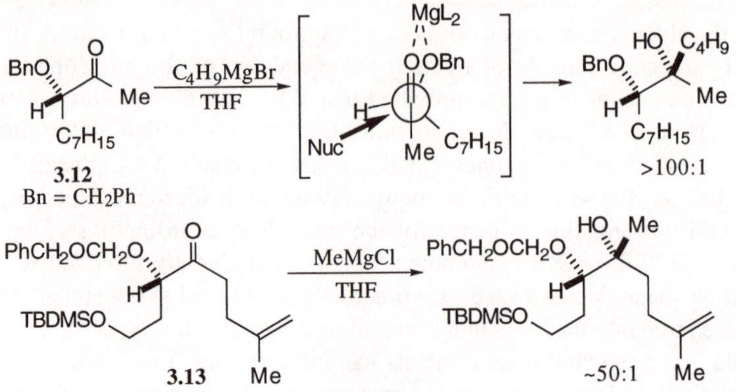

Fig. 3.5

3.13 can exhibit very high chelation controlled additions (Fig. 3.5).[8]

Chiral auxiliaries have been used to control the stereochemistry of additions to carbonyl compounds, and can be attached either to the carbonyl compound or to the nucleophile, with the former approach proving more productive.

One such approach involves the use of chiral acetals and related derivatives of α-ketoaldehydes. Two examples typical of this approach are illustrated in Fig. 3.6.[9,10] The chiral auxiliaries are derived from natural proline, and are easily removed to give the product alcohols **3.14** by acidic hydrolysis. Although only the (*S*)-enantiomer of the chiral auxiliary is readily available, either enantiomer of the final product **3.14** can be obtained simply by control of reaction variables. For **3.15** the absolute stereochemistry of the product depends on the order in which the two nucleophilic groups are introduced.

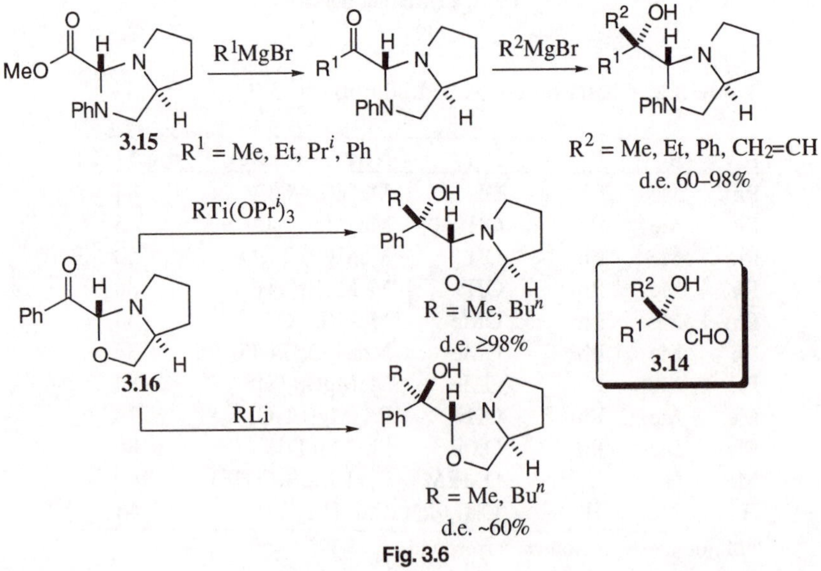

Fig. 3.6

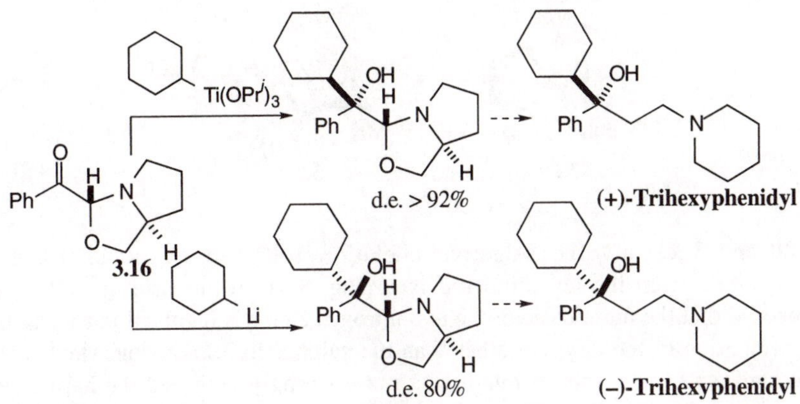

d.e. > 92%   (+)-**Trihexyphenidyl**

d.e. 80%   (−)-**Trihexyphenidyl**

Fig. 3.7

In the case of **3.16** choice of organometallic reagent can be used to determine the absolute stereochemistry of the product. This method was used to prepare both enantiomers of trihexyphenidyl (Fig. 3.7), used in treatment of parkinsonism.

Several chiral auxiliaries which use 1,3-oxathianes as the acetal unit have been successful in this area.[11] Typical reactions of the most convenient of these systems **3.17**, derived from natural pulegone, are shown in Fig. 3.8.[12] The product from **3.18** was used in a total synthesis of (*R*)-mevalolactone **3.19**, an important biogenetic precursor of terpenes and steroids.[13] It has also been shown that addition of ytterbium chloride causes reversal of stereoselectivity in addition of Grignard reagents to this type of oxathiane substrate.[14]

Transition state models which are consistent with the addition products of

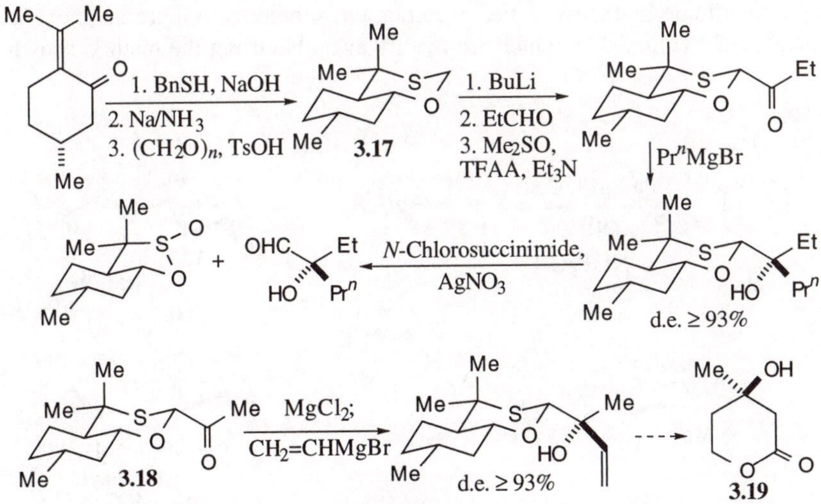

Fig. 3.8

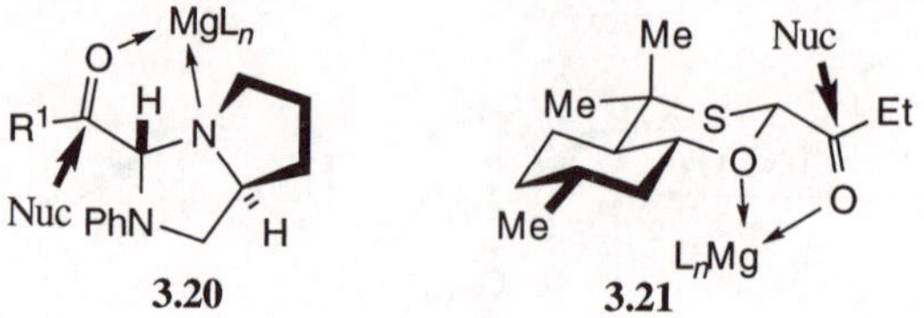

Fig. 3.9

3.20 and **3.21** involve selective chelation followed by attack of the nucleophile from the less hindered face (Fig. 3.9). In the case of **3.20** it is proposed that the more basic of the two nitrogen atoms is involved in chelation. Coordination to the oxygen rather than the sulphur in **3.21** is consistent with the hard acid (magnesium) interacting more strongly with and the hard base (ether) than the soft base (thioether).

The acetal derivatives discussed above possess a chiral centre adjacent to the carbonyl group, which makes an important contribution to the excellent stereoselectivity observed in these reactions. Examples of stereochemical control in this area which involve more remote chiral centres are relatively rare, but exceptional levels of stereoselectivity have been observed in additions to **3.22**.[15] The chiral auxiliary, 8-phenylmenthol **3.23**, can be prepared from pulegone (both enantiomers of which are available) as shown in Fig. 3.10.[16] This chiral auxiliary has also been used in the control of Diels–Alder reactions and conjugate additions to α,β-unsaturated esters (Chapter 6). Face blocking by the phenyl group is thought to be an important feature in all these reactions, and additions to **3.22** are consistent with reaction via this type of conformation (Fig. 3.10).

As stated earlier, attachment of a chiral auxiliary to the nucleophilic component is possible, although there are far fewer examples which provide high enantiomeric excess.[17] For example, enantiomerically pure sulphoxides such as **3.24** (Fig. 3.11), which are readily available using the method shown,

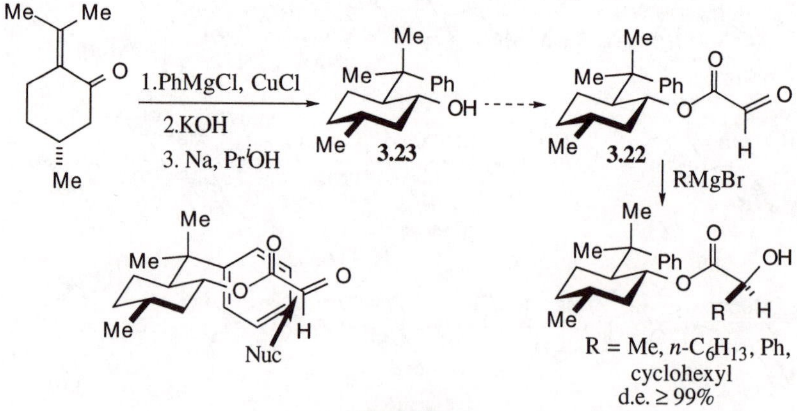

Fig. 3.10

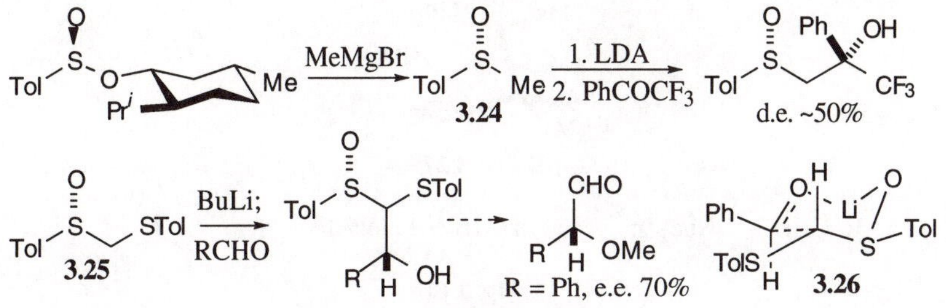

Fig. 3.11

add to carbonyl compounds with modest selectivity.[18] This selectivity can be increased by the incorporation of a sulphide group adjacent to the carbanionic centre (**3.25**).[19] The observed stereochemistry in this reaction is consistent with the chair-like transition state **3.26**.

In the preceding examples the chirality of the carbonyl compound is responsible for the stereoselectivity of the addition. Enantioselective synthesis should also be possible if a chiral reagent adds to a prochiral carbonyl compound, and powerful methods are emerging which can achieve this with high levels of stereoselectivity. These methods rely on changing the nature of the organometallic reagent by the addition of specific chiral ligands, often chelating ligands, which bind to the metal. Effectively, this converts an achiral nucleophile into a chiral reagent, which can now distinguish the prochiral faces of a carbonyl compound. This approach is illustrated by the examples shown in Fig. 3.12.[20,21]

In both of these examples the chiral chelating ligand is required in at least stoichiometric amounts. An extremely attractive alternative is the use of a chiral ligand which itself is a catalyst for the addition of the organometallic derivative to the carbonyl group.

This is demanding as the organometallic reagent must be relatively unreactive towards the carbonyl group unless combined with the catalyst ('ligand acceleration'), and the catalyst must possess the correct three-dimensional

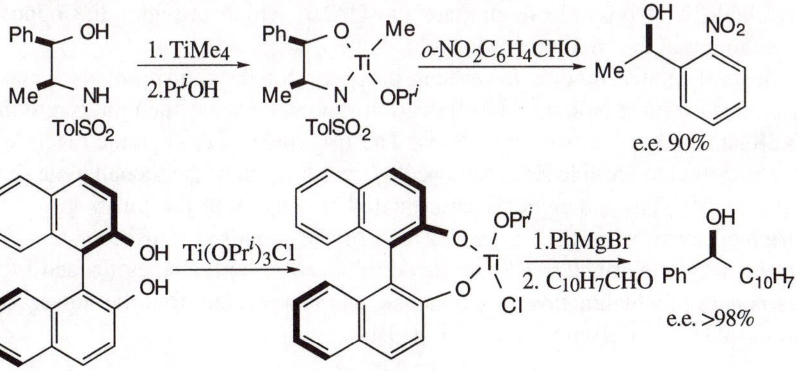

Fig. 3.12

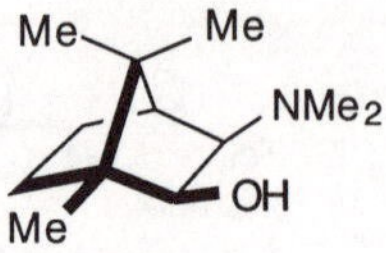

(–)-DAIB **3.27**

$$R-CHO \;+\; Nuc_2Zn \;+\; (-)\text{-DAIB} \;(\sim 2\ mol\%) \longrightarrow$$

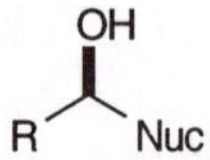

Fig. 3.13

Table 3.3 Additions of dialkylzinc reagents to aldehydes catalysed by **3.27**

R	Nuc	e.e. (%)
Ph	Me	91
Ph	Et	99
Ph	Bu^n	98
$p\text{-ClC}_6\text{H}_4$	Et	93
$p\text{-MeOC}_6\text{H}_4$	Et	93
2-Furyl	C_5H_{11}	>95
$(E)\text{-}C_6H_5CH{=}CH$	Et	96
$(E)\text{-}(Bu^n)_3SnCH{=}CH$	C_5H_{11}	85
$PhCH_2CH_2$	Et	90

structure to provide high enantiomeric excesses. In spite of these complexities remarkable advances have been made in this area using zinc organometallics and various catalysts.

The enantioselective addition of some dialkylzinc reagents to aldehydes can be catalysed by a range of chiral β-aminoalcohols. This is exemplified by the reactions of diethylzinc catalysed by (–)-3-*exo*-(dimethylamino)isoborneol ((–)-DAIB) **3.27** (Fig. 3.13 and Table 3.3).[22]

The catalyst **3.27** is prepared from camphor and as both enantiomers are available it is possible to prepare (+)-DAIB, which provides the opposite enantiomer of the addition product.

Several related models have been proposed for the transition state, one of which is outlined below.[23] The transition state structure is thought to resemble **3.28** and to involve two zinc atoms. The first mole of dialkylzinc reacts with the hydroxyl to provide Zn_A, followed by coordination of the second mole (Zn_B) (Fig. 3.14). The aldehyde is coordinated to Zn_A with the bulky group (R) oriented away from the congested area of the proposed transition state. The group which is transferred (Nuc) is coordinated to both zinc atoms and to the carbon atom which is undergoing attack, and is delivered from the 'lower' face of the aldehyde (as represented in Fig. 3.14).

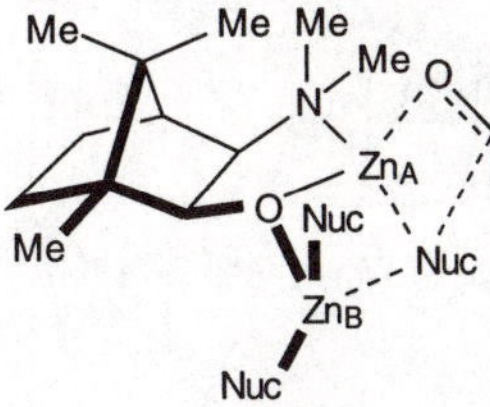

3.28
Fig. 3.14

A selection of the many chiral β-aminoalcohols and related systems which have been found to act as asymmetric catalysts in this type of addition is presented in Fig. 3.15.[24] The enantiomeric excess and absolute configuration of the product obtained with the particular catalyst in the addition of diethylzinc to benzaldehyde is shown below the structure of the catalyst (Fig. 3.15).

The scope and potential of this type of asymmetric catalysis have been widened by the development of methods for the preparation of various functional zinc derivatives and other catalyst systems. These are illustrated by the examples in Figs 3.16 and 3.17.

It is possible to transmetallate readily available vinylboranes **3.29** (Fig. 3.16) to produce organozinc reagents **3.30** which will undergo enantioselective addition to aldehydes in the presence of the previously used catalyst (−)-DAIB.[25] In this case the nucleophilic carbon unit is derived from the corresponding alkyne. This reaction sequence is sufficiently selective to allow the highly efficient intramolecular cyclization shown in Fig. 3.16, which was used in a short total synthesis of (*R*)-muscone using only 1 mole per cent of (+)-DAIB.[26]

Functional organozinc reagents can also be generated from the corresponding

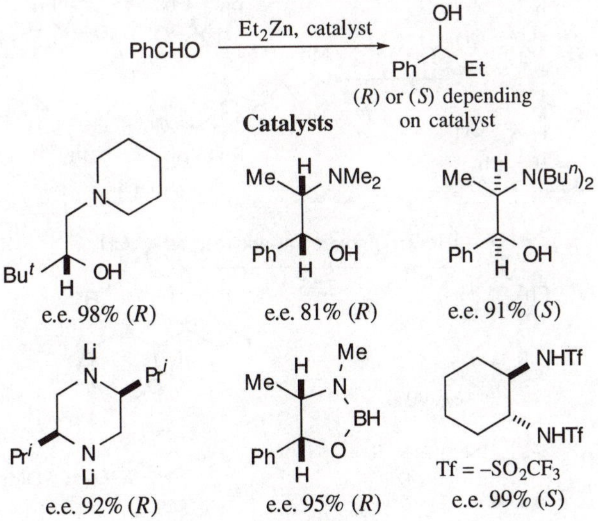

Catalysts

Fig. 3.15

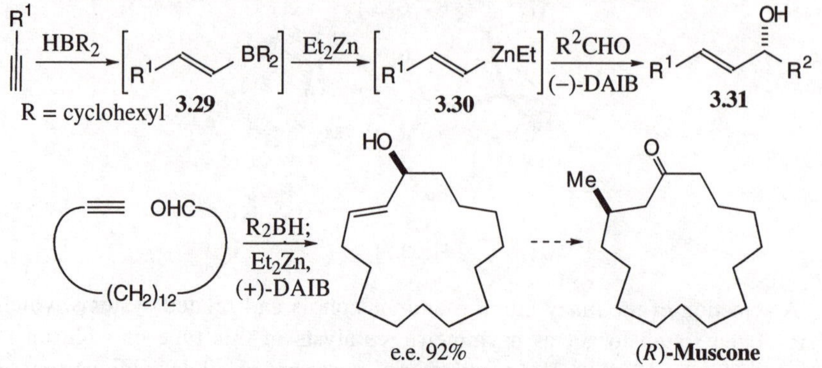

e.e. 92%   (*R*)-**Muscone**

Fig. 3.16

Grignard reagents by addition of zinc chloride, which establishes equilibrium 1 (Fig. 3.17), followed by dioxane, which precipitates the magnesium salt as a highly insoluble complex (reaction 2, Fig 3.17). This provides a 'salt free' solution of the organozinc reagent, which adds to aldehydes with good enantioselectivity in the presence of a catalyst derived from **3.32** and titanium tetra*iso*propoxide (formulated as **3.33**).[27]

From the foregoing illustrations it can be seen that these catalysed enantioselective additions have the potential to be important powerful techniques for asymmetric synthesis. A remarkable observation, which has been

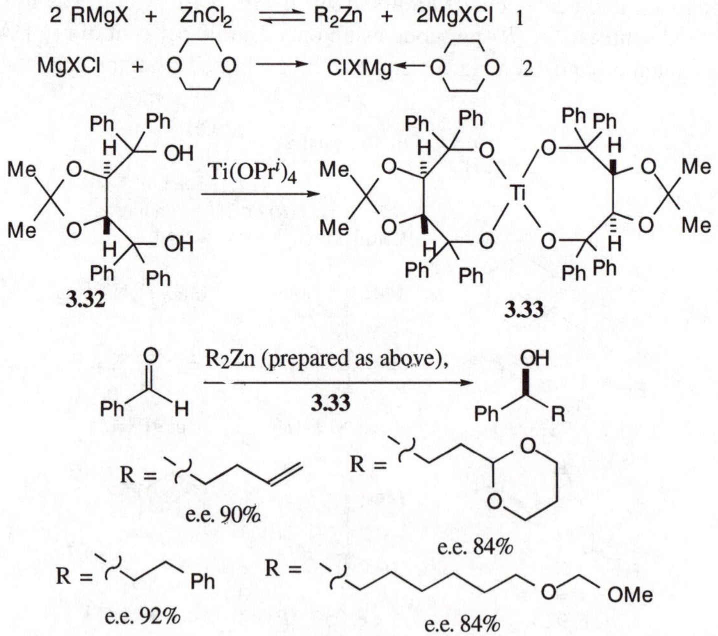

Fig. 3.17

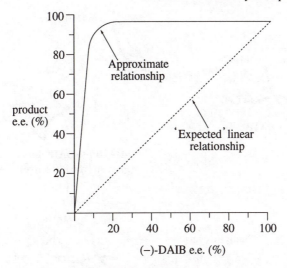

Fig. 3.18

studied for the diethylzinc/(−)-DAIB system, is that the catalyst does not necessarily have to be of high optical purity to provide products with high enantiomeric excesses. In a reaction mediated by an asymmetric catalyst, as the catalyst is the only chiral species involved in the transition state, it might reasonably be expected that the enantiomeric excess of the catalyst would limit the enantiomeric excess of the product. A catalyst of 50 per cent enantiomeric excess should give a product of no more than 50 per cent enantiomeric excess, and there should be a linear relationship between the enantiomeric excess of the catalyst and product. It can be seen from the graph in Fig. 3.18 that this is certainly not the case for the reaction between diethylzinc and benzaldehyde catalysed by (−)-DAIB. A catalyst of 25 per cent enantiomeric excess gives a product of greater than 90 per cent enantiomeric excess![28]

The origin of this 'chiral amplification' has been found to lie in the nature of the species produced when the catalyst DAIB reacts with the diethylzinc. These species are dimeric, rather than monomeric, and the dimers are not catalytically active themselves. The catalytically active species is the monomer, and the various dimers show different tendencies to dissociate. The case in which (−)-DAIB of 100 per cent enantiomeric excess is used is illustrated in Fig. 3.19. One catalytically *inactive* dimer (the (−)-DAIB/(−)-DAIB dimer) is formed, which dissociates into the monomer, and it is only the monomer which is capable of entering the catalytic cycle.

If (−)-DAIB of less than 100 per cent enantiomeric excess is used, three dimers are possible as both (−)-DAIB and (+)-DAIB can be added to the diethylzinc, and equilibrium is rapidly established between these species (Fig. 3.20). The (−)-DAIB/(−)-DAIB and the (+)-DAIB/(+)-DAIB dimers are enantiomeric, and the monomer from each of these will give opposite

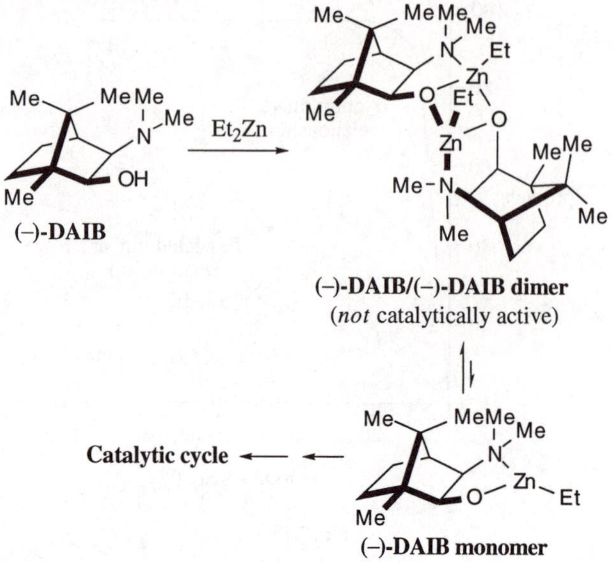

Fig. 3.19

asymmetric induction. However, the third dimer, derived from (–)-DAIB and (+)-DAIB, and labelled (–)-DAIB/(+)-DAIB dimer (Fig. 3.20) happens to be the most stable of the three dimers, and shows little tendency to dissociate under the reaction conditions. This dimer is a *meso*-compound, which is why there are three, rather than four, possible dimers.

Effectively, formation of this *meso*-dimer is irreversible under the reaction conditions, which means that at equilibrium all of the minor enantiomer of the catalyst ((+)-DAIB in this case) is used up in formation of the *meso*-dimer. The minor enantiomer of the catalyst is therefore 'reacted away' leaving the major enantiomer 'free' to function as normal (as in Fig. 3.19). This is illustrated schematically for a sample of (–)-DAIB of 50 per cent enantiomeric excess (75 per cent (–) and 25 per cent (+)) below (Fig. 3.20).

In the example shown (Fig. 3.20), at equilibrium there will be effectively none of the (+)-DAIB/(+)-DAIB dimer, the 25 mole per cent (+)-DAIB present is consumed by reaction with an equimolar amount of (–)-DAIB to give the stable, catalytically inactive *meso*-dimer. The remaining (–)-DAIB (50 mole per

(–)-DAIB + **(+)-DAIB**
(75%) (25%)

↓ Et_2Zn

(–)-DAIB/(–)-DAIB dimer (50%) + **(+)-DAIB/(+)-DAIB dimer** → **(–)-DAIB/(+)-DAIB *meso*-dimer** (50%)

Fig. 3.20

cent of the original total DAIB added) then functions as normal. In this example, using a catalyst of 50 per cent enantiomeric excess is in effect equivalent to using 50 per cent of the amount actually added (the rest is unavailable for reaction as part of the *meso*-dimer).

In general terms, this phenomenon of chiral amplification is likely to be of particular significance if the catalyst in question is not easy to obtain in enantiomerically pure form. For example, the starting material might be of low enantiomeric excess, or a resolution process might be necessary (in which case resolution to enantiomeric purity would not be required). In these cases the possibility of obtaining a product of high enantiomeric excess from a catalyst of moderate optical purity would be most attractive.

The nucleophilic species which have been covered so far have all been either alkyl, vinyl, or aromatic. The addition of allylmetals to carbonyl groups has received particular attention, and will be covered in the next part of this chapter. One of the reasons for the relative importance of the addition of allylmetals is that it provides an alternative route to aldol-type products (Fig. 3.21), which have many applications in organic synthesis (see Chapter 5 for a discussion of asymmetric aldol reactions).

There are many similarities between the aldol reaction and the addition of allylmetal reagents to aldehydes and ketones. Two new chiral centres are produced at the same time as the new C–C bond (assuming $R^2 \neq H$ in Fig. 3.21) so there are four possible stereoisomers of the product. To be consistent with the discussion of the aldol reaction itself (Chapter 5), analogous stereochemical descriptors will be used (Fig. 3.22, cf. Fig. 5.1). A complicating factor, which is not present in aldol reactions, is the possibility of reaction of the allylmetal reagent at either C-1 or C-3 (Fig. 3.21), but all the examples discussed here will concern allylmetal reagents which react at C-3.

If neither **3.34** nor **3.35** is chiral, then **3.36** and **3.37** are enantiomeric (the *syn*-racemate), as are **3.38** and **3.39** (the *anti*-racemate). The best that could be achieved would be the exclusive formation of either the *syn*- or the *anti*-racemate (without the use of a chiral catalyst or other chiral component). If either or both reactants **3.34** and **3.35** are chiral then in principle it should be possible to obtain one of the individual stereoisomers as the major or exclusive product. In this case the transition states which lead to the four possible product stereoisomers will be diastereoisomeric and not necessarily equal in energy.

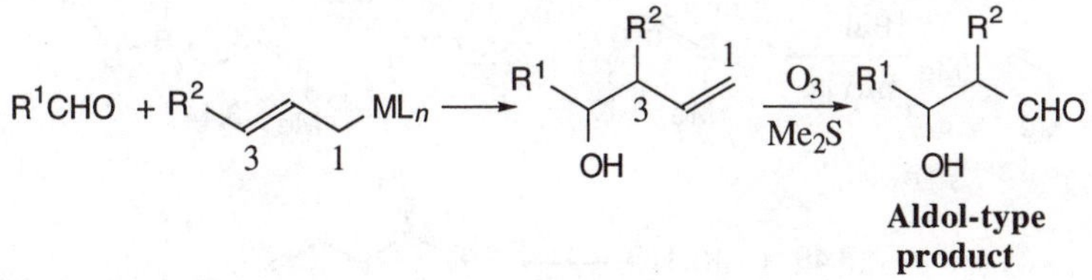

Fig. 3.21

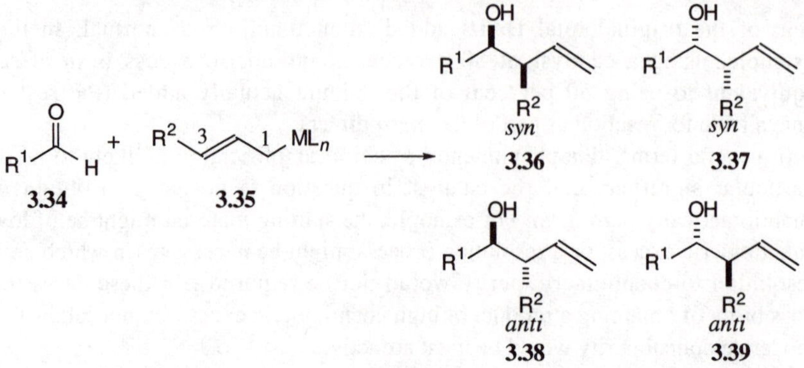

Fig. 3.22

The use of a chiral allylmetal reagent **3.35** which carries asymmetric ligands is an attractive approach to enantioselective synthesis in this area, and several examples of this type have been studied. Various metals have been studied (including silicon, tin, titanium, and boron), but the following discussion will be confined to boron reagents, as there are several readily prepared and relatively effective reagent types. Moreover, reasonably consistent transition state models are available to account for the observed stereoselectivity.

Convenient chiral boron reagents can be prepared from the appropriate allylmetal (Fig. 3.23) by reaction with tri*iso*propyl borate, followed by exchange of the remaining alkoxide ligands on boron with di*iso*propyl tartrate (Fig. 3.23). Representative examples of the addition of **3.40** to achiral aldehydes are provided in Table 3.4.[29]

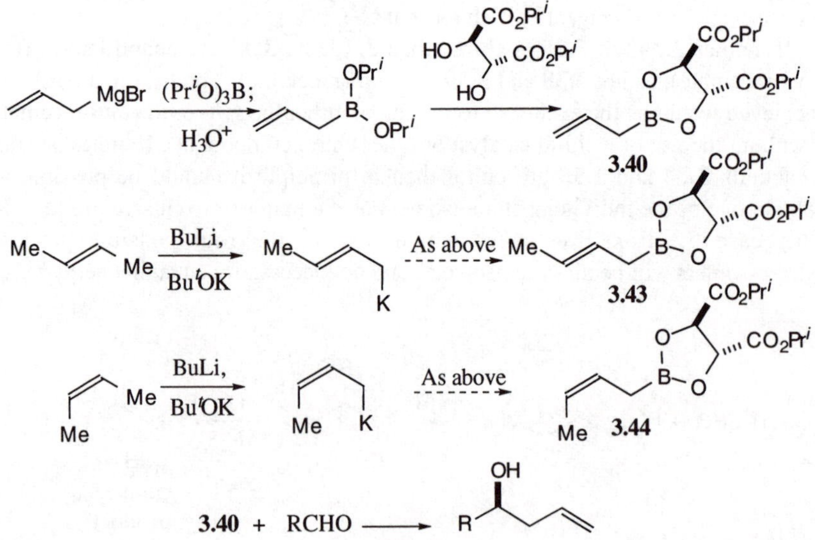

Fig. 3.23

Table 3.4 Additions of **3.40** to aldehydes

R	e.e. (%)	Yield (%)
n-C$_6$H$_{11}$	87	97
n-C$_{10}$H$_{19}$	86	69
Bu^t	86	78

The transition state model which has been proposed to account for the sense of stereoselectivity is illustrated in Fig. 3.24. The important aspects of this model include the requirement for complexation of the carbonyl group to the boron atom, and the resulting six-membered transition state which adopts a chair-like conformation. There are two possible chair conformations in which the aldehyde substituent (R) is equatorial (**3.41** and **3.42**). Lone pair–lone pair (and presumably dipole–dipole) repulsion is then responsible for raising the energy of chair transition state **3.42** relative to **3.41**. This is similar to the Zimmerman–Traxler transition state model which is used to account for the stereochemical relationships observed in aldol reactions (Chapter 5).

The (E)- and (Z)-propenyl reagents **3.43** and **3.44**, prepared as shown in Fig. 3.23, react with aldehydes to give the *anti*- and *syn*-products respectively, in moderate to good enantiomeric excess (Fig. 3.25).[30]

These stereochemical results (Fig. 3.25) can be understood by using the above transition state model, as illustrated in Fig. 3.26. In these transition state models, the framework is exactly as in **3.41** (Fig. 3.24). The (E)- and (Z)-

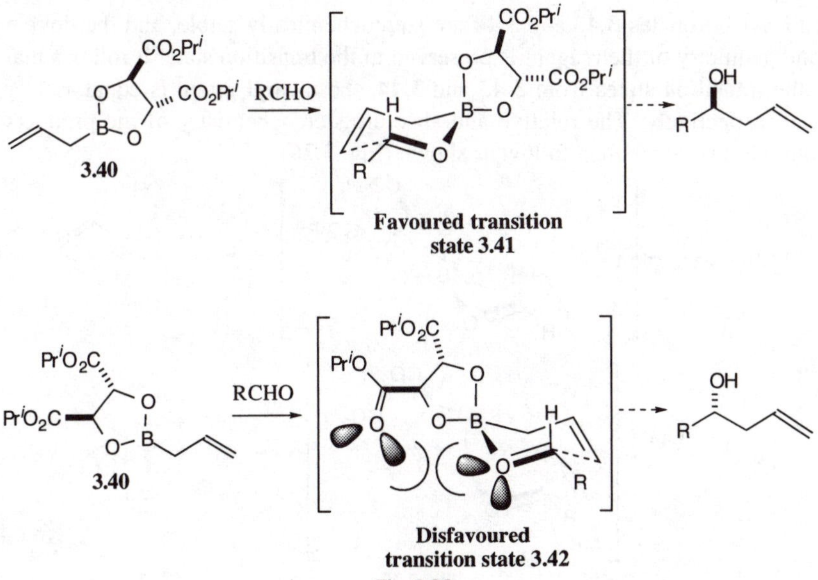

Favoured transition
state 3.41

Disfavoured
transition state 3.42

Fig. 3.24

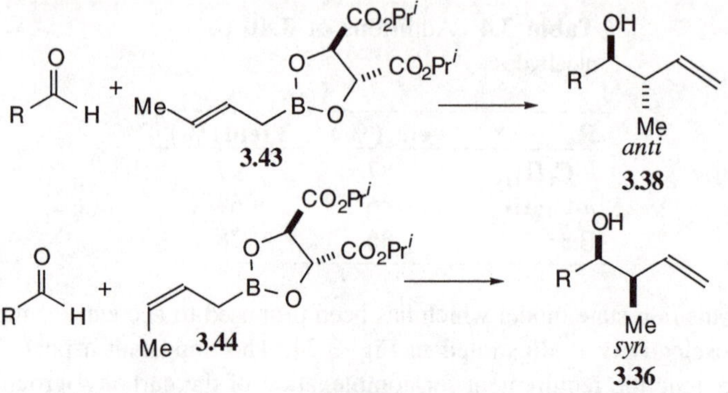

Fig. 3.25

Table 3.5 Additions of **3.43** and **3.44** to aldehydes

R	Reagent	*anti:syn*	e.e.(%)
$n\text{-}C_9H_{19}$	**3.43**	>99:1	88
$n\text{-}C_9H_{19}$	**3.44**	3:97	86
$TBSOCH_2CH_2$	**3.43**	≥98:2	85
$TBSOCH_2CH_2$	**3.44**	≥2:98	72
Bu^t	**3.43**	95:5	73
Bu^t	**3.44**	>1:99	70
$n\text{-}C_7H_{15}CH{=}CH$	**3.43**	>99:1	74
$n\text{-}C_7H_{15}CH{=}CH$	**3.44**	3:97	62

propenyl boronates **3.43** and **3.44** are stereochemically stable, and the double bond geometry of the reagent is preserved in the transition state. It follows that in the transition states from **3.43** and **3.44**, the methyl group is equatorial or axial respectively. The relative and absolute stereochemistry of the products from these reagents then follow as shown (Fig. 3.26).

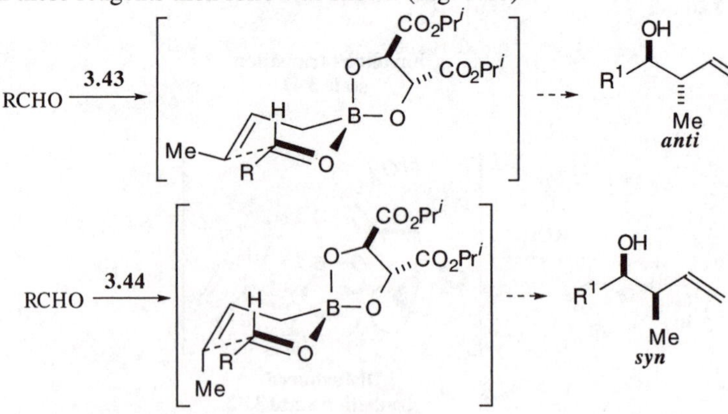

Fig. 3.26

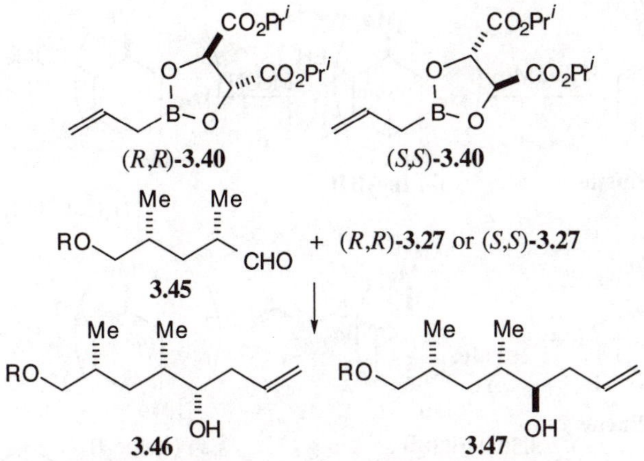

(R,R)-**3.40** (S,S)-**3.40**

3.45   + (R,R)-**3.27** or (S,S)-**3.27**

3.46   **3.47**

(R,R)-**3.40** '*matched pair*'   **3.46:3.47** = 92:8
(S,S)-**3.40** '*mismatched pair*'   **3.46:3.47** = 13:87

Fig. 3.27

The enantioselectivity of these reagents can be modest, but they are easy to prepare and to use, the sense of the asymmetric induction is predictable, and as both enantiomers of the tartrate ester are readily available, both enantiomers of the product are equally easy to obtain.

The diastereoselectivity of these additions can be increased by using a chiral non-racemic aldehyde. Both reactants being chiral, the possibility now exists for double asymmetric synthesis (see Chapter 2), a situation which is helped by the ready availability of both enantiomers of these chiral reagents. This is illustrated by the reactions shown in Fig. 3.27 with aldehyde **3.45**.

Asymmetric synthesis by the addition of allyl or crotyl units can be iterative,

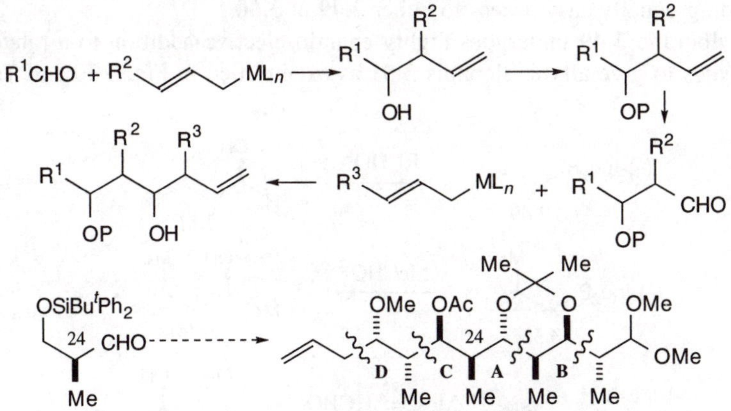

3.48

Order of C–C bond formation **A, B, C, D**;
each takes place with stereoselectivity ≥ 90%

Fig. 3.28

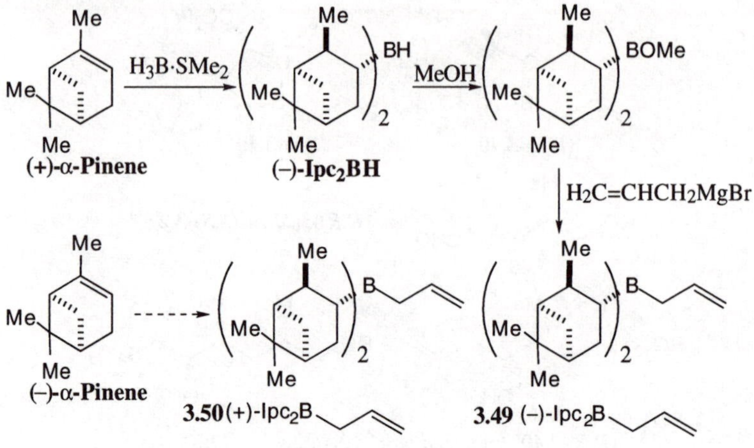

Fig. 3.29

simply by protection of the hydroxyl group followed by alkene cleavage to the protected aldehyde. This aldehyde than serves as the substrate for the next allylation or crotylation. This approach is illustrated in Fig. 3.28, and has been used in a total synthesis of **3.48** which represents the 'ansa' chain of the natural product rifamycin S.[31]

Asymmetric addition of allyl and related groups can be carried out with even higher enantioselectivity by using allylboranes with ligands other that tartrate, in particular isopinocampheyl (Ipc).[32] These ligands are easily attached to the boron by hydroboration of α-pinene (Fig. 3.29). The resulting diisopinocampheylborane (Ipc$_2$BH), which is discussed in some detail in Chapter 6 in the context of asymmetric hydroboration, is easily converted into the allyl derivative **3.49**.[33] Both enantiomers of α-pinene are available, providing equally easy access to either **3.49** or **3.50**.

Allylborane **3.49** undergoes highly enantioselective addition to a range of aldehydes to give allylic alcohols **3.51** as exemplified in Fig. 3.30 and Table

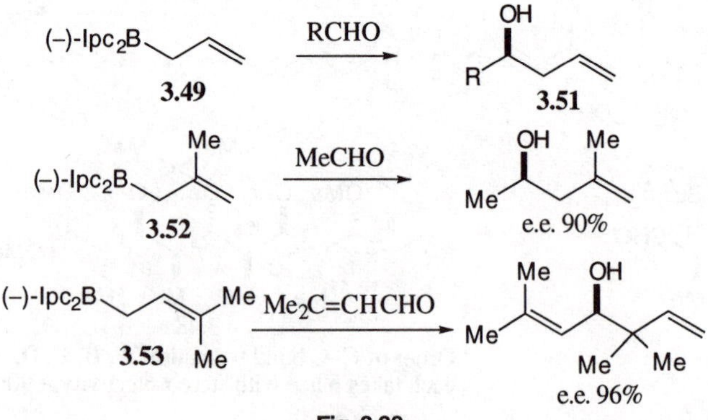

Fig. 3.30

Table 3.6 Additions of
3.49 to aldehydes

R	e.e.(%)	Yield(%)
Me	93	74
Et	86	71
Bu^n	87	72
Pr^i	90	86
Bu^t	83	88
Ph	96	81

3.6.[33] Other simple allyl derivatives such as **3.52** and **3.53** also react with high enantioselectivity in the same sense as **3.49**.[34]

Crotyl boranes and related reagents are equally easy to prepare, using either a method analogous to that used for the allyl reagents or occasionally by selective hydroboration of a diene (Fig. 3.31). The crotylpotassium reagents are prepared as shown in Fig. 3.23.

Reagents **3.54**, **3.55**,[35] **3.56**,[36] and **3.57**[37] all react with aldehydes with very high diastereoselectivity and enantioselectivity, in a consistent and predictable fashion. These aspects of these reagents are illustrated by their reaction with acetaldehyde (Fig. 3.32).

This addition is also consistent, predictable, and highly stereoselective when the aldehyde is chiral. Both allyl boranes[38] and crotyl boranes[39] react with a

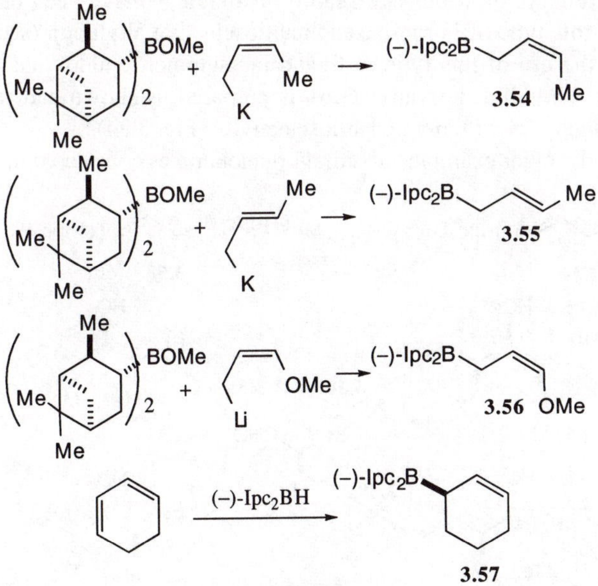

Fig. 3.31

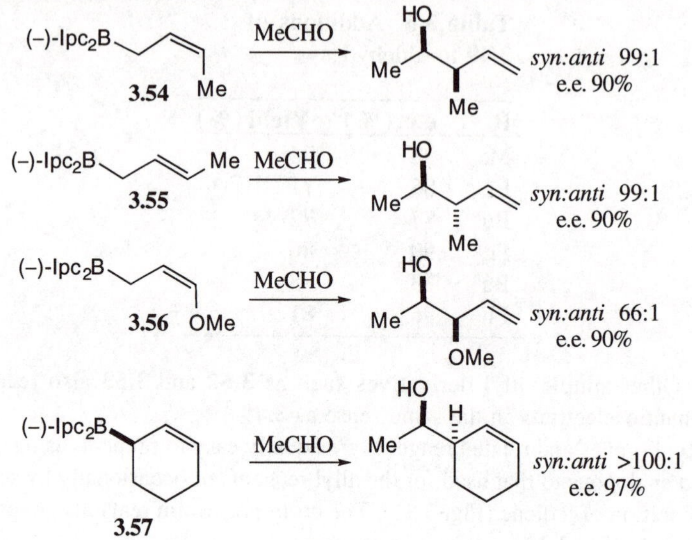

Fig. 3.32

range of α-chiral aldehydes, and in all cases the stereochemistry of the addition is controlled by the chirality of the borane. In this way it is possible to prepare all four stereoisomeric products from reaction of a chiral aldehyde and a crotyl borane simply by choice of reagent (Fig. 3.33), which demonstrates the power of this approach.

Chiral ligands based on disulphonyl derivatives of diamine **3.60** (Fig. 3.34), both enantiomers of which are readily available,[40] have been used as chiral ligands in this type of boron-based enantioselective allylation (see Chapters 5 and 6 for the use of this type of ligand in asymmetric aldol and Diels–Alder reactions). The allyl derivative **3.61** is prepared as shown and reacts with a range of aldehydes with high enantioselectivity (Fig. 3.34).[41]

As with the other examples of chiral organoboranes considered in this chapter,

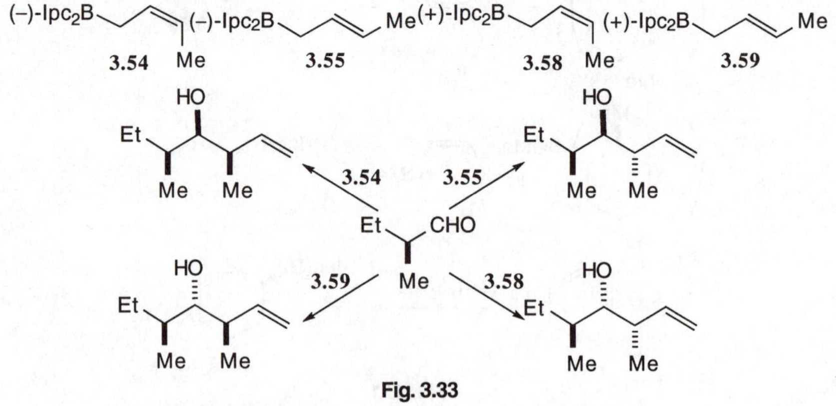

Fig. 3.33

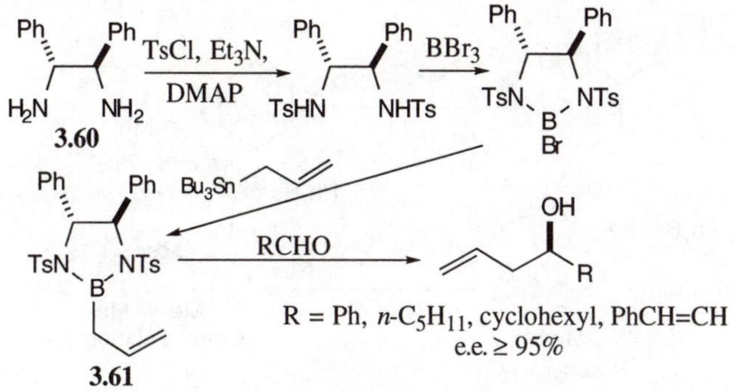

Fig. 3.34

the enantioselectivity of the reagent overrides the intrinsic facial selectivity of chiral aldehydes (Fig. 3.35), and the corresponding (*E*)-crotyl borane gives the *trans* product in high enantiomeric excess.

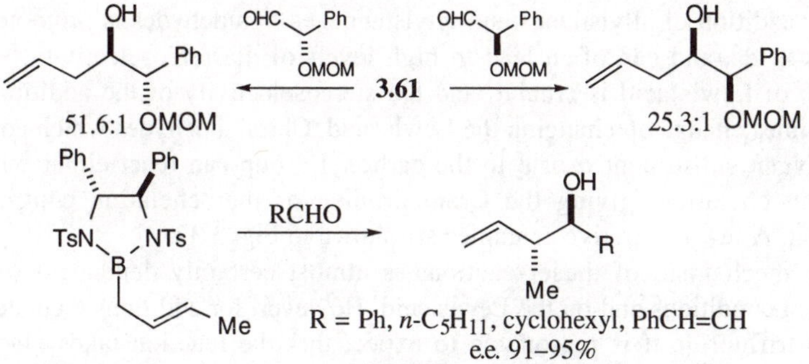

Fig. 3.35

The propynyl derivative **3.62** can be prepared as shown, and reacts with various aldehydes to produce allenyl alcohols **3.63**, which undergo a variety of potentially useful synthetic transformations (Fig. 3.36).[42]

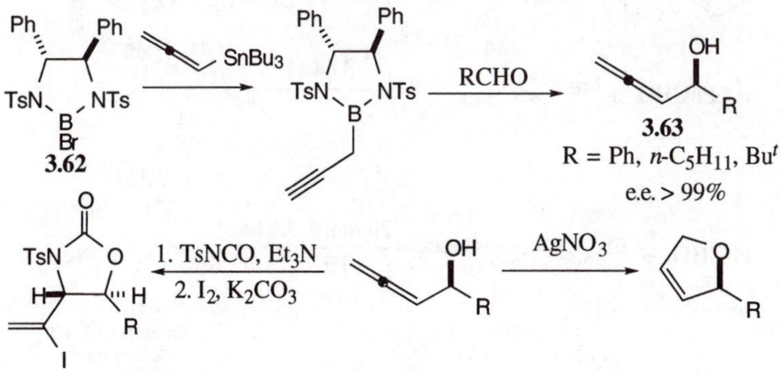

Fig. 3.36

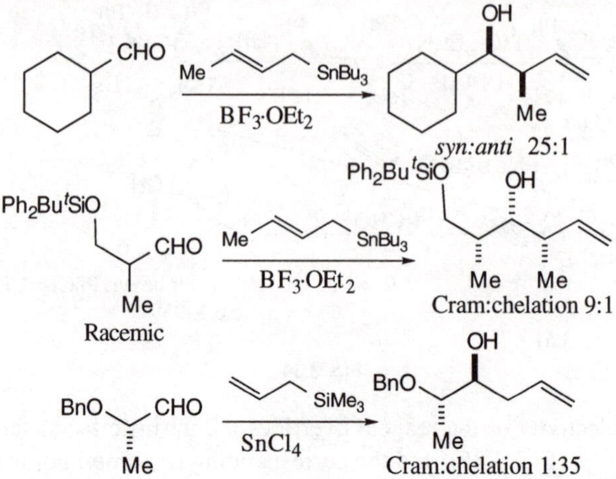

Fig. 3.37

The addition of allylsilanes and allylstannanes to aldehydes is promoted by Lewis acids, and can often lead to high levels of diastereoselectivity.[43] The choice of Lewis acid is crucial, and the stereoselectivity of the addition can sometimes change on changing the Lewis acid. Chiral aldehydes which contain an oxygen substituent α or β to the carbonyl group can react either with or without chelation, giving the Cram product or the 'chelation controlled' product. A few illustrative examples are shown in Fig. 3.37.

The mechanism of these reactions is almost certainly dependent on the reaction conditions and on the Lewis acid. However, for additions mediated by boron trifluoride it is reasonable to expect that the reaction takes place by complexation of the aldehyde prior to addition of the stannane. In this case the use of a chiral boron Lewis acid might be expected to result in asymmetric

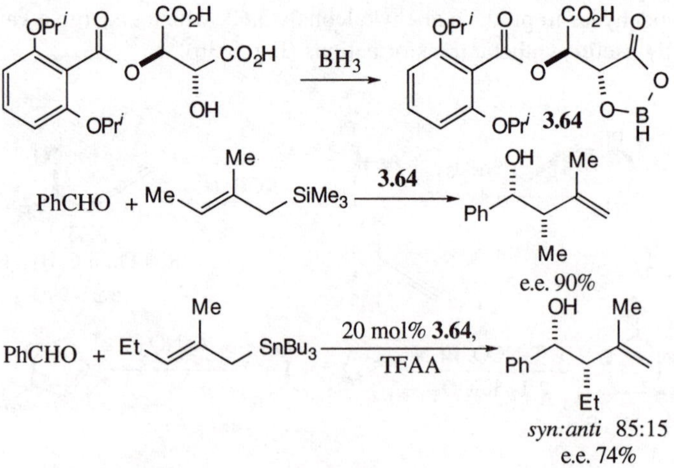

Fig. 3.38

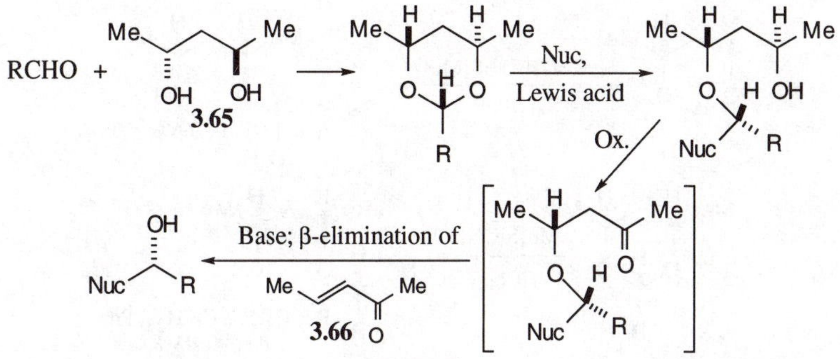

Fig. 3.39

addition to a prochiral aldehyde. This has been demonstrated for the chiral Lewis acid formulated as **3.64** and prepared as shown (Fig. 3.38).[44,45]

The carbonyl group of an aldehyde can easily be converted into an acetal, and such acetals will undergo reaction with certain types of nucleophile under the influence of a Lewis acid. If the diol used is chiral and the aldehyde prochiral, the possibility of asymmetric synthesis exists. This can be achieved with various diols,[46] but the most widely used is 2,4-dihydroxypentane **3.65**. The overall reaction scheme is outlined in Fig. 3.39. The chiral auxiliary is easily removed, but has to be destroyed by oxidation followed by base-catalysed reverse Michael reaction which causes elimination of **3.66**.

The stereoselectivity observed in such reactions of acetals of **3.65** is consistent with a model in which the acetal reacts in its most stable chair conformation. The Lewis acid coordinates to the oxygen adjacent to the axial methyl group, and the nucleophile then reacts with inversion at the acetal centre. Coordination to the other oxygen is disfavoured by steric repulsion due to the equatorial methyl group (Fig. 3.40).

In general, the nucleophilic species needs to be stable in the presence of the Lewis acid, and typical examples (Fig. 3.41) would include allylsilanes,[47] cyanotrimethylsilane,[48] alkynyltrimethylsilanes[49] and various organometallic reagents.[50].

From the selected examples provided in this chapter, it can be appreciated that enantioselective additions of carbon nucleophiles to carbonyl compounds are of great importance in asymmetric synthesis. A similar claim can be made for enantioselective substitution α to a carbonyl group, and some of the important methods for achieving such reactions are the topic of the following chapter.

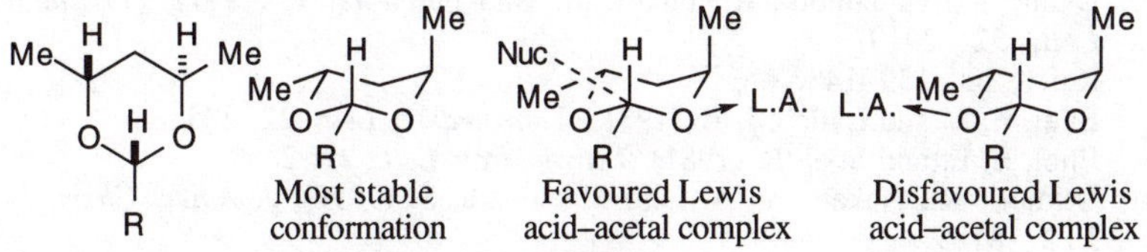

| Most stable conformation | Favoured Lewis acid–acetal complex | Disfavoured Lewis acid–acetal complex |

Fig. 3.40

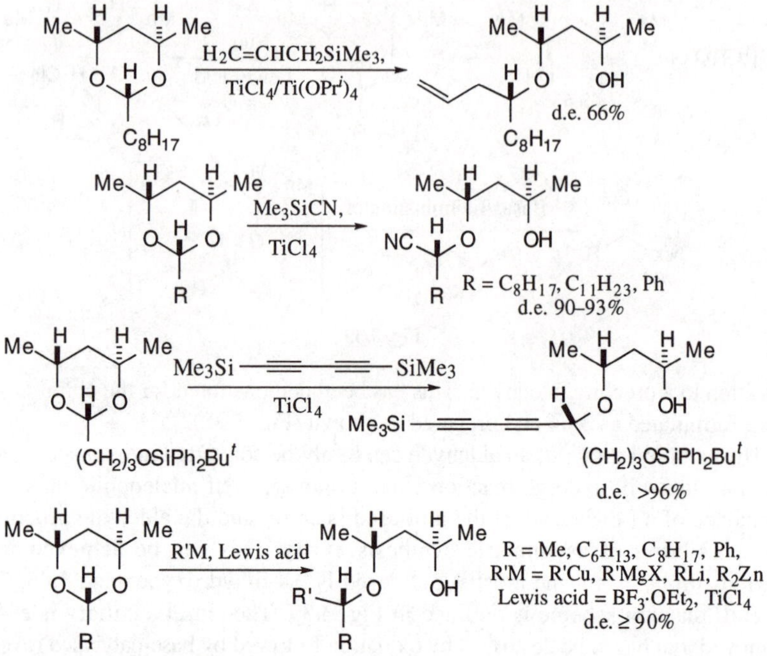

Fig. 3.41

References

1. Eliel, E. L. (1984), in *Asymmetric Synthesis*, (ed. J. D. Morrison), Vol. 2, pp. 125–156, Academic Press, New York, and references cited therein.
2. Anh, N. T. and Eisenstein, O. (1977), *Nouv. J. de Chim.*, **1**, 61; Chérest, M., Felkin, H., and Prudent, N. (1968), *Tetrahedron Lett.*, 2199.
3. Cram, D. J. and Elhafez, F. A. A. (1952), *J. Amer. Chem. Soc.*, **74**, 5828; Bassindale, A. (1984), in *The Third Dimension in Organic Chemistry*, pp. 206–207, Wiley, Chichester.
4. Reetz, M. T., Stanchev, S., and Haning, H. (1992), *Tetrahedron*, **48**, 6813.
5. Ref. 3; Reetz, M. T. (1982), *Top. Curr. Chem. Res.*, **106**, 1; Reetz, M. T. (1986), *Organotitanium Reagents in Organic Synthesis*, Springer, Berlin.
6. Taken from Table 1, ref. 1.
7. Chen, X., Hortelano, E. R., Eliel, E. L., and Frye, S. V. (1992), *J. Amer. Chem. Soc.*, **114**, 1778, and references cited therein.
8. Still, W. C. and McDonald, J. H. (1980), *Tetrahedron Lett.*, **21,** 1031.
9. Mukaiyama, T. (1981), *Tetrahedron*, **37**, 4111.
10. Ukaji, Y., Yamamoto, K., Fukui, M., and Fujisawa, T. (1991), *Tetrahedron Lett.*, **32**, 2919.
11. Ref.1, pp. 139–148.
12. Eliel, E. L. and Lynch, J. E. (1981), *Tetrahedron Lett.*, **22**, 2855.
13. Eliel, E. L. and Soai, S. (1981), *Tetrahedron Lett.*, **22**, 2859.
14. Utimoti, K., Nakamura, A., and Matsubara, S. (1990), *J. Amer. Chem. Soc.*, **112**, 8189.

15. Whitesell, J. K., Bhattacharya, A., and Henke, K. (1982), *J. Chem. Soc.,* *Chem. Commun.,* 988.
16. Corey, E.J. and Ensley, H. E. (1975), *J. Amer. Chem. Soc.,* **97**, 6908.
17. Solladié, G. (1984), in *Asymmetric Synthesis,* (ed. J. D. Morrison), Vol. 2, pp. 157–199, Academic Press, New York, and references cited therein.
18. Bravo, P., Frigerio, M., and Resnati, G. (1990), *J. Org. Chem.,* **55**, 4216.
19. Colombo, L., Gennari, C., Scolastico, C., Guanti, G., and Narisano, E. (1981), *J. Chem. Soc., Perkin Trans. 1,* 1278.
20. Reetz, M. T., Kükenhöner, T., and Weinig, P. (1986), *Tetrahedron Lett.,* **27**, 5711.
21. Seebach, D., Beck, A. K., Roggo, S., and Wonnacott, A. (1985), *Chem. Ber.,* **118**, 3673.
22. Kitamura, M., Suga, S., Kawai, K., and Noyori, R. (1986), *J. Amer. Chem. Soc.,* **108**, 6071.
23. For a discussion, see Noyori, R. and Kitamura, M. (1991), *Angew. Chem. Int. Ed. Engl.,* **30**, 49.
24. Taken from Table 5, ref. 23.
25. Oppolzer, W. and Radinov, R. N. (1992), *Helv. Chim. Acta,* **75**, 170.
26. Oppolzer, W. and Radinov, R. N. (1993), *J. Amer. Chem. Soc.,* **115**, 1593.
27. von dem Busche–Hünnefeld, J. and Seebach, D. (1992), *Tetrahedron,* **48**, 5719.
28. Kitamura, M., Okada, S., Suga, S., and Noyori, R. (1989), *J. Amer. Chem. Soc.,* **111**, 4028.
29. Roush, W. R., Hoong, L. K., Palmer, M. A. J., and Park, J. C. (1990), *J. Org. Chem.,* **55**, 4109.
30. Roush, W. R., Ando, K., Powers, K., Palkowitz, A. D., and Halterman, R. L. (1990), *J. Amer. Chem. Soc.,* **112**, 6339.
31. Roush, W. R., Palkowitz, A. D., and Ando, K. (1990), *J. Amer. Chem. Soc.,* **112**, 6348.
32. For a general account of asymmetric synthesis using boranes, see Brown, H. C. and Ramachandran, P. V. (1991), *Pure and Appl. Chem.,* **63**, 307.
33. Brown, H. C. and Jadhav, P. K. (1983), *J. Amer. Chem. Soc.,* **105**, 2092.
34. Brown, H. C., Jadhav, P. K., and Perumal, P. T. (1984), *Tetrahedron Lett.,* **25**, 5111; Brown, H. C. and Jadhav, P. K. (1984), *Tetrahedron Lett.,* **25**, 1215.
35. Brown, H. C. and Bhat, K. S. (1986), *J. Amer. Chem. Soc.,* **108**, 293.
36. Brown, H. C., Jadhav, P. K., and Bhat, K. S. (1988), *J. Amer. Chem. Soc.,* **110**, 1535.
37. Brown, H. C., Jadhav, P. K., and Bhat, K. S. (1985), *J. Amer. Chem. Soc.,* **107**, 2564.
38. Brown, H. C., Bhat, K. S., and Randad, R. S. (1987), *J. Org. Chem.,* **52**, 319.
39. Brown, H. C., Bhat, K. S., and Randad, R. S. (1987), *J. Org. Chem.,* **52**, 3701.
40. Pikul, S. and Corey, E. J. (1992), *Organic Syntheses,* **71**, 22.
41. Corey, E. J., Yu, C. M., and Kim, S. S. (1989), *J. Amer. Chem. Soc.,* **111**, 5495.
42. Corey, E. J. and Jones, G. B. (1991), *Tetrahedron Lett.,* **32**, 5713.
43. For a review, see Fleming, I. (1991), in *Comprehensive Organic Synthesis,* (ed. C. H. Heathcock), Vol. 2, pp. 595–628, Pergamon Press, Oxford.
44. Furuta, K., Mouri, M., and Yamamoto, H. (1991), *Synlett,* 561.
45. Marshall, J. A. and Tang, Y. (1992), *Synlett,* 653.
46. For a review, see Alexakis, P. and Mangeney, P. (1990), *Tetrahedron Asymmetry,* **1**, 477.

47. Johnson, W. S., Crackett, P. H., Elliot, J. D., Jagodinski, J. J., Lindell, S. D., and Natarajan, S. (1984), *Tetrahedron Lett.*, **25**, 3951.
48. Elliot, J. D., Cjoi, V. M. F., and Johnson, W. S. (1983), *J. Org. Chem.*, **48**, 2294.
49. Tabor, A. C., Holmes, A. B., and Baker, R. (1989), *J. Chem. Soc., Chem. Commun.*, 1025.
50. Alexakis, A., Mangeney, P., Ghribi, A., Marek, I., Sedrani, R., Guir, C., and Normant, J. F. (1988), *Pure and Appl. Chem.*, **60**, 49; Normant, J. F., Alexakis, A., Ghribi, A., and Mangeney, P. (1989), *Tetrahedron*, **45**, 507; Ghribi, A., Alexakis, A., and Normant, J. F. (1984), *Tetrahedron Lett.*, **25**, 3075; Ghribi, A., Alexakis, A., and Normant, J. F. (1984), *Tetrahedron Lett.*, **25**, 3083; Lindell, S. D., Elliot, J. D., and Johnson, W. S. (1984), *Tetrahedron Lett.*, **25**, 3947; Mori, A., Marueka, K., and Yamamoto, H. (1984), *Tetrahedron Lett.*, **25**, 4421; Alexakis, A., Marek, I., Mangeney, P., and Normant, J. F. (1990), *J. Amer. Chem. Soc.*, **112**, 8042.

4 α-Substitution using chiral enolates

Substitution at the α position of chiral enolates is a large and important area of asymmetric synthesis, which has been thoroughly reviewed.[1] This chapter is concerned mainly with the alkylation of enolates which are chiral by virtue of being attached to a chiral auxiliary (see Chapter 2) or an equivalent. Nevertheless the general principles involved can be extended to the reactions of other enolates as the concepts which are used to account for the stereoselectivity of alkylation of such enolates can be applied to the reactions of other enolates.

The alkylation of an enolate with high and predictable stereochemical control requires several conditions to be fulfilled. These will be discussed using the generalized structure **4.1** (Fig. 4.1) in which 'Aux' represents a chiral auxiliary. The metal 'counter-ion' of the enolates is represented by 'M'.

The first requirement for stereoselective enolate alkylation is that the enolate is formed in only one of the two possible geometries. The importance of this can be seen by considering the result of non-selective enolate formation (Fig. 4.1). Even if the chiral auxiliary directs completely to one face of the enolate, the diastereoselectivity of the reaction will only be as high as the selectivity of enolate formation.

Given that the chiral auxiliary in **4.1** directs reaction completely to the 'upper' face of the enolates, then enolate **4.2** will lead exclusively to product **4.3**, and enolate **4.4** to **4.5**. The products **4.3** and **4.5** are diastereoisomers and could be separated in principle, but if the selectivity was low this would render the asymmetric synthesis inefficient, and lower the yield of the desired stereoisomer.

Moreover, in the event of the diastereoisomers **4.3** and **4.5** being inseparable, removal of the chiral auxiliary would lead to a relatively low enantiomeric excess, directly proportional to the diastereoisomeric excess

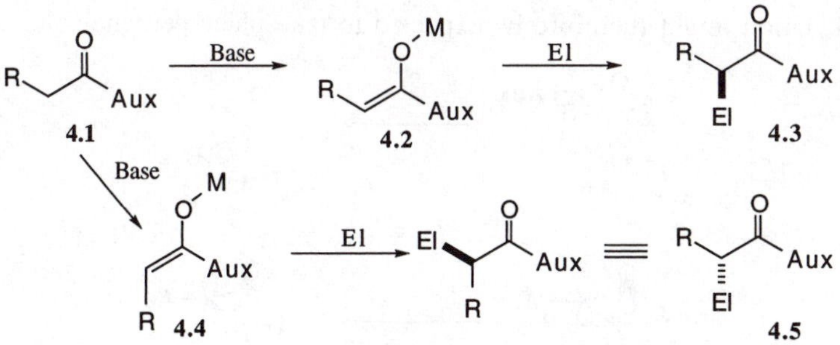

Aux directs exclusively to 'upper' face of enolate
El = electrophile
Fig. 4.1

obtained in the alkylation (assuming that removal does not racemize the product).

Factors which can influence the stereoselectivity of enolate formation include choice of base, solvent, temperature, counter-ion, and most importantly from the point of view of this chapter, the structure of the chiral auxiliary. Ideally, a given chiral auxiliary should produce the same enolate geometry, exclusively **4.2** or **4.4**, irrespective of the nature of R.

Before considering individual chiral auxiliaries, the stereoelectronic requirements for deprotonation and alkylation α to a carbonyl group will be addressed as these concepts can be useful in discussing the reasons for the stereoselectivity obtained using a particular chiral auxiliary.

Deprotonation α to a carbonyl group is thought to follow the path shown in Fig. 4.2[2] in which the breaking C–H bond lines up with the π orbital of the carbonyl group **4.6** to produce the enolate **4.7**. The corresponding orbital picture is also shown (**4.8** and **4.9**), and it can be seen that in this conformation **4.8** the C–H σ orbital which 'becomes' part of the enolate π system is at the point of maximum overlap with the π orbital of the carbonyl group. In this generalized case deprotonation must therefore lead specifically to the enolate **4.7** in which group A is *cis* to the oxygen atom, rather than the enolate which has A *trans* to the oxygen atom. It follows that the conformation of **4.6** during deprotonation is likely to be important in determining the geometry of the enolate, or the geometrical composition of the enolate mixture.

The reaction of an electrophile (El) with enolate **4.7** results in the formation of a new σ bond in place of the original C–H bond which was broken. This involves the conversion of the 'p orbital' at the C-terminus of the enolate π system back into a σ orbital. Arguments analogous to those used above lead to the expectation that formation of the new σ bond, like the cleavage of the C–H σ bond (Fig. 4.2), takes place with 'maximum orbital overlap' (Fig. 4.3). In principle this could take place on *either* face of the enolate, and if the enolate is not chiral then attack on either face is equally likely. One of the functions of the chiral auxiliary is to direct reaction to one face of the enolate, as shown in Fig. 4.3.

Alkylation might therefore be expected to take place perpendicular to the

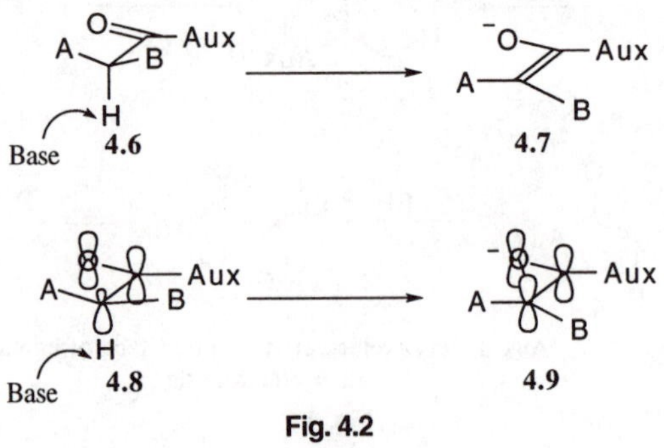

Fig. 4.2

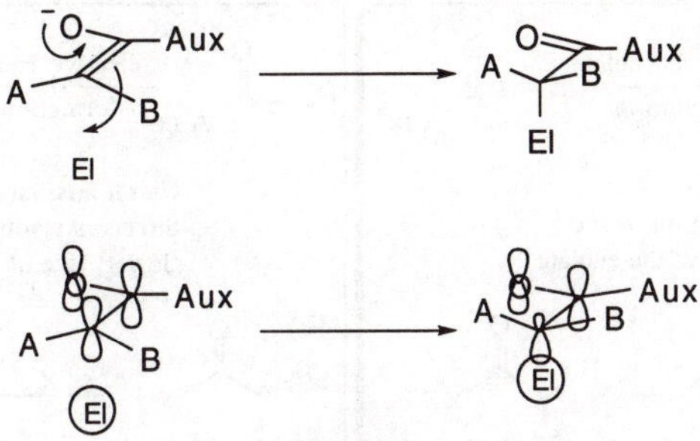

Fig. 4.3

plane of the enolate.[3] A more detailed analysis of the orbitals involved leads to the prediction that the electrophile should approach the enolate along a trajectory somewhat displaced from the 'vertical', due to an antibonding interaction between the orbital on the incoming electrophile and the oxygen p orbital in the HOMO of the enolate.[4] Two views of this are represented in Fig. 4.4. According to this idealized representation the electrophile's trajectory is displaced away from the 'perpendicular', *towards* the chiral auxiliary, which is likely to accentuate steric effects from this part of the molecule.

It is clear from the above that both deprotonation and alkylation α to a carbonyl group are likely to be subject to geometrical constraints as a result of stereoelectronic effects. These effects are important in understanding the stereochemistry of enolate alkylation (and aldol reactions) controlled by known chiral auxiliaries, and in the design of new ones.

Given that highly stereoselective enolate formation is possible, and that the alkylation of such an enolate has geometric constraints, all that remains in order to carry out a successful asymmetric synthesis is to use a chiral auxiliary which imposes a strong facial bias on the alkylation.

Detailed discussion of selected chiral auxiliaries follows, but the overall general strategy for the alkylation of an enolate controlled by a chiral auxiliary is shown in Fig. 4.5. In this idealized scheme it is supposed that one geometrical isomer of the enolate can be formed exclusively, and that the chiral auxiliaries direct all reaction to the 'upper' (Aux^a) or 'lower' (Aux^b) face of the enolate. Removal of the chiral auxiliary without loss of stereochemical integrity α to the carbonyl group would then provide either enantiomer of the alkylation product.

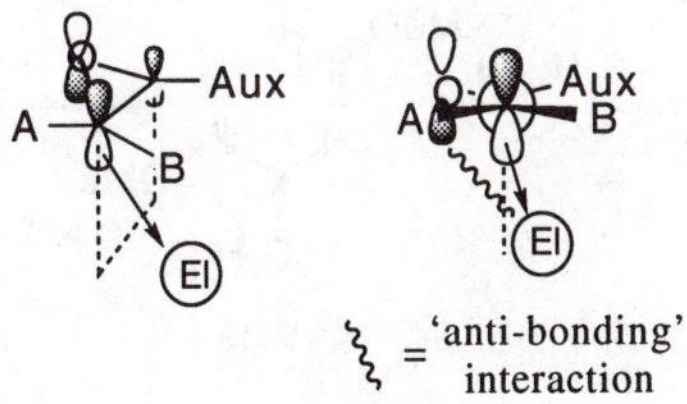

$\}$ = 'anti-bonding' interaction

Fig. 4.4

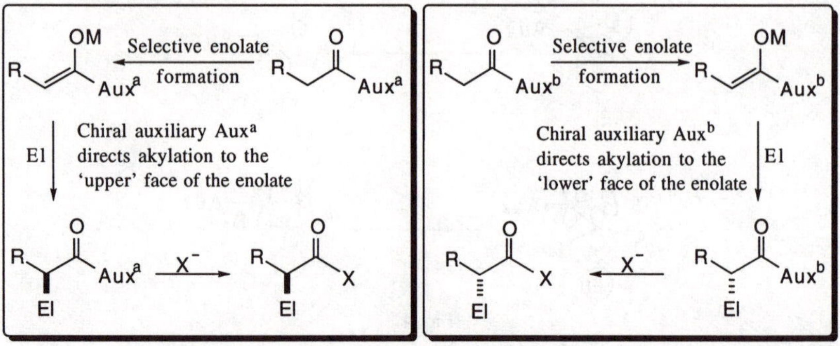

Fig. 4.5

A straightforward example of this type of asymmetric synthesis is the enantioselective synthesis of α-amino acids which uses (*S*)-valine, a naturally occurring α-amino acid, as the chiral auxiliary.[5] In this process (Fig. 4.6) the chiral auxiliary, (*S*)-valine **4.10**, is condensed with glycine **4.11**, the carbonyl compound which is to undergo stereoselective alkylation.

Treatment of the product **4.12** with trimethyloxonium tetrafluoroborate provides the bis-imino ether **4.13**, which is enolized with LDA at the less hindered position, the methylene of the glycine residue. The geometry of this enolate **4.14** is fixed by the six-membered ring, and the isopropyl group of the valine residue imposes a strong facial bias by hindering the 'lower' face of the enolate, thereby directing the electrophile to the 'upper' face. Hydrolysis then provides the esters of the desired alkylated glycine derivative **4.15**, and of the chiral auxiliary. In the example given, the diastereoselectivity of the alkylation

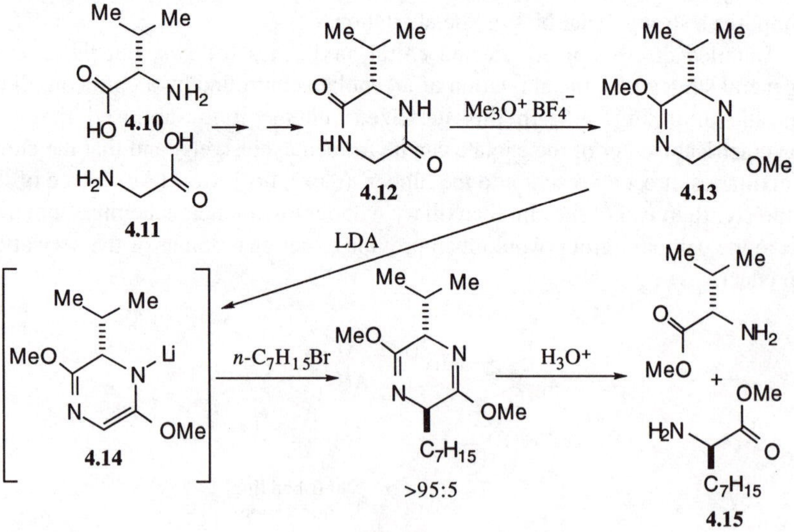

Fig. 4.6

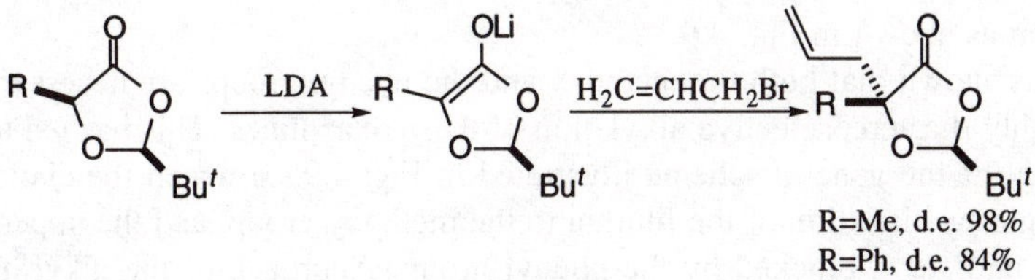

Fig. 4.7

is >95:5; other electrophiles react with similar stereoselectivities.

Another example of the highly diastereoselective alkylation of a chiral enolate whose geometry is fixed by a ring is the alkylation shown in Fig. 4.7.[6] In this case the ring is five-membered, and the facial bias is provided by the *t*-butyl group.

In both of the preceding examples only one enolate geometry is possible due to the constraints imposed by it being part of a ring. An alternative approach is to use chelation effects to constrain enolate geometry. Indeed, several important and generally useful methods for asymmetric synthesis via the alkylation of chiral enolates use just this approach. Here the chiral auxiliary constrains enolate geometry through chelation of the counter-ion as well as providing a high facial bias.

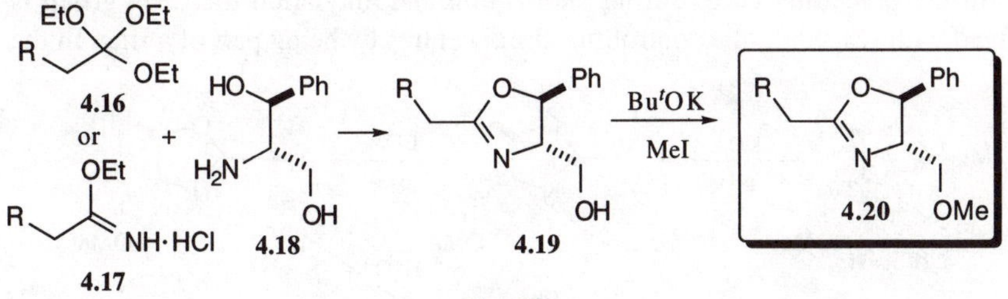

Fig. 4.8

Carboxylic acid derivatives **4.16** or **4.17** react with the commercially available amino diol **4.18** to produce chiral, non-racemic oxazolines of general structure **4.19** (Fig. 4.8), which are then methylated as shown. The resulting oxazolines **4.20** undergo several important reactions including highly diastereoselective alkylation.[7]

The level of stereoselectivity of enolate formation depends on the base used, with LDA providing the enolate in highest stereochemical purity.[8] The predominant isomer is thought to be that in which the group R^1 is *trans* to the

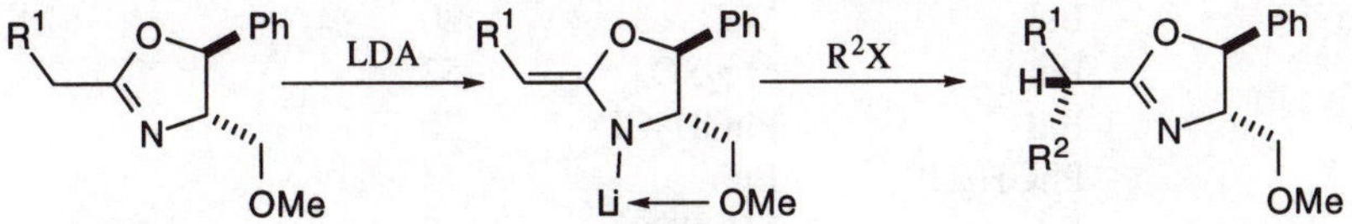

Fig. 4.9

nitrogen as shown in Fig. 4.9.

It was shown that both the methoxy and the phenyl groups are necessary for the highly diastereoselective alkylation of these oxazolines. This has led to the proposal of the general scheme illustrated in Fig. 4.9, in which the enolate is held rigid by chelation of the lithium to the methoxy group, and the upper face of this enolate is blocked by the phenyl group. Approach of the alkyl halide from the lower face results in alkylation with the stereochemistry shown (Fig. 4.9).[9]

A general route to enantiomerically enriched carboxylic acids is presented in Fig. 4.10, in which two alkyl groups are introduced by sequential alkylation. The order of their introduction determines the absolute configuration of the final product. From the selected examples given in Table 4.1 it can be seen that the enantiomeric excess can also depend on the order of alkylation. Removal of the chiral auxiliary by hydrolysis with acid is also illustrated in Fig. 4.10, and the enantiomeric excesses given in Table 4.1 refer to the disubstituted carboxylic acid obtained by hydrolysis.

This oxazoline methodology was used in the total synthesis of a pheromone of the European pine sawfly (Fig. 4.11). The chiral, non-racemic bromide **4.21** was prepared from the corresponding carboxylic acid. Further reactions provided the desired natural product.[10]

In the preceding case, during enolization and alkylation the C=N group is fixed with respect to the controlling chiral centres by being part of a ring. In the

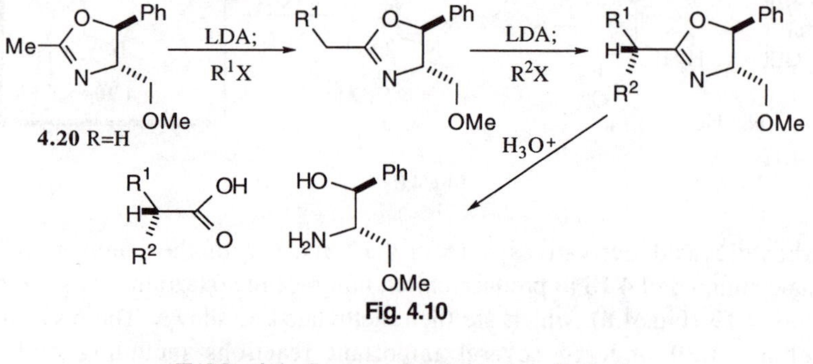

Fig. 4.10

Table 4.1 Stereoselectivity in alkylation of **4.20** R=H

R^1X	R^2X	e.e. (%)
MeI	BuI	66
BuI	MeI	20
BuI	Me_2SO_4	52
BuI	$PhCH_2Cl$	70
$PhCH_2Cl$	BuI	73
EtI	Me_2SO_4	70

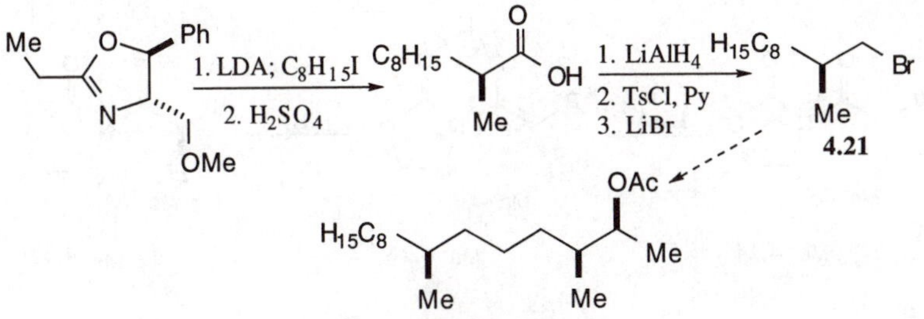

Sawfly pheromone

Fig. 4.11

more general case of deprotonation and reaction adjacent to a C=O group, incorporation of the C=O group into a ring is not possible. However, it *is* possible to incorporate these atoms in a 'temporary ring' by coordination of the metal ion of the enolate to the chiral auxiliary.

Enolates which behave in this way can be formed from acyl derivatives of the oxazolidinones **4.22** and **4.23** (Fig. 4.12) which are highly effective and useful chiral auxiliaries for the alkylation of enolates.[11] These chiral auxiliaries are easily prepared from readily available 1,2-aminoalcohols, and their anions react readily with acid chlorides to provide the acylated oxazolidinones needed for alkylation (Fig. 4.12).[12]

Treatment of the acylated oxazolidinones with base, usually LDA, produces a chelated enolate with >99:1 selectivity for the geometrical isomer shown in Fig. 4.13.

Taking account of the stereoelectronic requirement for deprotonation discussed previously (Fig. 4.2), and the likelihood that the group R will occupy a position which reduces steric interactions with the rest of the molecule, two conformations which retain the planarity of the N–C=O system are likely for deprotonation, **4.24** and **4.25** (Fig. 4.13).

Both of these would lead to the observed enolate geometry, although the

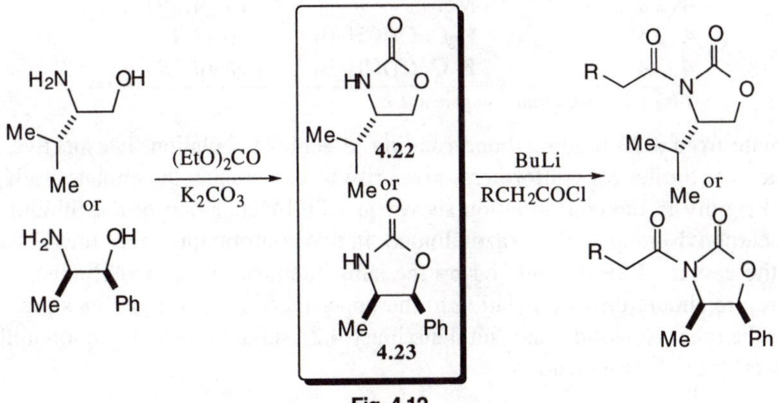

Fig. 4.12

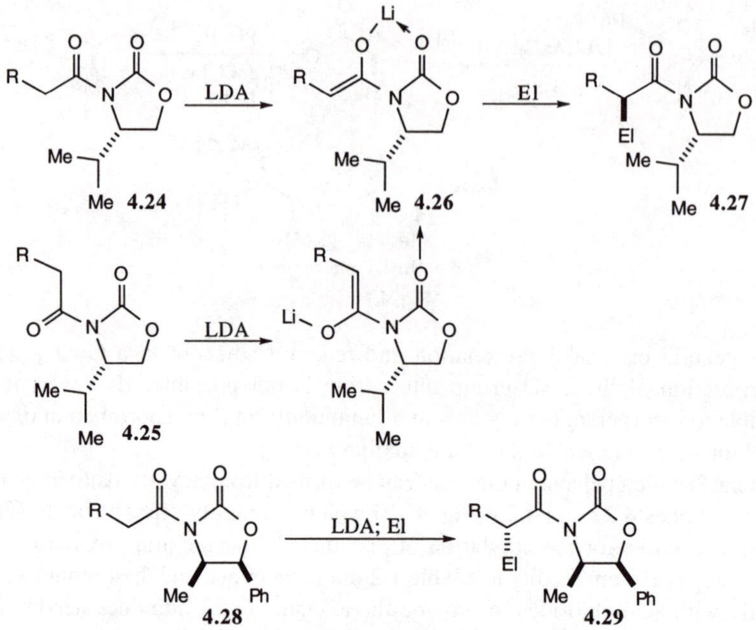

Fig. 4.13

Table 4.2 Stereoselectivity in the alkylation of oxazolidinones **4.24** and **4.28** R=Me

Oxazolidinone	El	d.e. (%)[a]
4.24	EtI	92 (**4.27**)
4.28	EtI	76 (**4.29**)
4.24	PhCH$_2$Br	98 (**4.27**)
4.28	PhCH$_2$Br	96 (**4.29**)
4.24	MeI	80 (**4.27**)
4.28	MeI	74 (**4.29**)
4.24	H$_2$C=CHCH$_2$Br	96 (**4.27**)
4.28	H$_2$C=CHCH$_2$Br	96 (**4.29**)

[a]Major diastereoisomer in parentheses.

enolate from **4.25** requires bond rotation to achieve chelation. Irrespective of which conformer (or conformers) gives rise to the enolate, the enolate itself is held rigidly in the conformation shown in **4.26** by chelation of the lithium to the carbonyl group of the oxazolidinone. In this conformation, the 'lower' face of the enolate **4.26** is hindered by the substituent on the oxazolidinone, and therefore should direct alkylation to the upper face. Similar arguments predict that the other oxazolidinone chiral auxiliary **4.23** should direct alkylation to the 'lower' face of the enolate.

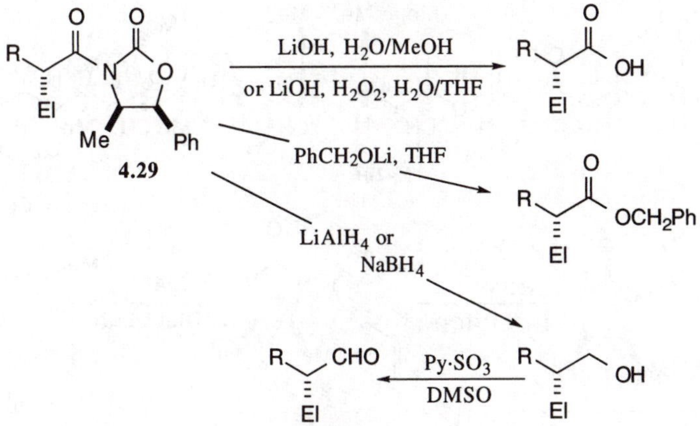

Fig. 4.14

In practice, both these predictions are upheld, and very high levels of diastereoselectivity are observed in the alkylation of this type of acylated oxazolidinone.[13] As expected, alkylation of **4.24** gives the stereochemistry shown in **4.27**, and **4.28** gives **4.29**.[14] Representative examples are provided in Table 4.2.

Not only is the alkylation of these acylated oxazolidinones highly diastereoselective, but the major diastereoisomer is usually easy to separate from the minor by conventional purification techniques, such as column chromatography or crystallization.

Removal of the oxazolidinone chiral auxiliary after alkylation and purification (if necessary) may be achieved by hydrolysis, alcoholysis, or reduction. In most cases reaction takes place at the desired carbonyl group, but in cases where the system is very hindered, reaction at the oxazolidinone carbonyl group is sometimes observed. In these cases hydrolysis using basic hydrogen peroxide often brings about the desired hydrolysis.[15] These possibilities are illustrated in Fig. 4.14. In all cases the chiral auxiliary is removed without loss of stereochemical purity in the product. Fig. 4.14 demonstrates how this oxazolidinone methodology can be used to prepare chiral, non-racemic carboxylic acids, esters, alcohols, and aldehydes.

The chemistry of oxazolidinones discussed so far, combined with aldol reactions controlled by the same chiral auxiliaries (see Chapter 5), has formed the cornerstone of several total syntheses of complex, naturally occurring organic compounds, such as the ionophore antibiotic X-206.[16] In addition to providing examples of the use of the oxazolidinone methodology discussed so far, this total synthesis will be used to illustrate the uses of asymmetric aldol reactions (Chapter 5) and of catalytic asymmetric epoxidation (Chapter 7). Indeed, it is the successful application of synthetic methods in such enantioselective syntheses that demonstrate their usefulness. If the methodology can provide sufficient material of high enantiomeric purity to allow the

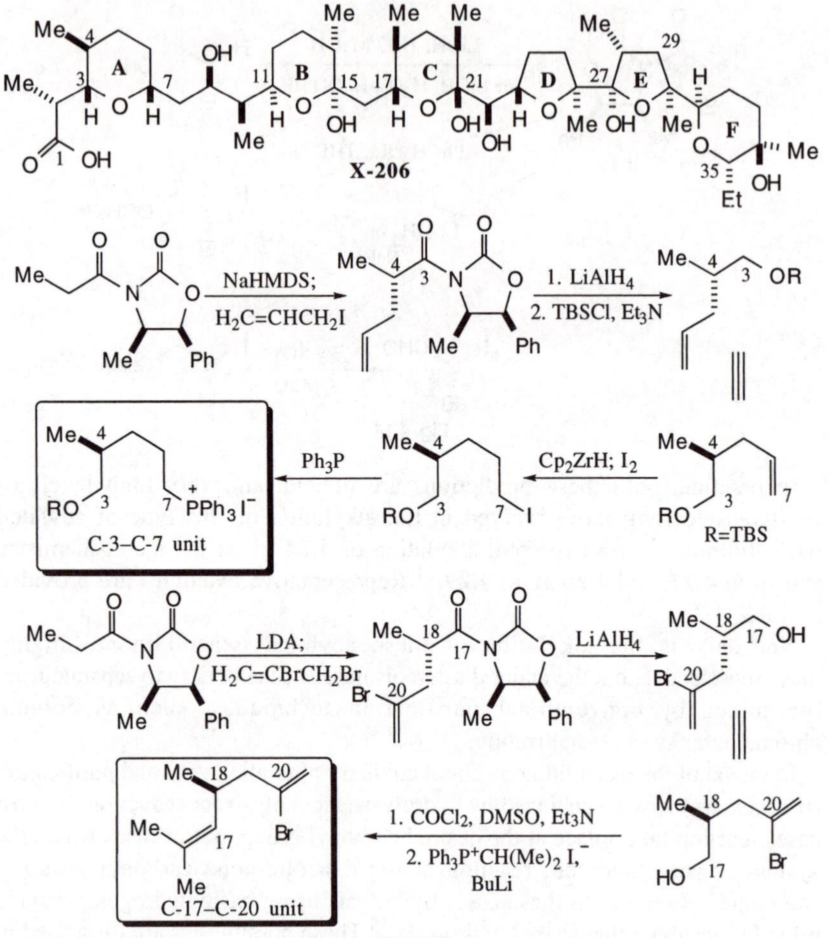

Fig. 4.15

completion of such a synthesis, then it should certainly be applicable to less complex targets.

The discussion of the complete synthesis of X-206 is beyond the scope of this volume, but the preparation of two important units in the synthesis which involves alkylation of the chiral auxiliary **4.23** is shown in Fig. 4.15.

In the examples discussed so far, the electrophile has been an alkyl halide, and while such alkylation represents a fundamental, stereoselective method for C–C bond formation, other electrophilic reagents can be used to useful effect. Three types of reactions of acylated oxazolidines which have been developed to a synthetically useful level are enolate hydroxylation, amination, and bromination.

Diastereoselective hydroxylation of these enolates is best accomplished by reaction of the sodium enolate of the appropriate acylated oxazolidinone with the

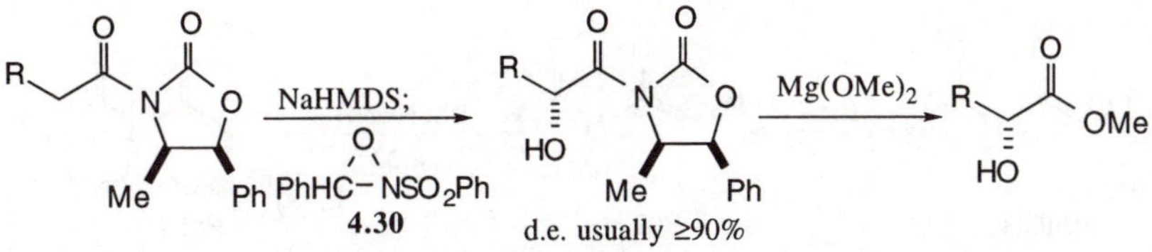

Fig. 4.16

oxaziridine **4.30**.[17] While the optimal experimental conditions for hydroxylation are somewhat different to those for alkylation, the reaction still provides an extremely effective and highly diastereoselective asymmetric synthesis of α-hydroxy acid derivatives (Fig. 4.16).

The bromination of acylated oxazolidinone was studied as part of a general approach to the synthesis of α-amino acids. In this case the boron enolate **4.31** was treated with NBS to provide the corresponding bromide **4.32**, the stereochemistry of which was analogous to that of enolate alkylation of these systems. Azide displacement of the bromide took place with inversion to provide azide precursors to protected α-amino acid derivatives (Fig. 4.17).[18]

An alternative route to α-amino acid derivatives would be the direct introduction of the nitrogen substituent by reaction of the enolate with an electrophilic aminating reagent. Two such reagents have proved successful, di-*t*-butyl azodicarboxylate **4.33**[19] and triisopropylbenzenesulphonyl azide **4.34**, the latter being the reagent of choice (Fig. 4.18).[17] In these amination reactions, consistently high diastereoselectivities are observed, and the chiral auxiliary can be removed either by hydrolysis or by ester exchange.

In the case of the azides, the products from these cleavage reactions, α-azido acids or esters, are excellent masked α-amino acid equivalents as the azide group is relatively stable, but is readily reduced to the desired amino function when required.

Enolization and subsequent azidation of this type of oxazolidine can be highly chemoselective as well as stereoselective. The oxazolidinone **4.35**, on

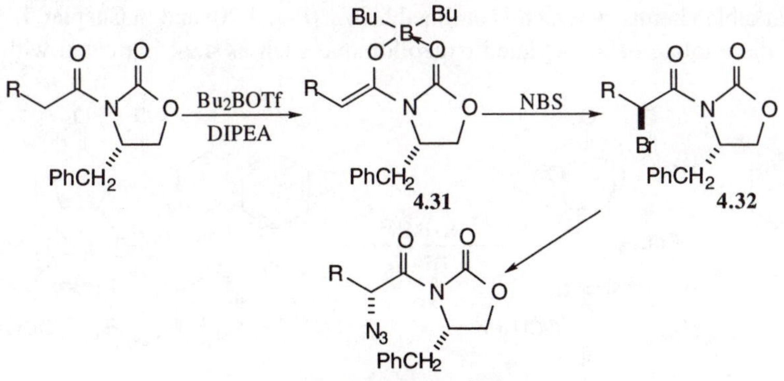

d.e. usually ~90%

Fig. 4.17

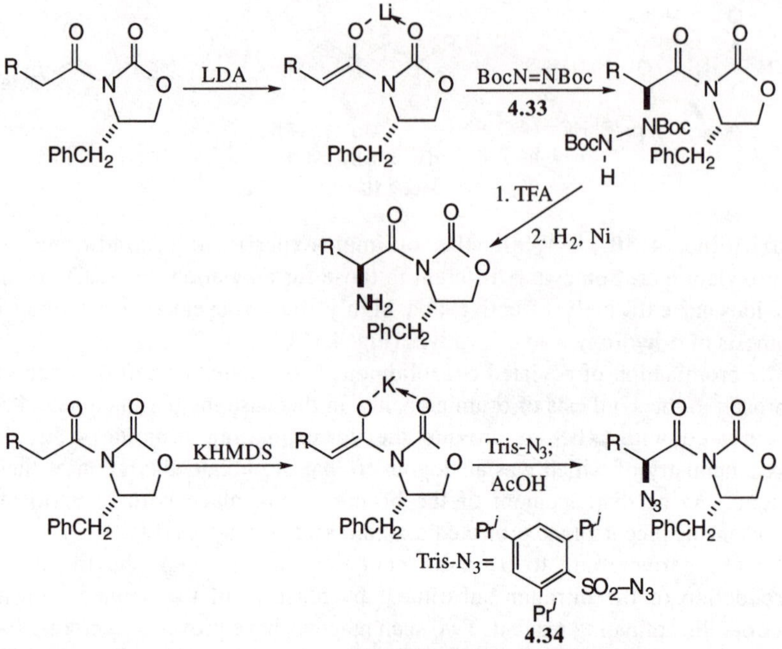

Fig. 4.18

treatment with two equivalents of KHMDS (one each for the NH and the methylene proton), followed by tri*iso*propylbenzenesulphonyl azide, gave the azido compound **4.36** in 85 per cent yield and with 97:3 stereoselectivity, in spite of the complexity of the substrate (Fig. 4.19).

One limitation of these acylated oxazolidinone chiral auxiliaries is that it is not possible to replace *both* methylene protons with substituents. The stereoselective alkylation of systems such as **4.37** (Aux = oxazolidinone chiral auxiliary, R^1, $R^2 \neq$ H) is not possible. Nevertheless the relative inertness of this proton (H*, Fig. 4.20) has been exploited and provides the basis for some remarkable chemistry which is outlined below (Fig. 4.20) and in Chapter 5.

If the enolate of an acylated oxazolidinone such as **4.24** is treated with an

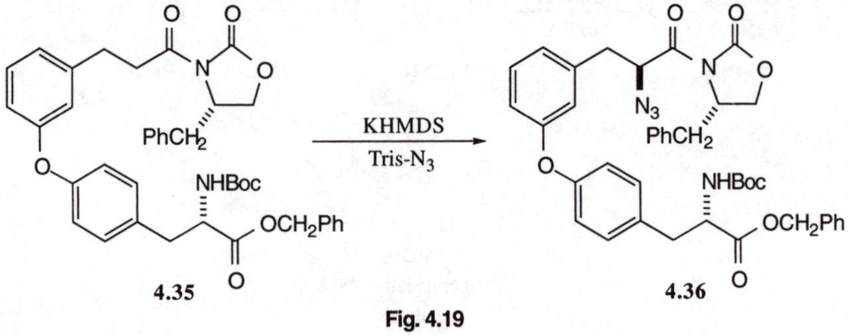

Fig. 4.19

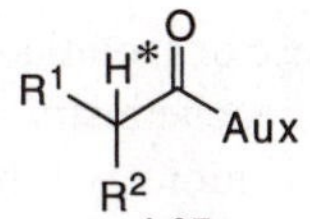

4.37

Aux = oxazolidinone chiral auxiliary

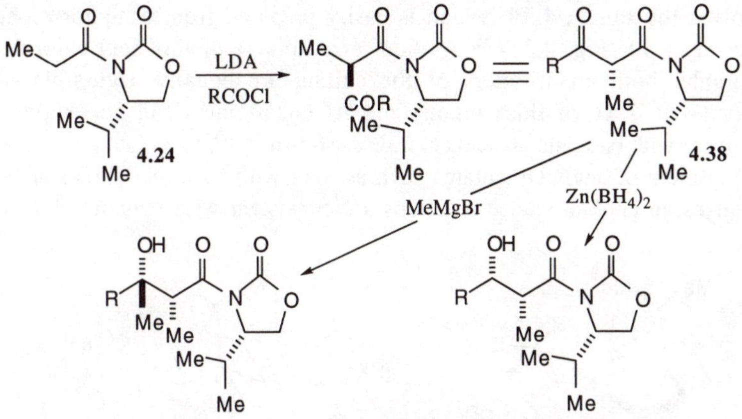

Fig. 4.20

acid chloride, the corresponding β-dicarbonyl compound **4.38** is produced.[20] These β-dicarbonyl compounds are relatively difficult to enolize and, remarkably, the chiral centre between the two carbonyl groups is configurationally stable under a range of conditions. This allows for further reactions to be carried out which would be impossible to achieve in normal β-dicarbonyl compounds without loss of stereochemical integrity. These reactions include reduction, nucleophilic addition, (Fig. 4.20), and aldol reactions[21] (Chapter 5).

The oxazolidinone chiral auxiliaries discussed in this chapter are clearly important in controlling the stereochemical outcome of enolate alkylation and

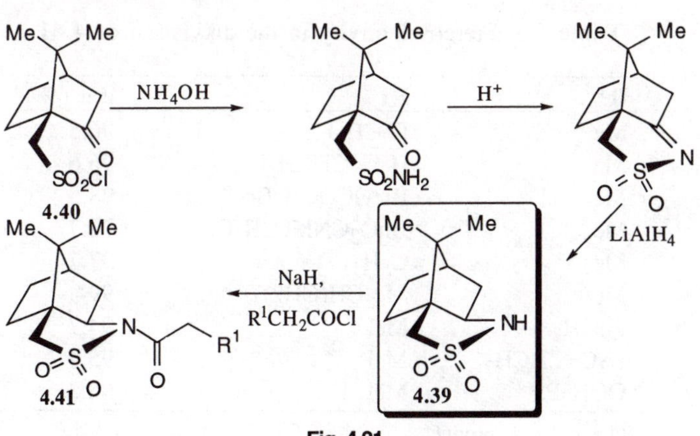

Fig. 4.21

related reactions. Moreover, these oxazolidinone chiral auxiliaries are also useful for controlling the absolute stereochemistry in aldol reactions and Diels–Alder cycloadditions, and these applications will be considered in Chapters 5 and 6 respectively.

A more recent example of enolate alkylation controlled by a chiral auxiliary involves the sultam **4.39**, which is easily prepared from camphorsulphonyl chloride **4.40** (Fig. 4.21).[22] As both enantiomers of this acid chloride are available, both enantiomers of the sultam are equally accessible. Acyl derivatives **4.41** of this sultam (and its enantiomer) can be obtained by treatment with base and an acid chloride as shown.

Treatment of acylated sultams such as **4.41** with base (NaHMDS or BuLi) produces an enolate whose reactions are consistent with structure **4.42** (Fig.

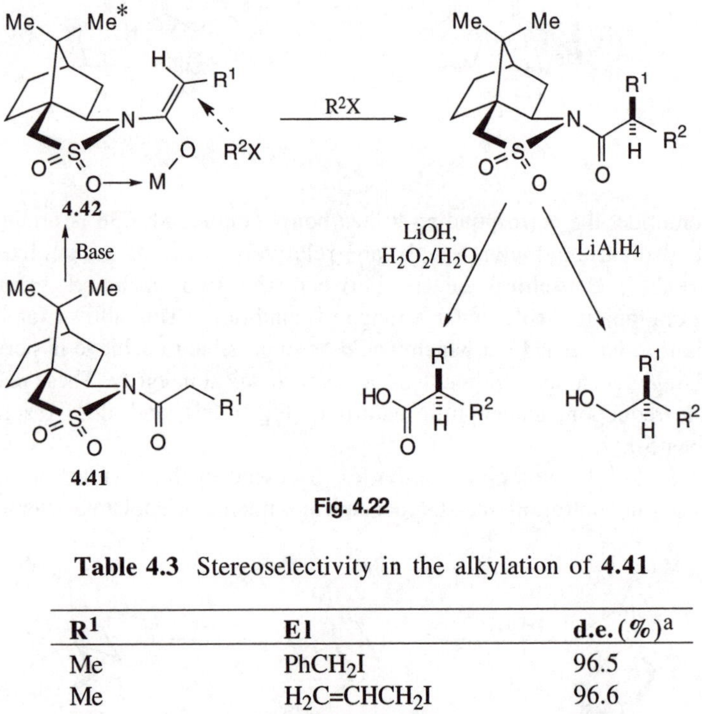

Fig. 4.22

Table 4.3 Stereoselectivity in the alkylation of **4.41**

R^1	El	d.e. (%)[a]
Me	$PhCH_2I$	96.5
Me	$H_2C=CHCH_2I$	96.6
Me	Bu^tOCOCH_2Br	98.5
Me	$BnO_2CNMeCH_2Cl$	72.7
Me	C_5H_{11}	97.7
Me	$Me_2CH(CH2)_3$	99
$PhCH_2$	MeI	94.7
$H_2C=CHCH_2$	MeI	95.4
OCH_2Ph	MeI	98.2

[a]d.e. of crude product.

4.22). Alkylation of this enolate then takes place from the lower face as the upper face is hindered by the methyl group Me* (**4.42** Fig. 4.22).[23] Some examples of this alkylation are given in Table 4.3 and in all cases shown the diastereoselectivity is high. The auxiliary can be removed either by hydrolysis to provide the alkylated carboxylic acid, or by reduction to the alcohol. In neither case is loss of stereochemical purity observed.

Interesting methodology which has been used to carry out highly stereoselective alkylations, and several other types of reactions, involves the use of the iron acyl **4.43** (Fig. 4.23).[24] This is available in both enantiomeric forms, although at the time of writing their cost effectively restricts the widespread use of these versatile systems. Complex **4.43** is essentially octahedral with all ligands occupying one site, except for the cyclopentadienyl (which occupies three), and the bond angles between the groups attached to the iron being approximately 90°. The preferred conformation of **4.43** is such that the oxygen of the acetyl group is *anti* to the carbon monoxide ligand, and a phenyl group of the triphenylphosphine ligand is shielding one face of the acetyl group, as represented in Fig. 4.23.

The acetyl group in **4.43** undergoes enolization on treatment with strong base, usually butyllithium, to produce an enolate which is very nucleophilic at carbon, and which reacts with a wide variety of electrophiles. The enolates **4.44** derived from these products **4.45** undergo highly diastereoselective alkylation with a variety of electrophiles (Fig. 4.23).[25]

The stereoselectivity of these alkylations is consistent with a model in which the oxygen atom of the enolate **4.44** is *anti* to the carbon monoxide ligand and the 'rear' face of the enolate is blocked by one of the phenyl groups of the

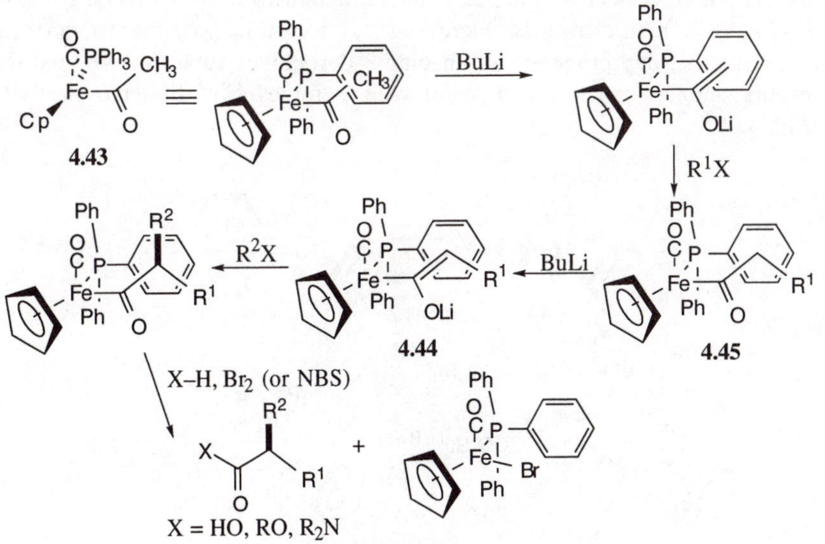

Fig. 4.23

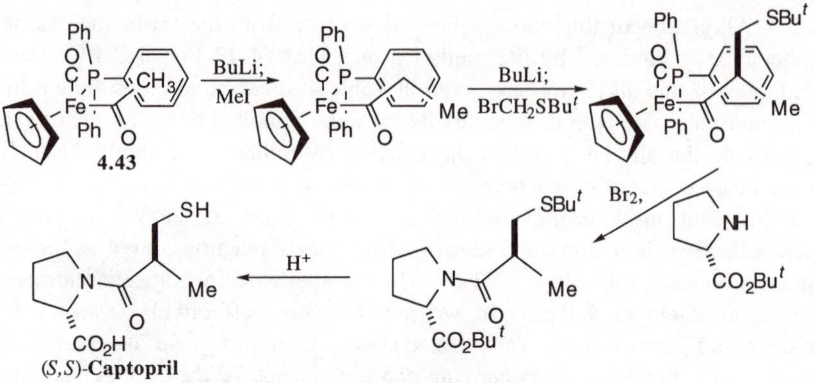

Fig. 4.24

triphenylphosphine, as in the carbonyl derivative itself. The level of diastereoselectivity of these alkylations is usually extremely high. The chiral auxiliary can be cleaved easily by various one-electron oxidants (bromine, NBS, cerium(IV), Fe(III)), with bromine or NBS being the reagents of choice. Deprotection in the presence of nucleophiles such as water, alcohols, or amines provides the corresponding carboxylic acid, ester, or amide respectively. A simple example of this strategy is provided by the synthesis of (*S*,*S*)-captopril **4.46** from the (*R*)-enantiomer of the iron acyl (Fig. 4.24).[26]

Deprotonation of iron acyls such as **4.43** and **4.45** requires a strong base, as the acidity of the protons adjacent to the carbonyl group is considerably less than that of a typical ketone. The origin of this decrease in acidity is related in part to the large charge separation in the ground state as represented in Fig. 4.25, and is consistent with the very low i.r. frequency of the carbonyl group of **4.43**. As a result of this decreased acidity it is possible to perform regioselective deprotonation of succinoyl derivatives such as **4.46**, and the resulting enolate undergoes alkylation with useful levels of diastereoselectivity (Fig. 4.25).[27]

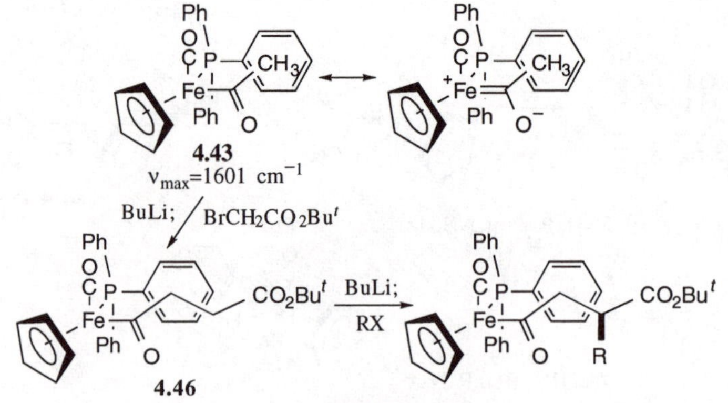

Fig. 4.25

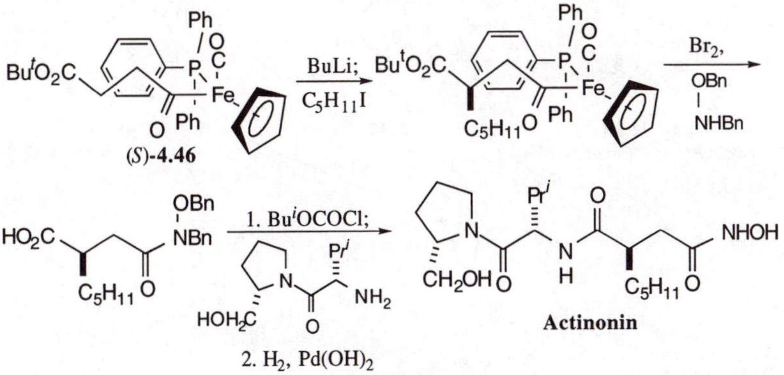

Fig. 4.26

The usefulness of this approach has been demonstrated by the asymmetric synthesis of actinonin, a compound with interesting anticollagenase activity (Fig. 4.26).[28]

The high diastereoselectivity observed in reactions of the enolates of acyl iron derivatives such as **4.43** and **4.45** has led to the development of asymmetric synthesis in which the chiral enolate reacts selectively with one enantiomer of a racemic electrophile (kinetic resolution, see Chapter 2). Racemic electrophiles which have been used in this approach include α-bromo esters and monosubstituted epoxides. Reaction with a racemic epoxide is illustrated in Fig. 4.27. The chiral enolate reacts much more rapidly with the (S)-enantiomer of the racemic epoxide in the presence of diethylaluminium chloride to give the lactone **4.47** after oxidative work-up.[29]

The ability to use racemic electrophiles and to react selectively with one enantiomer is clearly a strategy of significant potential in asymmetric synthesis, one which should find increasing applications as the scope of the chemistry is widened.

The chiral auxiliaries discussed so far have all been attached by acylation to the system undergoing reaction. The products from these reactions after chiral auxiliary removal will therefore be chiral, non-racemic carboxylic acid derivatives or alcohols, depending upon the method of chiral auxiliary removal. It would clearly be of considerable synthetic value to be able to achieve the enantioselective synthesis of *ketone* derivatives using chiral auxiliaries. Most of the rest of this chapter is devoted to this topic.

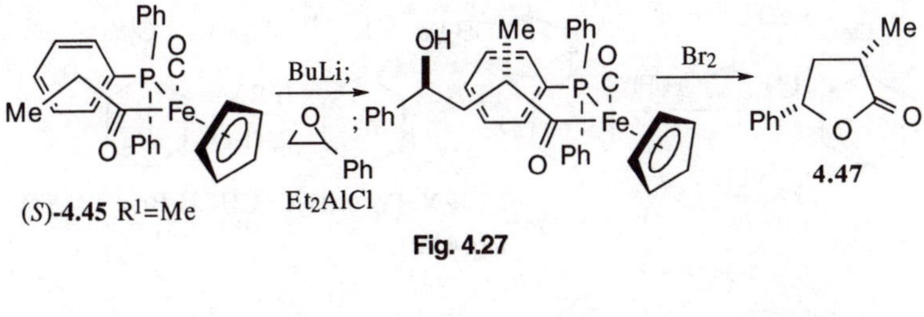

Fig. 4.27

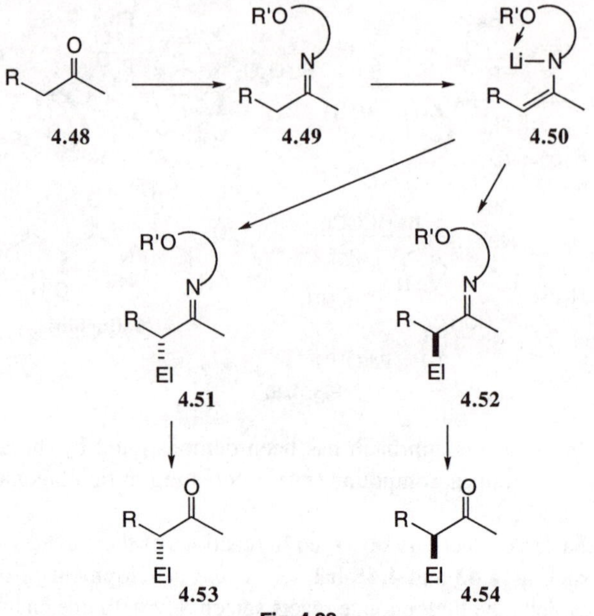

Fig. 4.28

The chiral auxiliaries which will be considered are all attached to the ketone through formation of a C=N bond from the C=O group, and are thought to control the stereochemistry of alkylation through chelation effects similar to those discussed previously in this chapter. The overall strategy is outlined in Fig. 4.28.

Reaction of the ketone **4.48** with a suitable chiral auxiliary produces **4.49**, which is treated with base to give the chelated lithium enolate **4.50**. If this enolate has a rigid conformation and the auxiliary provides a strong facial bias, then it would be expected that one of the two possible diastereoisomers **4.51** or **4.52** would predominate on reaction with an electrophile (El) (Fig. 4.28). Separation (if necessary) and removal of the chiral auxiliary would give either enantiomer **4.53** or **4.54**. As previously, there are stringent requirements which must be met to achieve a general, predictable, and practical method, and there are relatively few such methods currently available.

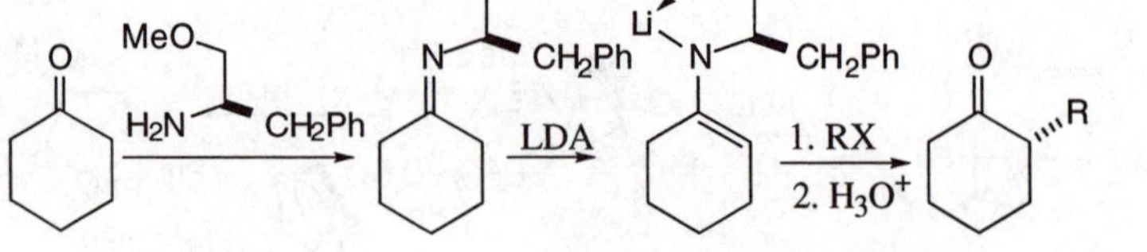

RX=PrI or H_2C=CHCH$_2$Br, e.e. >90%

Fig. 4.29

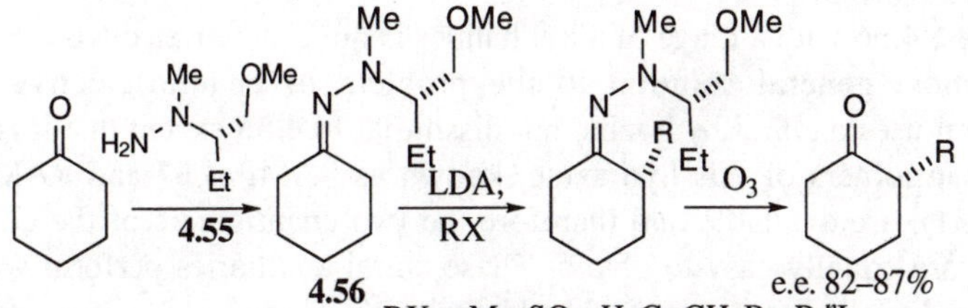

RX = Me_2SO_4, $H_2C{=}CH_2Br$, Pr^nI

Fig. 4.30

The alkylation of cycloalkanones has been carried out using a variety of acyclic, enantiomerically pure β-alkoxy amines as chiral auxiliaries.[30] A simple example of this approach is provided in Fig. 4.29, and although somewhat limited in scope, the enantiomeric excesses obtained can be high.

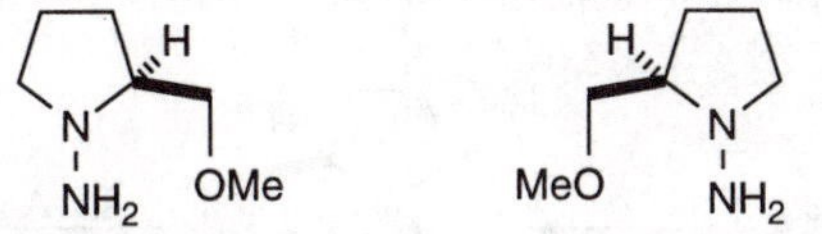

SAMP **4.57** RAMP **4.58**

Fig. 4.31

The reaction scheme shown in Fig. 4.30 uses the hydrazine **4.55** as the chiral auxiliary, and leads to alkylation products with useful levels of enantiomeric purity.[31] It is interesting that in this case, the asymmetric centre of the chiral auxiliary is one atom more distant from the site of alkylation than in the imine derivative illustrated in Fig. 4.28. In spite of this, the alkylation of

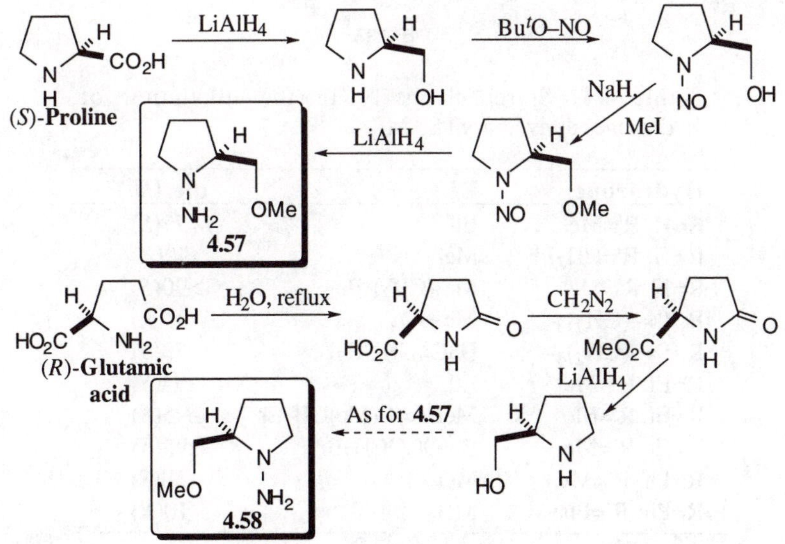

Fig. 4.32

hydrazone **4.56** with a range of alkyl halides is quite stereoselective.

The most general solution to the problem of enantioselective ketone alkylation uses a chiral hydrazine not dissimilar to **4.55** except that it is cyclic. Both enantiomers of this hydrazine, known as SAMP **4.57** and RAMP **4.58** (Fig. 4.31), are available and therefore the two enantiomers of the alkylation product are equally easy to obtain. These chiral auxiliaries perform well with acyclic and cyclic ketones, and with aldehydes.[32]

The hydrazines SAMP **4.57** and RAMP **4.58** may be prepared as outlined in Fig. 4.32, and both syntheses can be carried out on a relatively large scale (1 mole).[33] In the following discussion of this methodology SAMP **4.57** is used in most of the examples. Using RAMP **4.58** would simply produce enantiomeric products.

Reaction of SAMP with ketones (or aldehydes) provides the chiral, non-racemic hydrazones as mixtures of (*E*)- and (*Z*)-isomers (where appropriate)

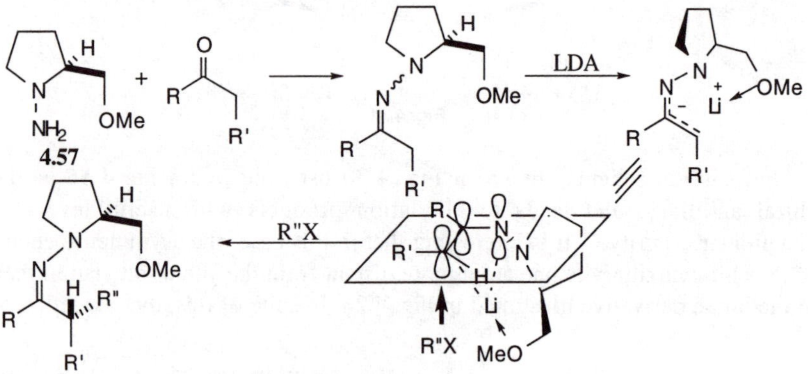

Fig. 4.33

Table 4.4 Stereoselectivity in the alkylation of hydrazones derived from SAMP

Hydrazone	El	d.e. (%)[a]
R=H, R'=Me	EtI	77(*S*)
R=H, R'=CH$_2$Ph	MeI	>90(*R*)
R=H, R'=Me	Me$_2$(CH$_2$)$_3$Br	>90(*S*)
R,R'=–(CH$_2$)$_3$–	Me$_2$SO$_4$	86(*R*)
R,R'=–(CH$_2$)$_4$–	H$_2$C=CHCH$_2$Br	73(*S*)
R=Et, R'=Me	EtI	94(*S*)
R=Et, R'=Me	MeHC=C(Me)CH$_2$Br	≥95(*S*)
R=Et, R'=Me	Bu^{t}OCOCH$_2$Br	≥95(*S*)
R=Ph, R'=Me	MeI	30(*S*)
R=Ph, R'=Ph	MeI	10(*R*)

[a]Absolute configuration of major enantiomer in parentheses.

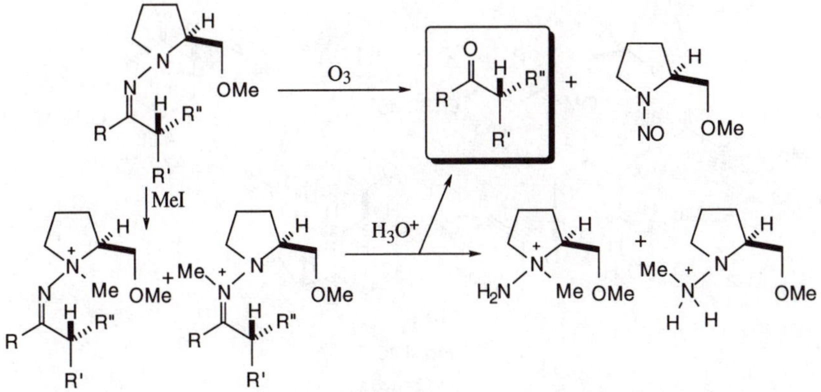

Fig. 4.34

which are then deprotonated (Fig. 4.33). The structure proposed for the lithiated SAMP hydrazone is illustrated in Fig. 4.33, and alkylation is thought to take place from the 'lower' face as shown by the arrow. Representative examples of the preparation of enantiomerically enriched carbonyl compounds by the alkylation of various types of SAMP hydrazones are given in Table 4.4.[34]

Detailed mechanistic studies support this overall scheme, and by comparison with related hydrazines it has been demonstrated that the methoxy group is important for high stereoselectivity to be obtained.[35] This is probably a chelation effect analogous to that observed in the oxazolines discussed previously in this chapter.

Removal of the chiral auxiliary is usually achieved either by *in situ* acid hydrolysis after reaction of the alkylated hydrazone with methyl iodide, or by ozonolysis (Fig. 4.34).[36] In both cases little or no loss of stereochemical purity is observed and consequently the enantiomeric excess of the product is the same as the diastereoisomeric excess of the alkylation reaction. The percentage enantiomeric excess given in Table 4.4 reflects, therefore, the level of stereoselectivity of the original alkylation of the SAMP hydrazone.

As with the other methods considered in this chapter, it is not surprising that this highly selective asymmetric synthesis of alkylated carbonyl compounds finds application in the enantioselective synthesis of natural products.

A simple example of such a synthesis is the enantioselective preparation of the defence substance used by 'daddy-long-legs' spiders (*Leioburnum vittatum and L. calcar*) **4.59**.[37] Both enantiomers were prepared using either SAMP or

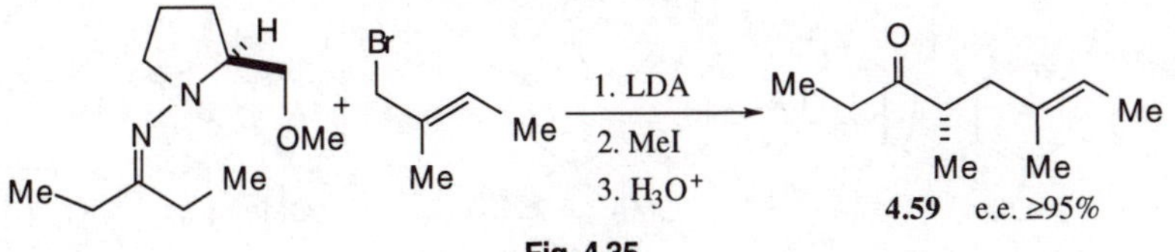

Fig. 4.35

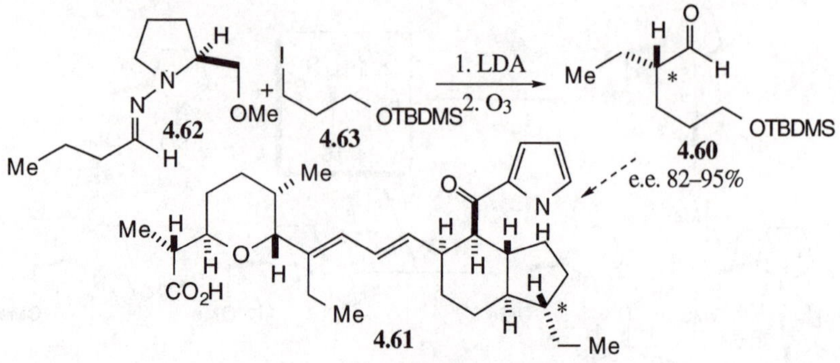

Fig. 4.36

RAMP in order to determine the absolute configuration of the natural substance. The sequence used is illustrated in Fig. 4.35.

This hydrazone alkylation methodology has also been used to prepare enantiomerically enriched units for the total synthesis of more complex natural products. For example, the aldehyde **4.60** (Fig. 4.36), used to construct the 'right-hand half' of the ionophore antibiotic X-14547A **4.61**, was prepared by alkylation of the SAMP hydrazone **4.62** with iodide **4.63**.[38]

α-Substitution using chiral enolates clearly provides a powerful approach to the asymmetric synthesis of a range of useful systems. Almost all the examples discussed so far involve the use of chiral auxiliaries attached to the enolate. It is possible to achieve enantioselective α-substitution using prochiral enolates in relatively few cases, and this chapter concludes with selected examples although these do not fall strictly under its title. Reaction of a prochiral enolate with a chiral, non-racemic electrophilic reagent provides a general strategy, which is particularly successful for hydroxylation and amination.

In both these cases the reagent is based on the camphor framework. For asymmetric hydroxylation, oxaziridines such as **4.64**, **4.65**, and **4.66** are

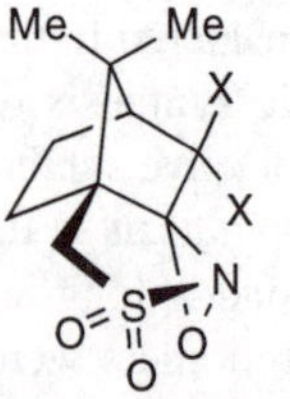

4.64 X = H **4.65** X = Cl **4.66** X = OMe

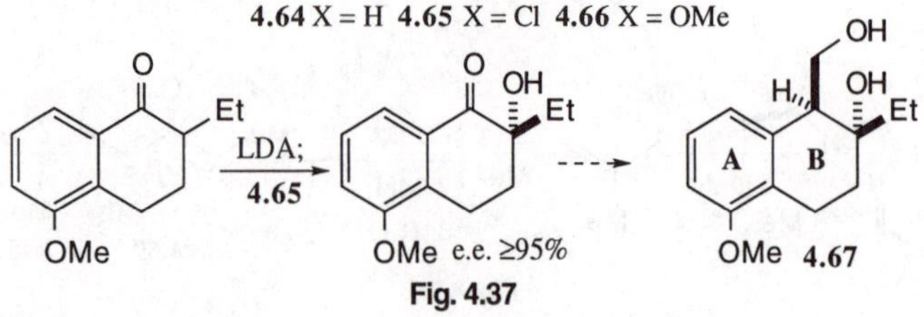

Fig. 4.37

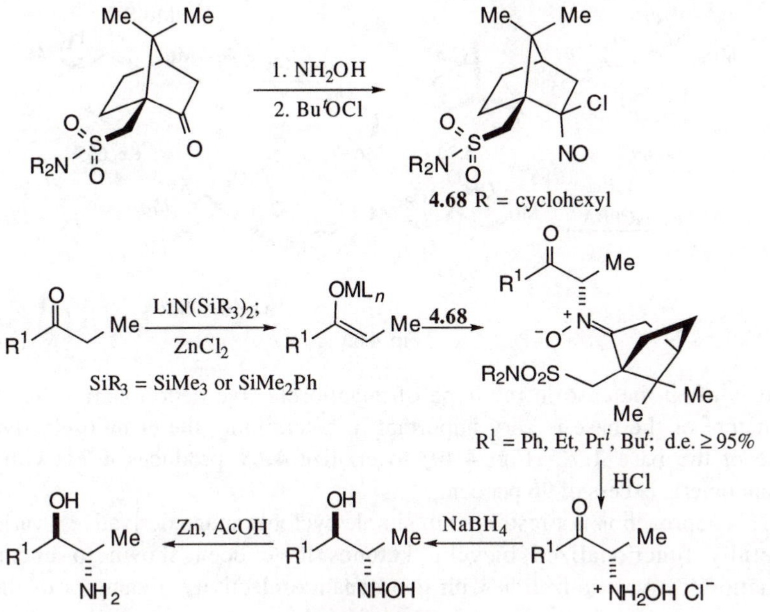

Fig. 4.38

particularly effective and both enantiomers are readily available.[39] This approach has been used in a synthesis of **4.67**, a unit representing the AB-ring of aklavinone (Fig. 4.37).[40]

The enantioselective aminating reagent **4.68** is easily prepared as shown, and undergoes highly stereoselective reaction with various metal enolates, the products of which can be converted into *anti*-1,2-aminoalcohols (Fig. 4.38).[41]

Another opportunity for achieving asymmetric α-substitution is to generate a chiral enolate from a prochiral (or in some cases racemic) ketone by using a chiral, non-racemic base. At present the highest enantiomeric excesses are usually obtained by *in situ* trapping of the enolate as the silyl enol ether followed by subsequent reaction. Although in its relatively early stages, there are signs that in the appropriate circumstances this approach can be an effective and efficient method for asymmetric synthesis,[42] as illustrated by the following selected examples.

Treatment of ketone **4.69** with the base **4.70** and chlorotrimethylsilane produces the enol ether **4.71** in reasonable enantiomeric excess (Fig. 4.39).[43] In effect the chiral base is discriminating between H_A and H_B, because the transition states for the removal of these protons are diastereoisomeric, and in this case removal of H_A is more rapid. Conversion of silyl enol ether **4.71** into the corresponding α-hydroxy ketone takes place without lowering the enantiomeric purity, and this ketone could then be converted into aeginetolide (Fig. 4.39).

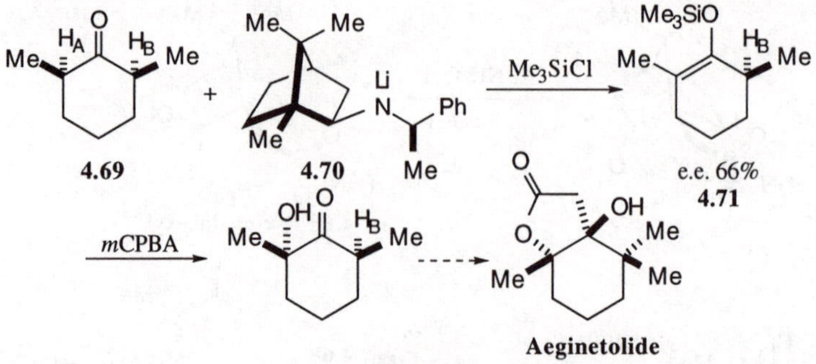

Fig. 4.39

It is often the case in this type of enantioselective deprotonation that the structure of the base is very important in determining the enantioselectivity. Use of the base **4.72** (Fig. 4.40) to enolize **4.69** produces **4.71** with an enantiomeric excess of 96 per cent.[44]

This approach is not restricted to simple cyclohexanone derivatives; various usefully functionalized bicyclic ketones have been shown to undergo enantioselective enolization with good enantioselectivity. Examples of these

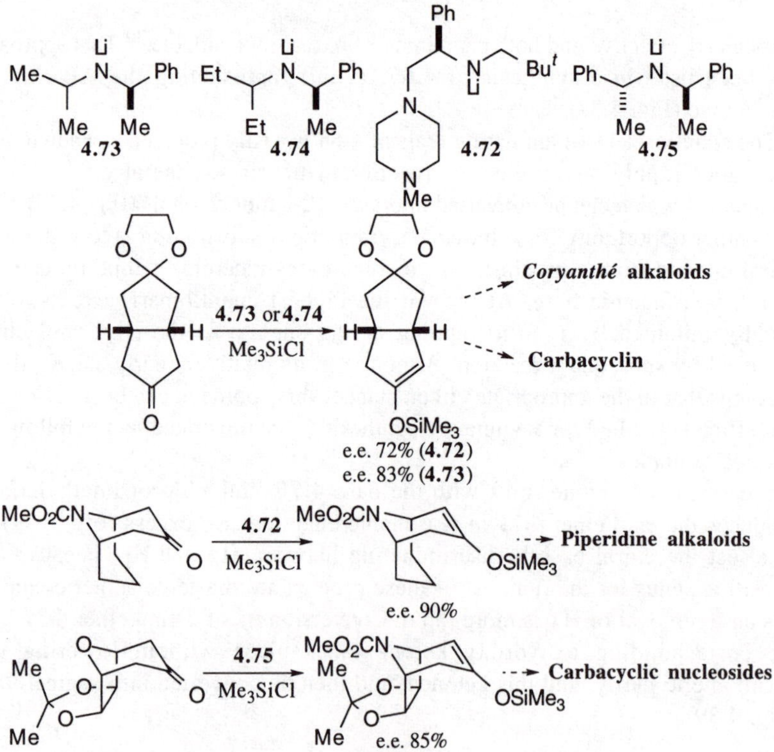

Fig. 4.40

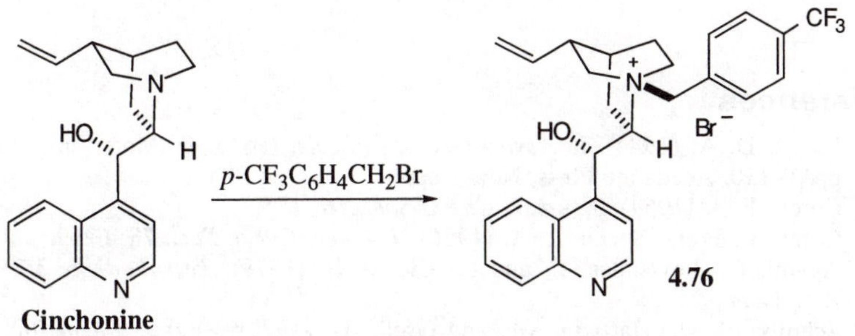

Cinchonine

Fig. 4.41

systems and the various bases used are provided in Fig. 4.40. The products from these reactions are precursors to a range of synthetic targets including the *Coryanthé* alkaloids,[45] prostaglandin analogues,[46] piperidine alkaloids,[47] and carbacyclic nucleosides.[48]

Enantioselective α-substitution using asymmetric catalysis is rather difficult to achieve. Nevertheless, some remarkable results have been achieved by the use of chiral, non-racemic quaternary ammonium salts such as **4.76** as phase transfer catalysts. Catalyst **4.76** is easily prepared from commercially available cinchonine (Fig. 4.41).[49]

Alkylation of the indanone **4.77** with methyl chloride under phase transfer conditions in the presence of 10 mole per cent of **4.76** gives the methylated product in very high enantiomeric excess (Fig. 4.42).[49] Closely related phase transfer catalysts have also been used to effect the enantioselective alkylation of glycine imine derivatives, although the enantiomeric excesses are somewhat lower (42–66 per cent).[50]

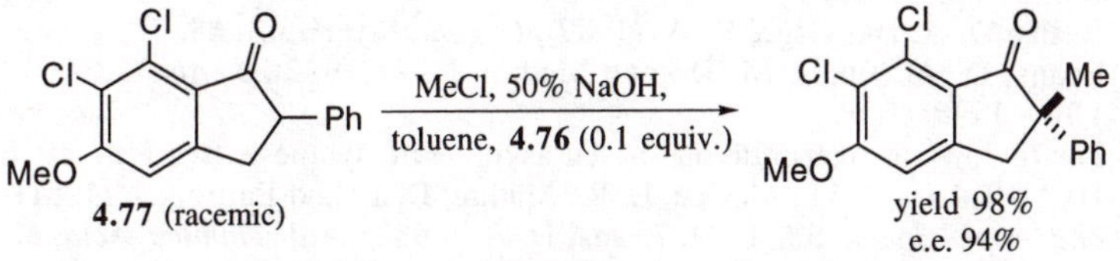

yield 98%
e.e. 94%

Fig. 4.42

The α-substitutions considered in this chapter create a single new chiral centre, adjacent to the carbonyl group, and provide powerful methods with real potential for use in asymmetric synthesis. The enantioselective reaction of enolates with prochiral electrophiles could result in the formation of products with two new chiral centres. Asymmetric versions of the aldol reaction constitute the single most important group of this type of reaction. These enantioselective aldol reactions provide extremely important methods for asymmetric synthesis and form the substance of the next chapter.

References

1. Evans, D. A. (1984), in *Asymmetric Synthesis*, (ed. J. D. Morrison), Vol. 3, pp. 1–110, Academic Press, New York.
2. Corey, E. J. (1954), *J. Amer. Chem. Soc.*, **76**, 175.
3. Corey, E. J. and Sneen, R. A. (1956), *J. Amer. Chem. Soc.*, **78**, 6269.
4. Agami, C., Levisalles, J., and Lo Cicero, B. (1978), *Tetrahedron*, **35**, 961; Ref. 1, pp. 25–26.
5. Schöllkopf, U., Hartwig, W., and Groth, U. (1979), *Angew. Chem. Int. Ed. Engl.*, **18**, 863; Schöllkopf, U., Hartwig, W., Popischil, K.-H., and Kehne, H. (1981), *Synthesis*, 966; Schöllkopf, U., Groth, U., Westphalen, K., and Deng, C. (1981), *Synthesis*, 969; Schöllkopf, U. and Groth, U. (1981), *Angew. Chem. Int. Ed. Engl.*, **20**, 977; Schöllkopf, U., Groth, U., and Deng, C. (1981), *Angew. Chem. Int. Ed. Engl.*, **20**, 798; Schöllkopf, U., Hartwig, W., Groth, U., and Westphalen, K. (1981), *Liebigs Ann. Chem.*, 696; Schöllkopf, U., Groth, U., and Hartwig, W. (1981), *Liebigs Ann. Chem.*, 2407.
6. Seebach, D., Boes, M., Naef, R., and Schweizer, W. B. (1983), *J. Amer. Chem. Soc.*, **105**, 5390; Frater, G., Muller, U., and Gunther, W. (1981), *Tetrahedron Lett.*, **22**, 4221.
7. For a review of the uses of oxazolines for asymmetric synthesis, see Lutomski, K. A. and Meyers, A. I. (1984), in *Asymmetric Synthesis*, (ed. J. D. Morrison), Vol. 3, pp. 213–274., Academic Press, New York.
8. Meyers, A. I., Knaus, G., Kamata, K., and Ford, M. E. (1976), *J. Amer. Chem. Soc.*, **98**, 567.
9. Meyers, A. I., Mazzu, A., and Whitten, C. E. (1977), *Heterocycles*, **6**, 971; Meyers, A. I., Snyder, E. S., and Ackerman, J. J. H. (1978), *J. Amer. Chem. Soc.*, **100**, 8186; Hoobler, M. A., Bergbreiter, D. E., and Newcomb, M. (1978), *J. Amer. Chem. Soc.*, **100**, 8182.
10. Byström, S., Högberg, H.-U., and Norin, T. (1981), *Tetrahedron*, **37**, 2249.
11. Ref. 1, pp. 87–90.
12. Evans, D. A. and Gage, D. A. (1989), *Organic Syntheses*, **68**, 77.
13. Evans, D. A., Ennis, M. D., and Mathre, D. J. (1982), *J. Amer. Chem. Soc.*, **104**, 1737.
14. For reviews of oxazolidinone-based asymmetric synthesis, see Ref. 11; Evans, D. A., Takacs, J. M., McGee, L. R., Mathre, D. J., and Bartroli, J. (1981), *Pure and Appl. Chem.*, **53**, 1109; Evans, D. A. (1982), *Aldrichimica Acta*, **53**, 23.
15. Evans, D. A., Britton, T. C., and Ellman, J. A. (1987), *Tetrahedron Lett.*, **28**, 6141.
16. Evans, D. A., Bender, S. L., and Morris, J. (1988), *J. Amer. Chem. Soc.*, **110**, 2506.
17. Evans, D. A., Morrisey, M. M., and Dorow, R. L.(1985), *J. Amer. Chem. Soc.*, **107**, 4346.
18. Evans, D. A., Britton, T. C., Dorow, R. L., and Ellman, J. A.(1990), *J. Amer. Chem. Soc.*, **112**, 4011.
19. Evans, D. A., Britton, T. C., Dorow, R. L., and Dellaria, J. F. (1988), *Tetrahedron*, **44**, 5525.
20. Evans, D. A., Ennis, M. D., and Le, T. (1984), *J. Amer. Chem. Soc.*, **106**, 1154.
21. Evans, D. A., Clark, J. S., Metternich, R., Novack, V. J., and Sheppard, G. S. (1990), *J. Amer. Chem. Soc.*, **112**, 866.

22. Oppolzer, W., Chapuis, C., and Bernardinelli, G. (1984), *Helv. Chim. Acta*, **67**, 1397; Davis, F. A., Towson, J. C., Weismiller, M. C., Lal, S., and Carroll, P. J. (1988), *J. Amer. Chem. Soc.*, **110**, 8477.
23. Oppolzer, W., Moretti, R., and Thomi, S. (1989), *Tetrahedron Lett.*, **30**, 5603.
24. Davies, S. G. (1990), *Aldrichimica Acta*, **23**, 31.
25. Baird, G. J., Bandy, G. J., Davies, S. G., and Prout, K. (1983), *J. Chem. Soc., Chem. Commun.*, 1202; Brown, S. L., Davies, S. G., Foster, D. F., Seeman, J. I., and Warner, P. (1986), *Tetrahedron Lett.*, **27**, 623.
26. Bashiardes, G. and Davies, S. G. (1987), *Tetrahedron Lett.*, **28**, 5563.
27. Bashiardes, G., Collingwood, S. P., Davies, S. G., and Preston, S. C. (1989), *J. Chem. Soc., Perkin Trans. 1*, 1162.
28. Bashiardes, G. and Davies, S. G. (1988), *Tetrahedron Lett.*, **29**, 6509.
29. Davies, S. G., Polywka, R., and Warner, P. (1990), *Tetrahedron*, **46**, 4847.
30. Meyers, A. I., Williams, D. R., and Druelinger (1976), *J. Amer. Chem. Soc.*, **98**, 3032.
31. Enders, D. and Eichenauer, H. (1976), *Angew. Chem. Int. Ed. Engl.*, **15**, 549.
32. Enders, D. (1984), in *Asymmetric Synthesis*, (ed. J. D. Morrison), Vol. 3, pp. 275–339, Academic Press, New York.
33. Enders, D. and Eichenauer, H. (1979), *Chem. Ber.*, **112**, 2933. For an alternative avoiding nitrosamines as intermediates, see Enders, D., Fey, P., and Kipphardt, H. (1987), *Organic Syntheses*, **65**, 173.
34. Compiled from examples given in Ref. 32.
35. Ref. 32, pp. 318–322.
36. Enders, D., Kipphardt, H., and Fey, P. (1987), *Organic Syntheses*, **65**, 183, and references cited therein.
37. Enders, D. and Baus, U. (1983), *Liebigs Ann. Chem.*, 1439.
38. Nicolaou, K. C., Papahatjis, D. P., Claremon, D. A., and Dolle, R. E. (1981), *J. Amer. Chem. Soc.*, **103**, 6967; Nicolaou, K. C., Claremon, D. A., Papahatjis, D. P., and Magolda, R. L. (1981), *J. Amer. Chem. Soc.*, **103**, 6969.
39. Davis, F. A., Kumar, A., and Chen, B. C. (1991), *J. Org. Chem.*, **56**, 1143; Davis, F. A., Sheppard, A. C., Chen, B. C., and Serajul Haque, M. J. (1990), *J. Amer. Chem. Soc.*, **112**, 6679; Davis, F. A., Ulatowski, T. G., and Serajul Haque, M. J. (1987), *J. Org. Chem.*, **52**, 5288; Mergelsberg, I, Gala, D., Scherer, D., DiBenedetto, D., and Tanner, M. (1992), *Tetrahedron Lett.*, **33**, 161.
40. Davis, F. A. and Kumar, A. (1991), *Tetrahedron Lett.*, **32**, 7671.
41. Oppolzer, W., Tamura, O., Sundarababu, G., and Signer, M. (1992), *J. Amer. Chem. Soc.*, **114**, 5900.
42. For a review, see Cox, P. J. and Simpkins, N. S. (1991), *Tetrahedron Asymmetry*, **2**, 1.
43. Cain, C. M. and Simpkins, N. S. (1987), *Tetrahedron Lett.*, **28**, 3723.
44. Kim, H., Shirai, R., Kawasaki, H., Nakajima, M., and Koga, K. (1990), *Heterocycles*, **30**, 307.
45. Leonard, J., Ouali, D., and Rahman, S. K. (1990), *Tetrahedron Lett.*, **31**, 739; Leonard, J., Hewitt, J. D., Ouali, D., Rahman, S. K., Simpson, S. J., and Newton, R. F. (1990), *Tetrahedron Asymmetry*, **1**, 699; Leonard, J., Ouali, D., and Rahman, S. K. (1992), *J. Chem. Soc., Perkin Trans. 1*, 1203.
46. Izawa, H., Shirai, R., Kawasaki, H., Kim, H., and Koga, K. (1989), *Tetrahedron Lett.*, **30**, 7221.

47. Momose, T., Toyooka, N., Seki, S., and Hirai, Y. (1990), *Chem. Pharm. Bull.*, **38**, 2072.
48. Cox, P. J. and Simpkins, N. S. (1991), *Synlett*, 321; Bunn, B. J., Cox, P. J., and Simpkins, N. S. (1993), *Tetrahedron*, **49**, 207.
49. Dolling, U.-H., Davis, P., and Grabowski, E. J. (1984), *J. Amer. Chem. Soc.*, **106**, 446; Hughes, D. L., Dolling, U.-H., Ryan, K. M., Schoenewaldt, E. F., and Grabowski, E. J. (1987), *J. Org. Chem.*, **52**, 4745.
50. O'Donnell, M. J., Bennett, W. D., and Wu, S. (1989), *J. Amer. Chem. Soc.*, **111**, 2353.

5 Asymmetric aldol reactions

The addition of an enolate to a ketone or aldehyde, often referred to as an aldol reaction, has a long pedigree, having been studied for many years from both the synthetic and mechanistic points of view.[1] Throughout this volume, the term 'aldol reaction' will be used to denote the formation of a β-hydroxycarbonyl system **5.2**, thereby distinguishing it from an 'aldol condensation' which refers to the formation of an α,β-unsaturated carbonyl compound **5.3** by β-elimination of an intermediate β-hydroxycarbonyl system. These relationships and definitions are illustrated in Fig. 5.1.

The aldol reaction as defined above is of great value in asymmetric synthesis, given good methods for stereochemical control. Two common types of synthetic units are combined by C–C bond formation, with the simultaneous formation of two new chiral centres. Before discussing some selected examples of this type of reaction and their applications in synthesis, it is important to consider the problems associated with the generation of these two new chiral centres.

Currently, most of the methods for asymmetric aldol reactions which are of general practical application use aldehydes as the electrophilic carbonyl component, and therefore most of the following discussion will concentrate on these reactions. In Fig. 5.2 the consequences of performing an aldol reaction in which there is no stereochemical control are illustrated, along with some of the jargon associated with this area of synthesis.

This general situation is analogous to that already encountered in additions to aldehydes (Chapter 3, Fig. 3.1 and Fig. 3.22). As mentioned in Chapter 3, there are many similarities between the addition of crotyl boranes to aldehydes and aldol reactions such as those depicted in Fig. 5.2.

When the stereochemical nomenclature shown in Fig. 5.2 is used it is important to draw the 'backbone' of the aldol product in the extended conformation shown, and only then to apply the '*syn*' (OH and R' on the same 'side') and '*anti*' (these groups on opposite 'sides') descriptors.

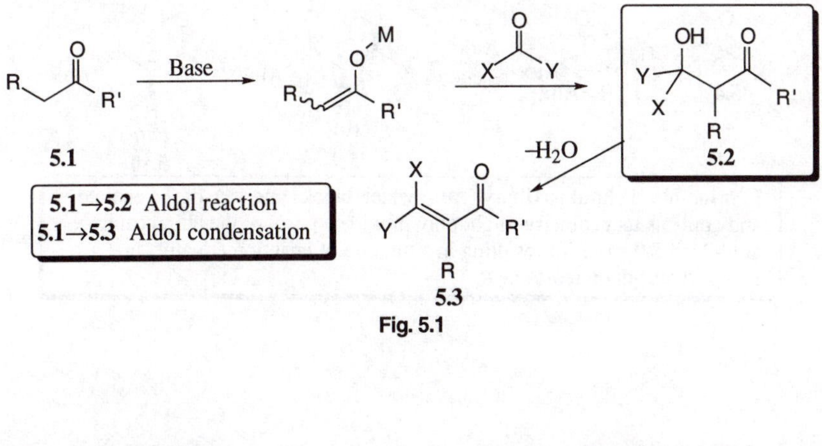

5.1

| 5.1→5.2 Aldol reaction |
| 5.1→5.3 Aldol condensation |

−H₂O

5.2

5.3

Fig. 5.1

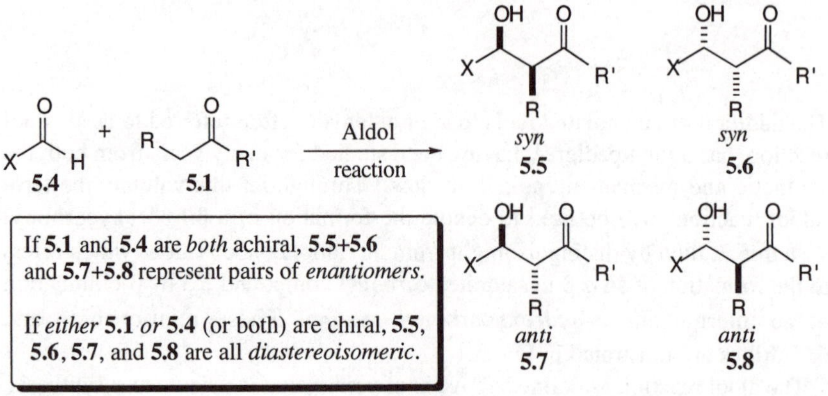

If **5.1** and **5.4** are *both* achiral, **5.5+5.6** and **5.7+5.8** represent pairs of *enantiomers.*

If *either* **5.1** *or* **5.4** (or both) are chiral, **5.5**, **5.6**, **5.7**, and **5.8** are all *diastereoisomeric.*

Fig. 5.2

Consideration of Fig. 5.2 suggests that stereochemical control in the aldol reaction is likely to be a challenging problem. For example, take the case of a (hypothetical) chiral auxiliary which directs attack of the electrophilic aldehyde **5.4** to the 'upper' face of the enolate derived from **5.1** (R'=Aux) and controls the enolate geometry (Fig. 5.3), requirements which suffice for asymmetric enolate alkylation (Chapter 4). In principle this need only result in control of the asymmetric centre adjacent to the carbonyl (C-2) of the product (just as in enolate alkylation). The question of the chiral centre which derives from the aldehyde **5.4** (C-3) remains, as shown in Fig. 5.3. In effect, the chirality at *C-2* depends upon which face of the *enolate* is attacked by the *aldehyde*, and that at *C-3* depends upon which face of the *aldehyde* is attacked by the *enolate.*

Fortunately, the problem is not quite as daunting as it might appear. For many types of aldol reactions a relationship exists between the geometry of the enolate and the relative stereochemistry of the new chiral centres. Often, one enolate geometry provides the *syn*-isomer, and the other the *anti*-isomer. Moreover, this relationship is reasonably consistent and arises because of mechanistic similarities between the various protocols for aldol reactions. In the best cases this can lead to predictability and high stereochemical control. Before discussing this relationship, as was the case in enolate alkylation (Chapter 4),

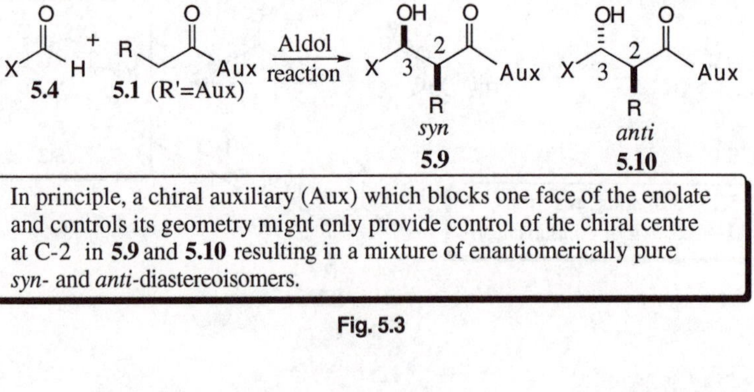

In principle, a chiral auxiliary (Aux) which blocks one face of the enolate and controls its geometry might only provide control of the chiral centre at C-2 in **5.9** and **5.10** resulting in a mixture of enantiomerically pure *syn-* and *anti*-diastereoisomers.

Fig. 5.3

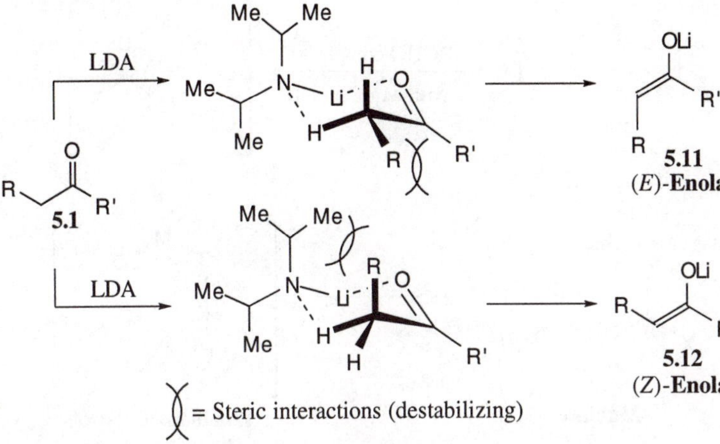

Fig. 5.4

some consideration must be given to the control of enolate geometry.

Enolization with hindered amide bases is often discussed in terms of deprotonation via a six-membered cyclic transition state, and although a simple account of this rationalization will be presented here, the area is still one of active research and debate.[2]

Two possible cyclic transition states for deprotonation and their relationship to enolate geometry are shown in Fig. 5.4, in which LDA is taken as the representative hindered amide base. In both transition states there are potentially destabilizing steric interactions, and which is favoured depends on which of these interactions is dominant.

The stereochemical notation which will be used for the two possible enolate geometries is illustrated in Fig. 5.5.[3] In this scheme, the two substituents at C-2 of the enolate are assigned relative priorities as normal in the (E)/(Z) stereochemical description of alkenes. For C-1, the highest priority is always given to 'OM', irrespective of the other substituent. The two enolates **5.11** and **5.12** have been named using this convention (Fig. 5.4).

The stereochemistry of the enolate has an important bearing on the stereoselectivity of its aldol reactions, along with several other variables including the nature of R', the metal, the solvent, and the precise reaction conditions. Given all these variables it might be expected to be unlikely that hard and fast 'rules' could be elaborated for the prediction of the outcome of an aldol reaction. Fortunately, a great body of work already exists, and from this it

Fig. 5.5

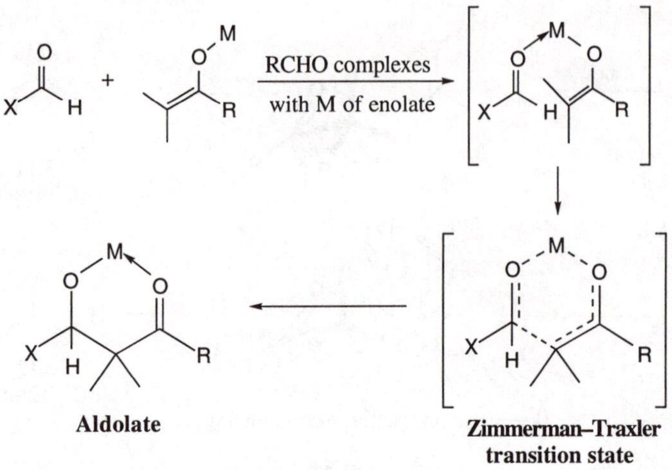

Fig. 5.6

is often possible to make a reasonable prediction in most cases. Moreover, several useful generalizations *can* be made, based *inter alia* on transition state models for aldol reactions.

The most widely encountered type of transition state model is based on the 'Zimmerman–Traxler' transition state, which accounts for the stereochemical outcome of aldol reactions of several types of metal enolates (especially lithium, boron, magnesium, and zinc).[4] An idealized representation of an aldol reaction based on this model is shown in Fig. 5.6.

In this model, the aldehyde complexes with the metal of the enolate, and aldol reaction proceeds via a six-membered transition state. This type of transition state is often represented as possessing a chair conformation by analogy with cyclohexanes, and is similar to that proposed for the addition of allyl boranes and crotyl boranes to aldehydes (Chapter 3, Fig. 3.24 and Fig. 3.26). Despite the fact that the geometry of such a chair transition state must be very different from a normal cyclohexane, this analogy proves to be most useful in accounting for structure/stereochemistry trends in many aldol reactions.

For many enolates, it is found that the (*E*)-isomer gives an *anti*-aldol product, and the (*Z*)-isomer a *syn*-aldol product. This can be understood by analysis in terms of the Zimmerman–Traxler model for the possible transition states. In the following analyses 'chair' conformations are used to represent the transition states, the X substituent of the aldehyde is in the more stable 'equatorial' orientation, and 'Aux' represents a generalized chiral auxiliary. The reaction of an (*E*)-enolate is illustrated in Fig. 5.7.

An enolate with this geometry allows for the possibility of the group R as well as X from the aldehyde to be equatorial. Which of the two possible *anti*-aldol products is obtained will depend on which face of the enolate reacts preferentially, which in turn usually depends on the chiral auxiliary (Aux).

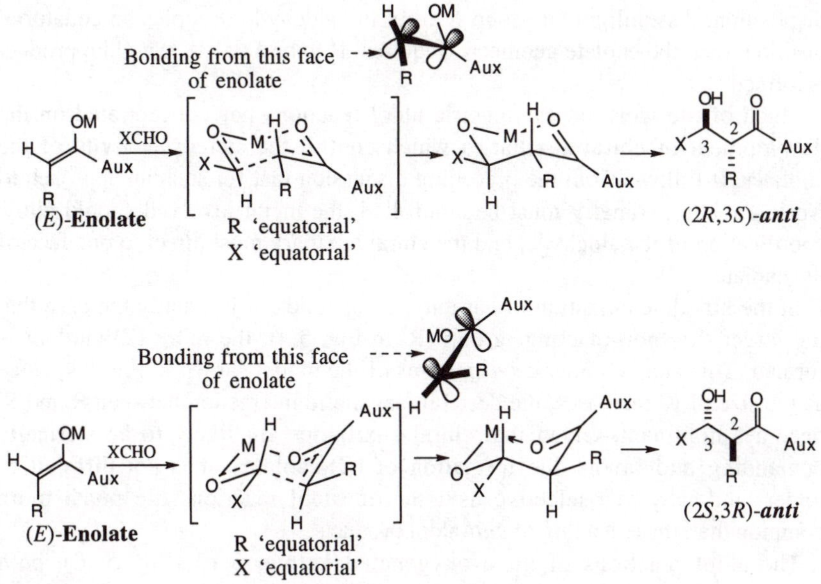

Fig. 5.7

It can be seen from the equivalent analysis for an aldol reaction of a (Z)-enolate (Fig. 5.8) that both R and X cannot both adopt an 'equatorial'

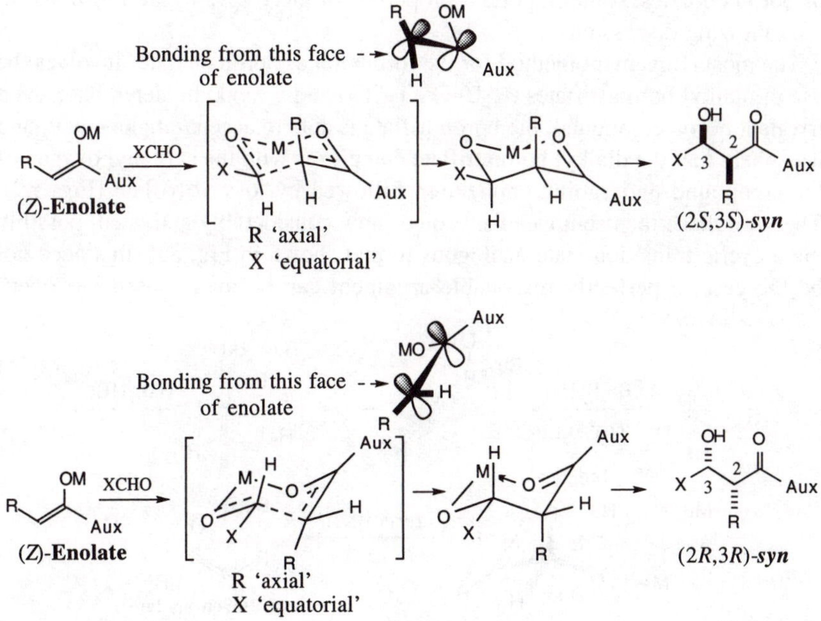

Fig. 5.8

disposition. Assuming that group X from the aldehyde occupies an equatorial position, then the enolate geometry 'requires' R to be axial. A *syn*-aldol product is formed.

Much of the work on asymmetric aldol reactions has concentrated on the development of chiral auxiliaries which control the stereoselectivity of the enolates. It follows from the preceding discussion that for such an approach to work, enolate geometry must be controlled, the metal involved should allow coordination of the aldehyde, and the chiral auxiliary must direct to one face of the enolate.

In the kinetic enolization of carbonyl compounds, it is usually the case that the larger the 'non-reacting' group (R' in Fig. 5.4), the more (Z)-enolate is formed. This can be understood in terms of the model shown in Fig. 5.4, since as the size of R' increases, the destabilizing steric interaction between R and R' becomes dominant. Given that chiral auxiliaries are likely to be sterically demanding and favour the formation of (Z)-enolates, it is not difficult to understand why *syn*-selective asymmetric aldol reactions are much more common than those leading to *anti*-aldol products.

The aldol reactions of the α-oxygenated ketone **5.13** (Fig. 5.10), both enantiomers of which have been thoroughly studied and applied to natural product synthesis, will be used to illustrate some general trends and mechanistic concepts in the area of asymmetric aldol reactions.[5] The boron enolates of **5.13** are the most stereoselective, usually reacting via the (Z)-enolate. As a considerable part of the following chapter will be concerned with the chemistry of boron enolates, some important properties of these species are illustrated in the following discussion.

The most convenient method for the formation of boron enolates involves the use of dialkyl boron triflates ($R_2B-OSO_2CF_3$) and a weak, hindered base. As a trivalent boron compound, the boron triflate is able to accept *one* lone pair, and it is likely that the dialkyl boron triflate complexes with the carbonyl oxygen of the compound undergoing enolization followed by loss of triflate (Fig. 5.9). The weak base then removes the proton, now considerably activated, possibly via a cyclic transition state analogous to that shown in Fig. 5.4 (this need not be the case, a perfectly reasonable argument can be made based on 'open'

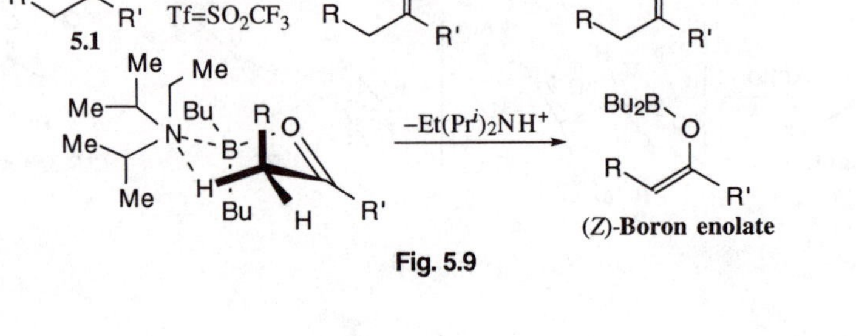

Fig. 5.9

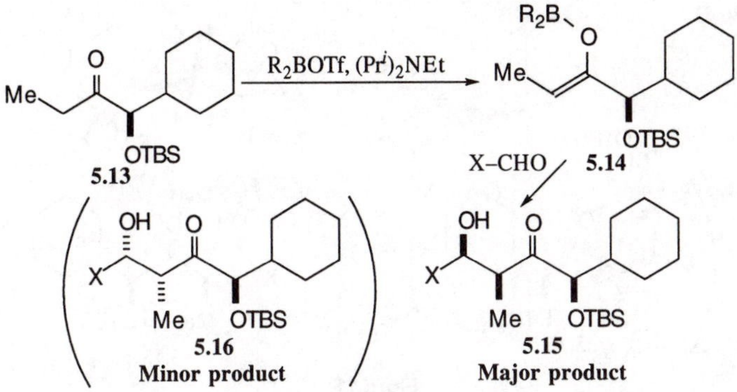

Fig. 5.10

Table 5.1 Stereoselectivity in aldol reactions of **5.13**

X	R	5.15:5.16
Ph	Bu^n	97.5:2.5
Ph	Cyclopentyl	98.7:1.3
Et	Bu^n	98:2
Et	Cyclopentyl	>99:1
BnOCH$_2$CH$_2$	Bu^n	96.5:3.5
BnOCH$_2$CH$_2$	Cyclopentyl	99:1
Pr^i	Bu^n	>99:1
Pr^i	Cyclopentyl	No reaction

transition states for this deprotonation).

In the case of ketone **5.13**, the (Z)-boron enolate **5.14** is formed with high stereoselectivity and reacts with aldehydes to give *syn*-aldol products with the stereochemistry as shown in Fig. 5.10. Selected examples are given in Table 5.1.[6,7]

To understand the stereochemical outcome of these reactions it is important to consider the likely conformation of the chiral auxiliary in the transition state as well as the geometry of the enolate. An analysis for enolate **5.14** is presented in Fig. 5.11. In the first step the aldehyde complexes with the enolate (cf. Fig. 5.6). For a boron enolate, this completes the coordination sphere around boron, which means that in the transition state the chiral auxiliary will not be held rigidly by chelation (in contrast to many cases of enolate alkylation, Chapter 4). As a result the chiral auxiliary adopts a conformation relative to the rest of the system which minimizes the energy involved. In the case of **5.14** this is thought to be a conformation in which the dipole–dipole repulsion between the two oxygen atoms of the enolate is minimized, represented by the antiperiplanar arrangement shown in Fig. 5.11. In adopting this conformation, the cyclohexyl group blocks approach of the aldehyde from the 'upper' face as shown. Addition

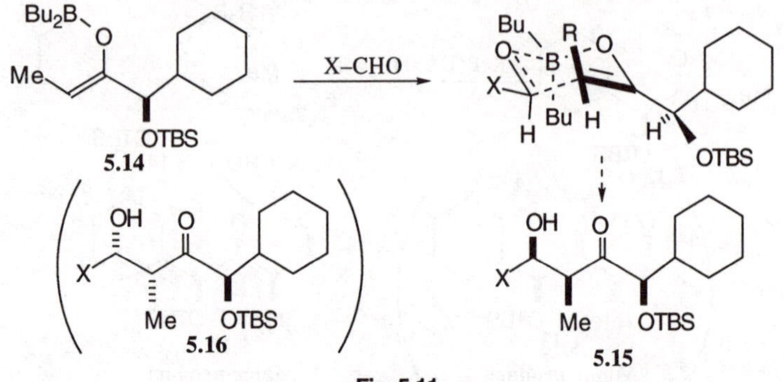

Fig. 5.11

to give the minor product **5.16** requires the aldehyde to approach from the same face as this bulky group.

The ketone **5.13** is readily available from mandelic acid **5.17** as shown in Fig. 5.12, and as both enantiomers of mandelic acid are readily available the enantiomer of **5.13** is equally easy to obtain.[7] One drawback with the use of **5.13** and its enantiomer is that the stereochemical control element, the α-oxygenated ketone unit, is necessarily destroyed in the process of its removal (Fig. 5.12).

Nevertheless, the high levels of stereochemical control and the ready availability of the ketones have allowed the synthesis of several complex natural products using this methodology, including 6-deoxyerythronolide B **5.18** (Fig. 5.13).[8]

The oxazolidinones **5.19** and **5.20** (R=Pr^i or Bn) (Fig. 5.14) have been used as excellent chiral auxiliaries for enolate alkylation (see Chapter 4) and it is not surprising to find that they also provide high levels of stereochemical control in aldol reactions.[9] However, there are important differences between the alkylation and aldol reactions of enolates derived from these oxazolidinones, which will be highlighted in the following discussion. The process of acylation and removal of these oxazolidinone chiral auxiliaries, as discussed in Chapter 4, is normally

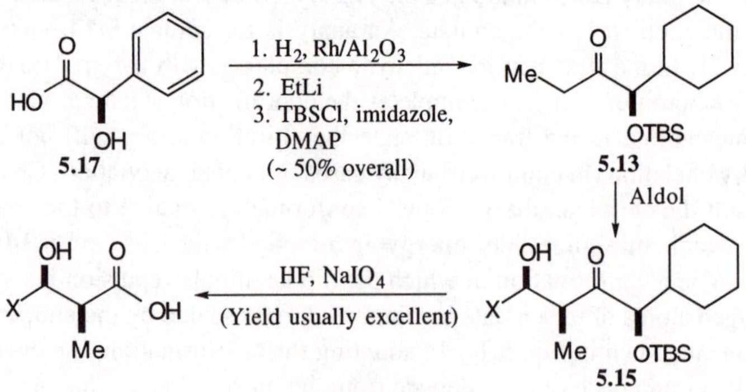

Fig. 5.12

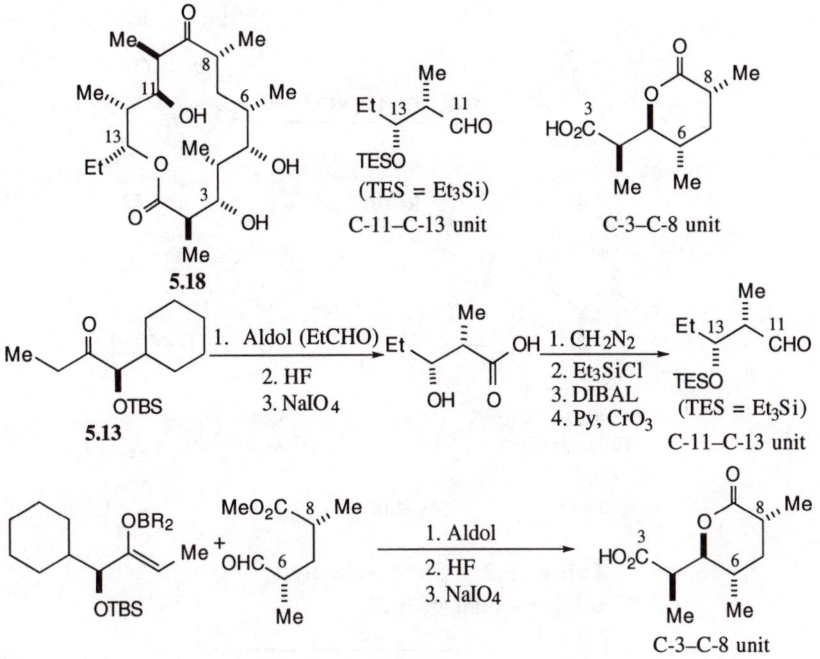

Fig. 5.13

efficient, and the diastereoselectivity in the reaction of enolates derived from the acylated oxazolidinones is usually excellent. The lithium enolates, which are so effective in alkylation and related reactions, give poor selectivity in aldol reactions, but the corresponding boron enolates provide exceptional levels of stereochemical control.

These boron enolates are prepared as described earlier in this chapter and are almost always exclusively (Z)-isomers (Fig. 5.15).[10] Aldol reaction with a wide range of aldehydes takes place with high *syn*-selectivity and with high facial selectivity, the product being essentially a single *syn*-isomer (Table 5.2).[11] Acyl derivatives of the chiral auxiliary **5.19** react to produce *syn*-aldol products of opposite absolute configuration with equally high diastereoselectivity, making either enantiomer of the *syn*-aldol equally available.

These aldol reactions could be viewed as reaction of the enolate with an electrophile (the aldehyde), and as such the sense of asymmetric induction might

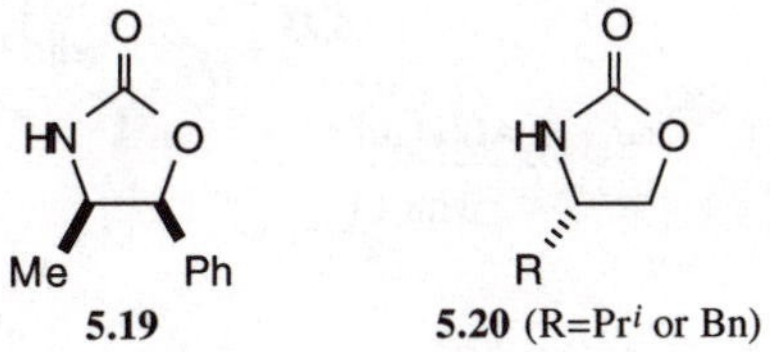

Fig. 5.14

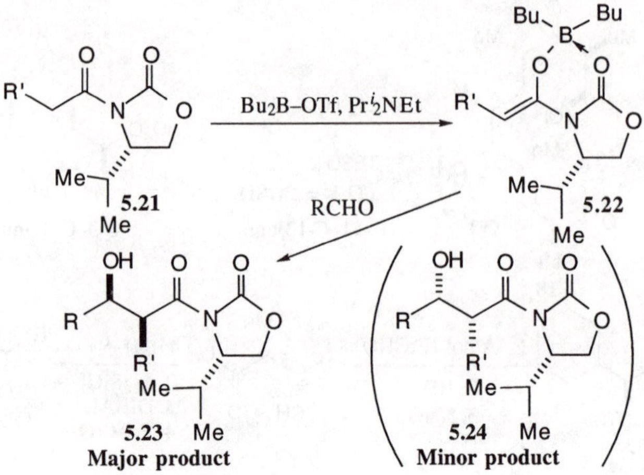

Fig. 5.15

Table 5.2   Stereoselectivity in aldol reactions of **5.21**

R	R′	5.23:5.24
Ph	Me	>99.8:0.2
Me	MeS	99.6:0.4
Pr^i	MeS	98.4:1.6
Bu^n	Me	99.3:0.7
Pr^i	Me	99.8:0.2

be expected to be the same as for alkylation of these enolates. Using the same oxazolidinone chiral auxiliary, the chirality induced at C-2 in an alkylation is *opposite* to that in the aldol reaction of the boron enolate (Fig. 5.16).

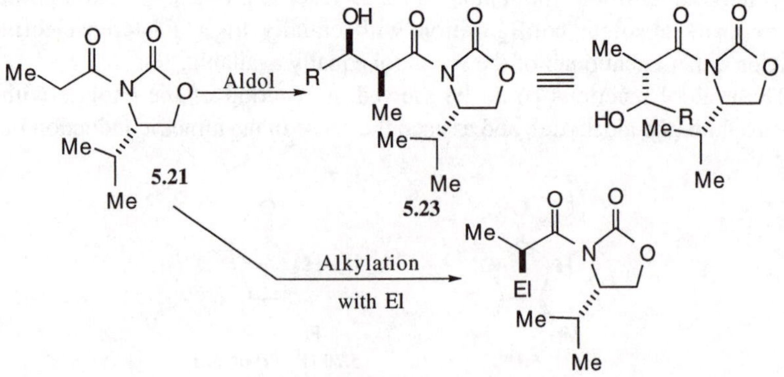

Fig. 5.16

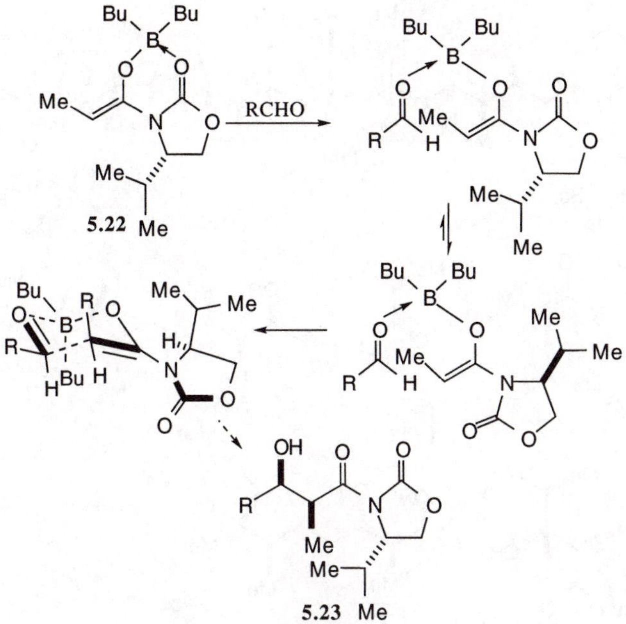

Fig. 5.17

The origin of this difference can be understood by considering the mechanism of the aldol reaction of boron enolates, as discussed above (Fig. 5.11). The boron enolates themselves exist as chelated species exemplified by **5.22**. For the aldol reaction to take place the O–B chelation must be broken, to allow the aldehyde to coordinate with the boron (cf. Fig. 5.6) as a trivalent boron derivative can only complex with one 'extra' ligand. The chiral auxiliary is now free to adjust its conformation and it is proposed that it rotates through 180° to minimize dipole–dipole repulsions (cf. Fig. 5.11 and the related discussion). If this happens then the isopropyl group now blocks the 'upper' face of the enolate (Fig. 5.17), thereby directing the electrophile (the aldehyde) to the lower face.[9]

It is interesting to note that although the controlling units are quite different in **5.14** and **5.22**, the same principles can be used to understand the diastereoselectivity found in the respective aldol reactions. The chair-like transition state is 'held together' by the boron, and the conformation of the chiral auxiliary (and consequently the facial selectivity of the enolate) is 'controlled' by the minimization of dipole–dipole repulsions between the oxygen of the enolate and an oxygen atom of the chiral auxiliary.

As discussed in Chapter 4, these oxazolidinone chiral auxiliaries possess features which make them particularly attractive; in effect they fulfil almost all of the criteria which were laid down in Chapter 2 (Table 5.1) for chiral auxiliaries. Consequently, this is one of the most widely used chiral auxiliaries for enolate and aldol reactions. Two simple examples which illustrate the use of

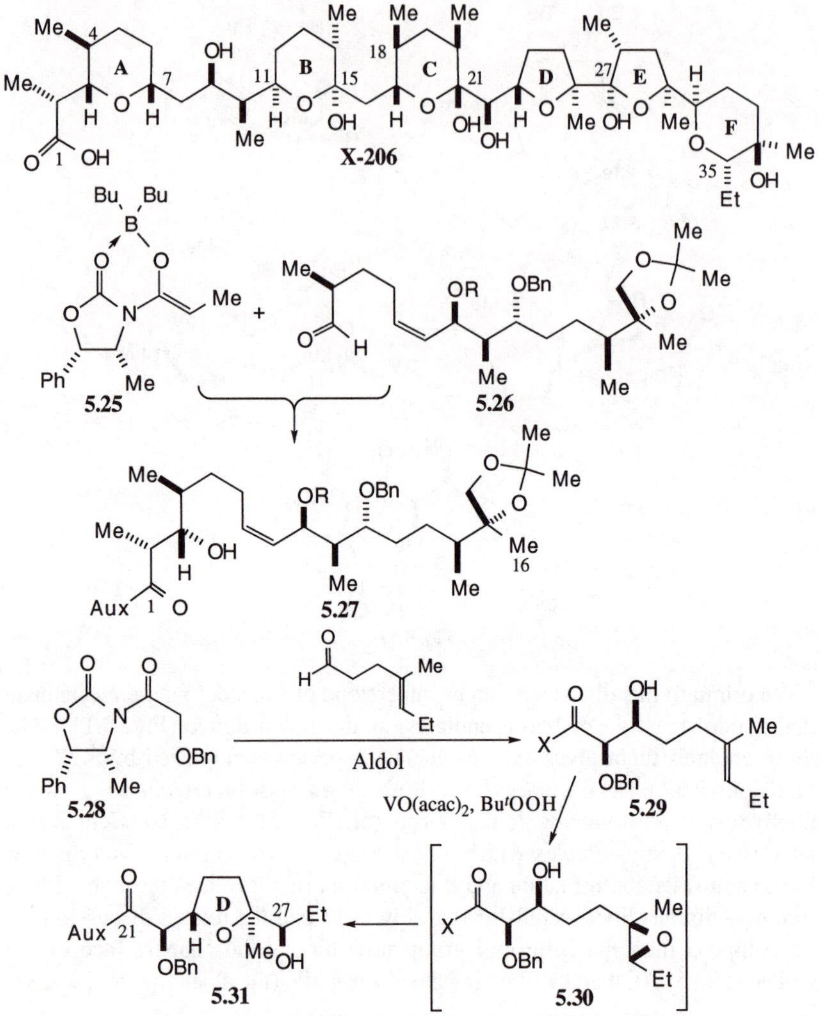

Fig. 5.18

this enolate chemistry are shown in Fig. 5.18, and are taken from the aforementioned total synthesis of X-206 (see Chapter 4).[12]

The examples of asymmetric aldol reactions shown in Fig. 5.18 illustrate two aspects of the power of this approach to the convergent synthesis of complex molecules. The first example, the reaction of the boron enolate **5.25** with aldehyde **5.26**, was chosen to emphasize the fact that the stereochemistry of the two chiral centres which are created in the aldol reactions of these enolates is determined solely by the choice of auxiliary, and that this effect overrides any 'inherent diastereoselectivity' of the aldehyde (due to chirality present in this component). It is also important to note that aldehyde **5.25** is itself structurally complex and contains a number of potentially sensitive functional groups (*cis*-alkene, ketal, tertiary ether, protected allylic alcohol) which are all compatible with the aldol reaction. The product **5.27** was isolated in 97 per cent yield as a single diastereoisomer.

The other example (Fig. 5.18) demonstrates that the aldol reaction of even quite simple components can provide access to relatively complicated systems in a very short sequence. Aldol reaction of the boron enolate of **5.28** with the aldehyde shown, provided **5.29** in 84 per cent yield. Product **5.29** is a bis-homoallylic alcohol (HO–CHR–CH$_2$–CH$_2$–CR=CR$_2$) and it is known that epoxidation of this type of alcohol using the reagents shown is highly diastereoselective. The intermediate epoxide **5.30** cyclizes under the reaction conditions.

In this way the tetrahydrofuran **5.31** was prepared in 89 per cent yield, the diastereoselectivity of the epoxidation step being ~95:5. Tetrahydrofuran **5.31** possesses four chiral centres (ignoring the chiral auxiliary) and there are 16 possible stereoisomers, *one* of which was prepared in a highly stereocontrolled fashion. Clearly the sequential use of two highly stereoselective reactions, each of which provides two new chiral centres, is an extremely powerful strategy for asymmetric synthesis.

The preparation and remarkable stability of β-dicarbonyl compounds (e.g. **5.32**) derived from these oxazolidinone chiral auxiliaries were discussed in Chapter 4 (Figure 4.20). In this case these β-dicarbonyl compounds were

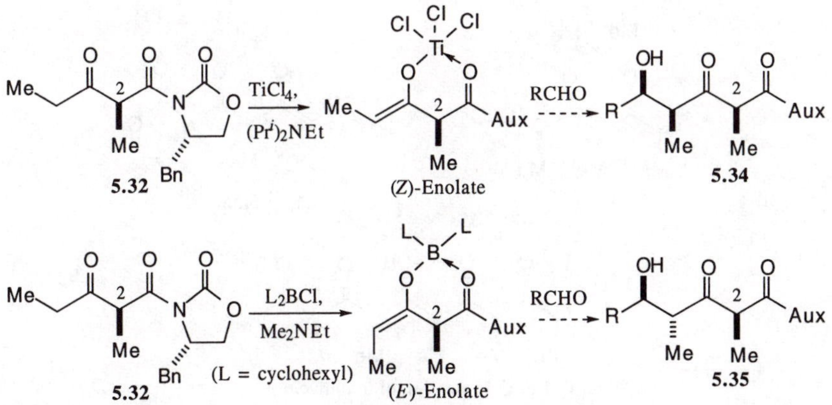

Fig. 5.19

prepared by the acylation of the corresponding enolate. A complementary route to such systems, which produces the other diastereoisomer at C-2, involves aldol reaction followed by oxidation of the secondary alcohol to the ketone (**5.32**, Fig. 5.19).[13]

Systems such as **5.32** exhibit remarkable reactivity under aldol conditions, as can be seen from Fig. 5.19. As discussed in Chapter 4, the proton at C-2 is inert under the reaction conditions, and aldol reaction occurs at the methylene group. Moreover, *either* of the possible *syn*-aldol products **5.33** or **5.34**, and one of the possible *anti*-products **5.35** can be obtained simply by choice of Lewis acid![14] For most of the examples represented by Fig. 5.19 which produce *syn*-isomers, the stereoselectivity is 96:4 or greater. The stereoselectivity in the reactions which result in *anti*-aldol products is somewhat lower (80:20 to 92:8), but still impressive and useful. Discussion of detailed transition state models for these reactions lies beyond the scope of this volume, and indeed an appropriate model which accounts for these and related reactions is unavailable at present. Nevertheless, the general correlation of enolate geometry and aldol stereochemistry still holds in that (Z)- and (E)-enolates give *syn*-and *anti*-aldol products respectively.

A remarkable example of this enolate chemistry is found in the synthesis of the C-1 to C-11 portion of the polyether antibiotic Ionomycin A (Fig. 5.20) by reaction of the tin(II) enolate **5.36** with aldehyde **5.37** (Fig. 5.20), itself prepared from the tin(II) enolate of **5.32** (Fig. 5.19).[13]

The discussion of asymmetric aldol chemistry has concentrated on oxazolidinone chiral auxiliaries as these are one of the most generally useful auxiliaries which are easily available in quantity. Their usefulness has been demonstrated in several total syntheses, some of which have been touched on in this chapter, and have been used successfully in many different laboratories. Nevertheless, these chiral auxiliaries are not the perfect solution. For example, only recently has it become possible to use boron enolates of acylated oxazolidinones to prepare *anti*-aldol products in high diastereoisomeric excess.

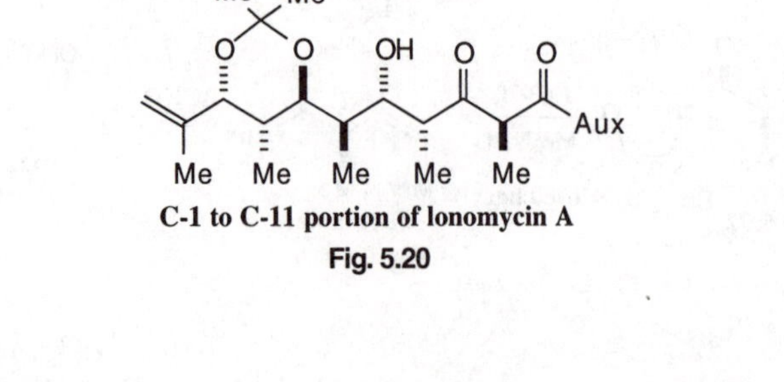

C-1 to C-11 portion of Ionomycin A

Fig. 5.20

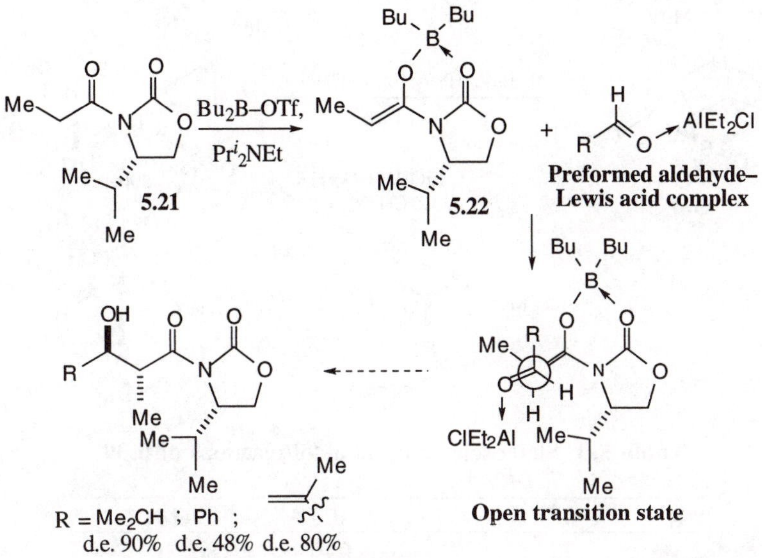

Fig. 5.21

This can been achieved by using reaction conditions which cause the aldehyde component of the aldol reaction to remain coordinated to a Lewis acid throughout the reaction. This allows the boron enolate to remain chelated during the aldol reaction. The reaction no longer proceeds via a Zimmermann–Traxler type transition state. An open transition state such as that shown in Fig. 5.21 is consistent with the results obtained using a sterically demanding Lewis acid.[15]

There are many other chiral auxiliaries which have been used for asymmetric aldol reactions, but a comprehensive review of these is outside the scope of this volume, and only two more will be discussed. Neither of these has yet achieved widespread use, but it is probably just a matter of time before they will be used as a matter of course in organic synthesis laboratories.

The first of these is based upon the sultam **5.38** (Fig. 5.22) and its enantiomer, both of which are readily available from either enantiomer of camphorsulphonyl chloride and easily acylated, as outlined in Chapter 4.[16]

The dialkylboron enolates of the acylated sultams **5.39** undergo highly diastereoselective aldol reactions with a range of aldehydes, providing products

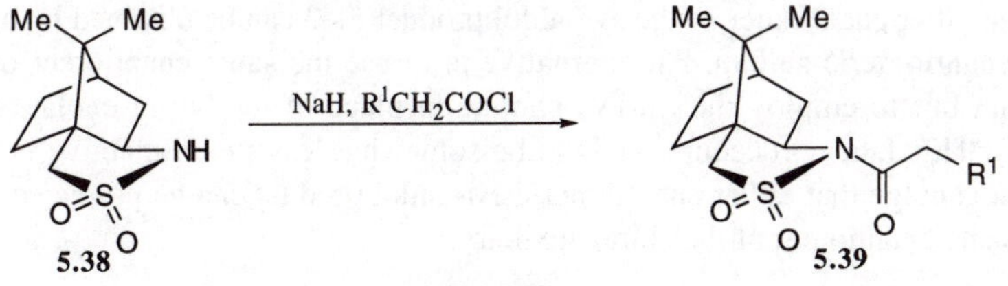

Fig. 5.22

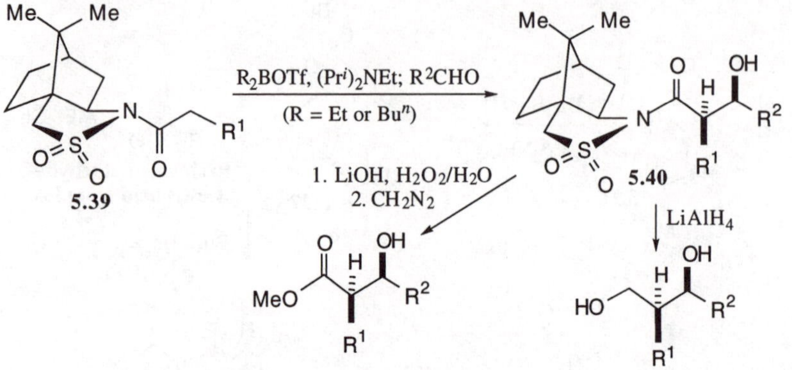

Fig. 5.23

Table 5.3 Stereoselectivity in aldol reactions of **5.39**

R^1	R^2	d.e.[a]	Yield (%)
Me	Ph	99:1	80
Me	Me	>99:1	69
Me	Pr^i	97:3	71
Me	MeCH=CH	>99:1	54
Ph	Pr^i	98.9:1.1	66
Bu^n	Ph	>98:2	64

[a]Ratio of *syn*-isomers; *anti*-isomers not detected.

which can be crystallized easily to >99 per cent purity. As with the oxazolidinone boron enolates, the major product is the *syn*-diastereoisomer and the auxiliary can be removed without loss of stereochemical purity (Fig. 5.23).[17]

The rationalization for the stereochemical outcome of these aldol reactions relies upon the principles which have been used previously in this chapter. A single (Z)-enolate **5.41** is formed, followed by aldol reaction via a six-membered transition state (Fig. 5.24). The facial selectivity is consistent with reaction of the enolate through a conformation which minimizes dipole–dipole repulsion between the SO_2 group and the enolate oxygen, with the 'upper' face of the enolate effectively shielded by the auxiliary, leading to the transition state **5.42**.

The other enantiomer of the *syn*-aldol product **5.43** can be obtained by use of the enantiomeric sultam. An alternative is to use the same enantiomer of the sultam but to employ the tin(IV) enolate in place of the boron enolate (Fig. 5.25). This latter procedure tends to be somewhat less stereoselective but has the advantage that either enantiomeric *syn*-aldol product can be produced from the same enantiomer of the chiral auxiliary.[17]

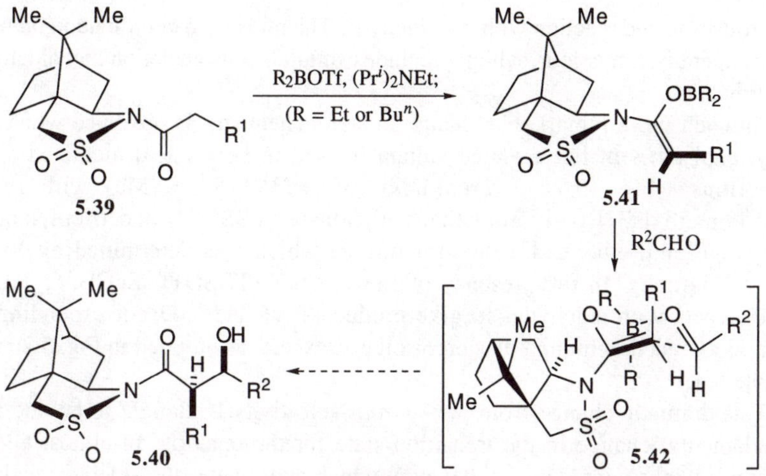

Fig. 5.24

This change in stereoselectivity on changing from a boron enolate to a tin(IV) enolate is thought to be related to the ability of tin(IV) to coordinate to an oxygen of the SO_2 group in addition to coordinating the aldehyde, with the result that the aldol reaction takes place via a conformation similar to **5.44** (Fig. 5.25).

In this conformation the enolate has rotated by 180° with respect to the sultam when compared to the boron enolate. This results in the opposite face of the enolate being blocked in the transition state as the tin(IV) can adopt (distorted) octahedral coordination, thereby maintaining enolate chelation during

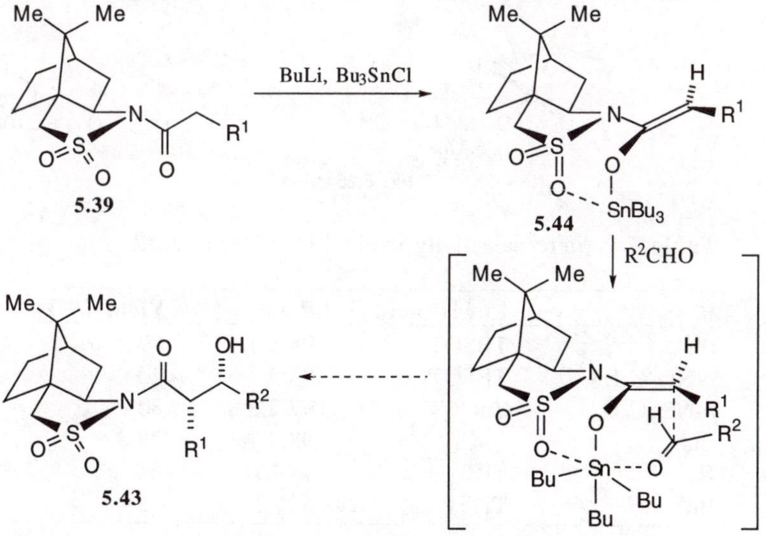

Fig. 5.25

coordination and reaction with the aldehyde. The maximum coordination number of trivalent boron is four, which precludes simultaneous chelation and aldehyde coordination.

An even more remarkable change in stereochemistry is observed when the silyl enol ether of the acylated sultam is used in Lewis acid mediated aldol reactions (Fig. 5.26). Treatment of **5.39** (R^1 = Me) with *tert*-butyldimethylsilyl trifluoromethanesulphonate (TBSOTf) and triethylamine gives the enol ether **5.45**, the structure of which was determined by X-ray crystallography. In the presence of Lewis acids (TBSOTf, ZnCl$_2$, or TiCl$_4$) **5.45** reacts with aldehydes to give products **5.46** and **5.47** corresponding to *anti* aldol reaction in high diastereoisomeric excess, as outlined in Fig. 5.26 and Table 5.4.[18]

This dramatic change from *syn*- to *anti*-selectivity is thought to be due to a fundamental change in the transition state for the reaction. In almost all the examples encountered so far, the atom which coordinates the aldehyde and the 'counter-ion' of the enolate have been one and the same, resulting in a cyclic

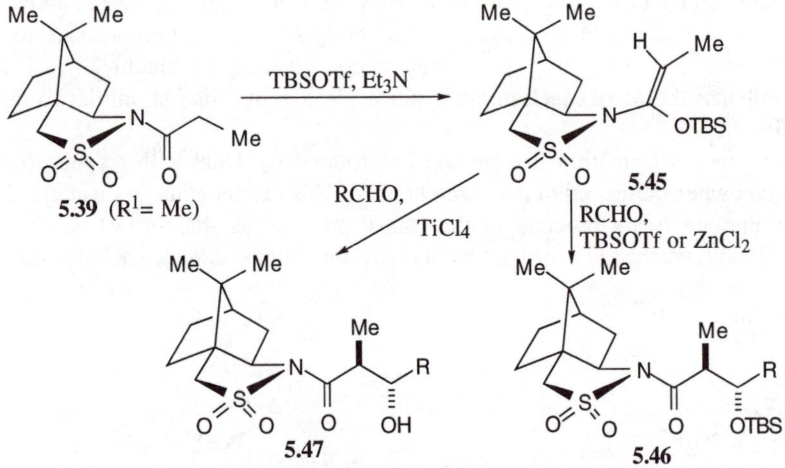

Fig. 5.26

Table 5.4 Stereoselectivity in aldol reactions of **5.39**

R	Lewis acid	d.e.[a]	Yield (%)
Ph	TBSOTf	98.5:1.5	89
p-NO$_2$C$_6$H$_4$	TBSOTf	95:5	<20
p-NO$_2$C$_6$H$_4$	ZnCl$_2$	97.2:2.8	80
Bu^i	ZnCl$_2$	98.9:1.1	78
Et	TiCl$_4$	>99:1	64
Bu^i	TiCl$_4$	>99:1	78

[a]Ratio of *anti*-isomer:*syn*-isomer

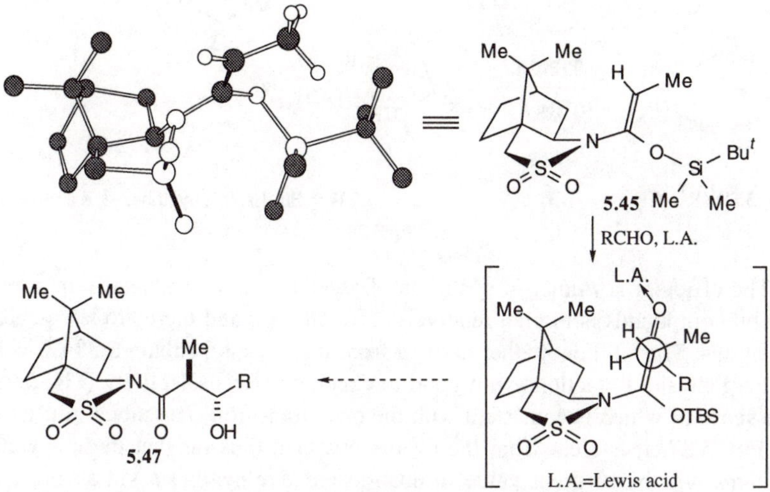

Fig. 5.27

transition state. For silyl enol ethers such as **5.45** this is no longer the case.

The 'counter-ion' (silicon) and the Lewis acid are different, and the transition state need not be cyclic. Indeed the aldol reactions of such silyl enol ethers are thought to proceed via acyclic or open transition states. The situation is analogous to that discussed for the aldol reactions with boron enolates derived from acylated oxazolidines in the presence of a Lewis acid (Fig. 5.21).

Analysis of the stereochemical outcome of reactions with open transition states will almost always be more difficult than those with cyclic transition states. A system with an open transition state has in principle many more degrees of freedom than a similar reaction with a cyclic transition state. A simplistic analogy may be made between these two situations and that of the conformations of cyclic and acyclic alkanes. Cyclohexane has two equivalent chair conformations which will be populated under normal conditions, whereas *n*-hexane has numerous low energy conformations which will be populated under equivalent conditions. Nevertheless, the extremely high level of stereoselectivity observed in the aldol reactions of **5.45** suggests that the open transition state might be highly ordered. Given that the structure of **5.45** is known in detail from X-ray crystallography, a reasonable model has been proposed for these reactions, and is presented in Fig. 5.27.[18]

A representation of the conformation of **5.45** derived from the X-ray crystal structure is shown in Fig. 5.27, as well as the proposed model for the open transition state for the Lewis acid mediated aldol reactions. In this model the coordinated aldehyde approaches the enol ether from the 'back' face, the other face of the double bond being blocked by the chiral auxiliary and the TBS group, with the R group of the aldehyde as far as possible from the bulk of the chiral auxiliary. This leads to an *anti*-aldol with the observed diastereoselectivity **5.47**.

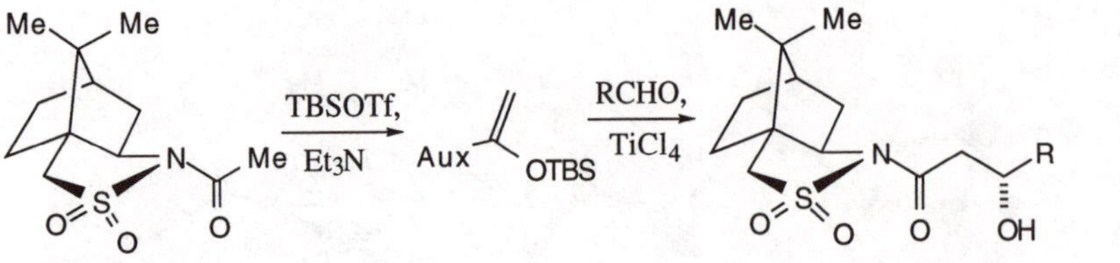

5.39 (R^1= H)

R = Ph, Et, Pr^i, cyclohexyl; d.e. 82–90%

Fig. 5.28

The efficient asymmetric synthesis of 'acetate' aldol products from simple acetyl compounds is often a relatively difficult task, and there are few general solutions. The silyl enol ether derived from the *N*-acetyl sultam **5.39** (R^1 = H) undergoes aldol reactions with good to excellent stereoselectivity (Fig. 5.28), the sense of which is consistent with the open transition state model illustrated in Fig. 5.27. It is clear from the results obtained thus far that these acylated sultams will be of great value in enantioselective synthesis via asymmetric aldol reactions.

Given the high levels of stereochemical control that can be achieved in enolate alkylation using chiral iron acyl compounds (Chapter 4), it is not surprising to find that such systems such as **5.48** (Fig. 5.29) are also useful in

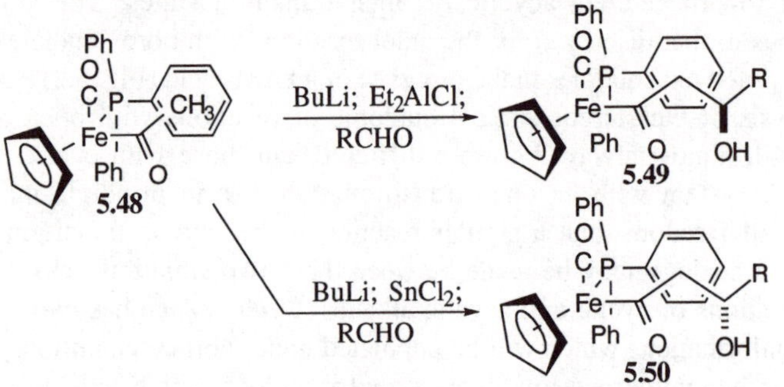

Fig. 5.29

Table 5.5   Stereoselectivity in aldol reactions of **5.48**

R	Lewis acid	d.e. (%)[a]
Me	Et$_2$AlCl	92(**4.49**)
Et	Et$_2$AlCl	>99(**4.49**)
Pr^i	Et$_2$AlCl	>99(**4.49**)
Et	SnCl$_2$	84(**4.50**)
Pr^i	SnCl$_2$	83(**4.50**)
Ph	SnCl$_2$	86(**4.50**)

[a]Major product in parentheses.

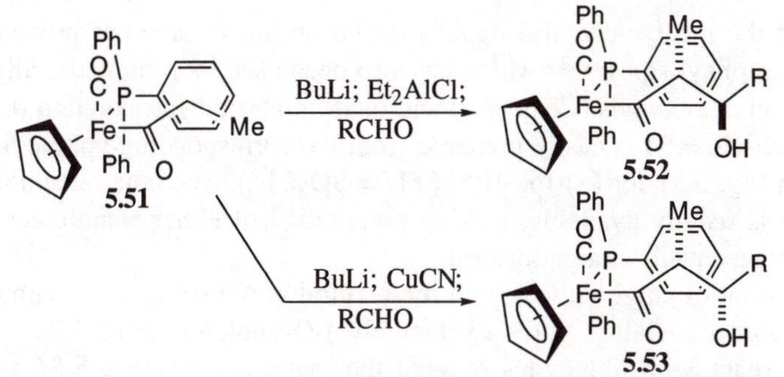

R = Me, Et, Pr^i, Bu^t. d.e. 97–100%

Fig. 5.30

asymmetric aldol reactions. In order to achieve high levels of stereoselectivity it is necessary to transmetallate from lithium, and depending upon the Lewis acid used *either* of the two possible diastereoisomers **5.49** or **5.50** can be made the major product (Fig. 5.29 and Table 5.5).[19,20]

The corresponding iron propionyl derivative **5.51** (Fig. 5.30) also undergoes highly diastereoselective aldol reactions, and again the stereochemistry of the product depends on the reagent used in the transmetallation step.[21] For a given enantiomer of the chiral auxiliary, the absolute stereochemistry at C-2 is independent of the transmetallation reagent. In effect, this reagent determines the absolute stereochemistry at C-3. In this way it is possible to prepare either the *anti*- or *syn*-aldol product **5.52** or **5.53** (Fig. 5.30).

The examples of asymmetric aldol reactions discussed so far have concentrated on the use of chiral auxiliaries to control the absolute stereochemistry of the products. Given that in many cases the reaction involves a cyclic transition state which includes a boron atom from the enolate, it might be expected that asymmetric synthesis could be achieved by using chiral ligands attached to this boron atom. In the simplest case where both enolate and aldehyde are achiral, this would amount to 'reagent control' of the reaction. This approach is attractive in that it would allow the asymmetric synthesis of aldol products from achiral ketones and aldehydes without the need for a 'normal' chiral auxiliary. In effect, the chiral auxiliary is attached to the ketone as a consequence of enolization using a chiral boron triflate, and is removed from the product during work-up.

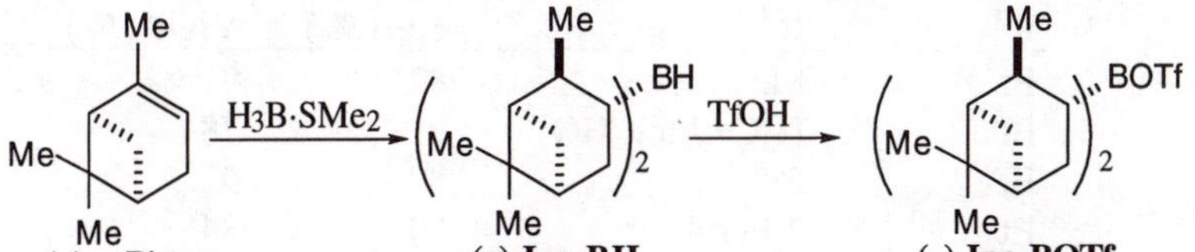

Fig. 5.31

One of the most successful ligands for boron in this area has proved to be isopinocampheyl (Ipc),[22,23] which has also been used in asymmetric allylation of carbonyl compounds (Chapter 3) and in asymmetric hydroboration (Chapter 6). The triflate **5.54** is easily prepared from the corresponding borane **5.55**, as shown in Fig. 5.31 for (−)-Ipc$_2$BOTf (Tf = SO$_2$CF$_3$). The other enantiomer of α-pinene is readily available, making preparation of either enantiomer of the boron triflate equally straightforward.

Enolization of ethyl ketones with these chiral boron triflates, as with simple dialkyl boron triflates, gives exclusively (Z)-enolates (Fig. 5.32). These enolates react with aldehydes to give the *syn*-aldol product **5.56** in good enantiomeric excess (Fig. 5.32 and Table 5.6). If the ethyl ketone is unsymmetrical, then enolization is usually selective for the less hindered substituent; in the cases shown in Fig. 5.32 this is always the ethyl group.

A transition state model which is consistent with these and related aldol reactions, and with molecular modelling studies has been developed and is illustrated in Fig. 5.33.[24] Of the various important chair transition states, that which would lead to the observed products **5.56** appears to be the most stable. The transition state which would lead to the enantiomeric products is destabilized by repulsion between a methyl group (Me*) of the Ipc ligand and the group R^1 of the enolate.

It is also possible to perform asymmetric aldol reactions of methyl ketones using this approach, but as is often the case with the related acetate aldol reactions, the enantioselectivity is somewhat lower. In addition, the stereoselectivity is opposite to that observed in reactions of ethyl ketones, for the same enantiomer of the chiral ligand. Nevertheless, the level of

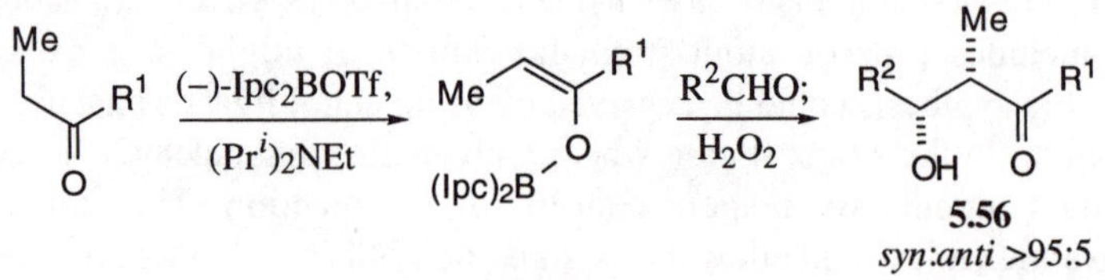

Fig. 5.32

Table 5.6 Stereoselectivity in aldol reactions of ethyl ketones using (−)-(Ipc)$_2$BOTf

R^1	R^2	e.e. (%)	Yield (%)
Et	Me	82	91
Et	H$_2$C=(Me)CHO	91	78
Et	Pr^n	80	92
Et	2-Furyl	80	84
Pr^i	H$_2$C=(Me)CHO	88	99
Bu^{i}CH$_2$	H$_2$C=(Me)CHO	86	79

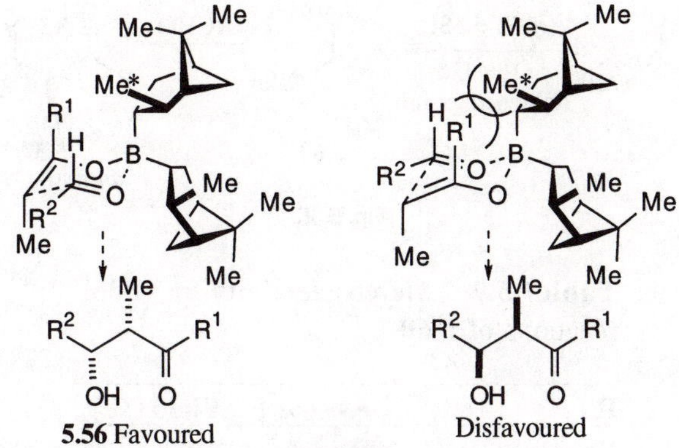

5.56 Favoured   Disfavoured

Fig. 5.33

enantioselectivity can be high enough to be useful in practice, as in the synthesis of **5.57** (Fig. 5.34), a precursor for a total synthesis of the marine natural product swinholide.[25]

Several other chiral boron reagents have been investigated for use in asymmetric aldol reactions,[26] but the only one which will be discussed here is **5.58** (Fig. 5.35).[27] This chiral reagent is analogous to that used in carbonyl allylation (Chapter 3) and Diels–Alder reactions (Chapter 6). Both enantiomers of the chiral diamine upon which this reagent is based are available, allowing the straightforward asymmetric synthesis of either enantiomer of the product.

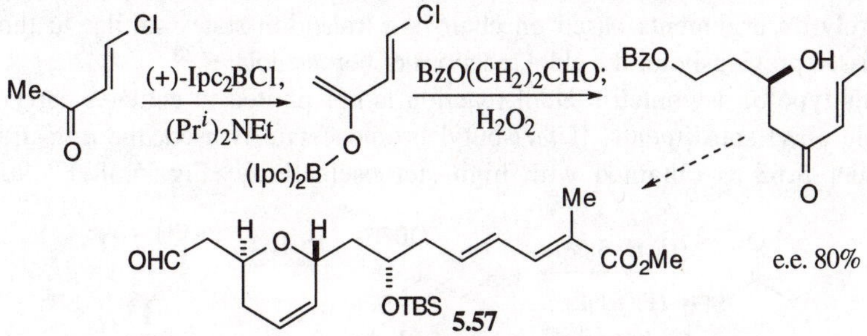

e.e. 80%

Fig. 5. 34

Control of the reaction conditions and the enol structure results in control of the relative stereochemistry of the products. If the *tert*-butyl propionyl or related

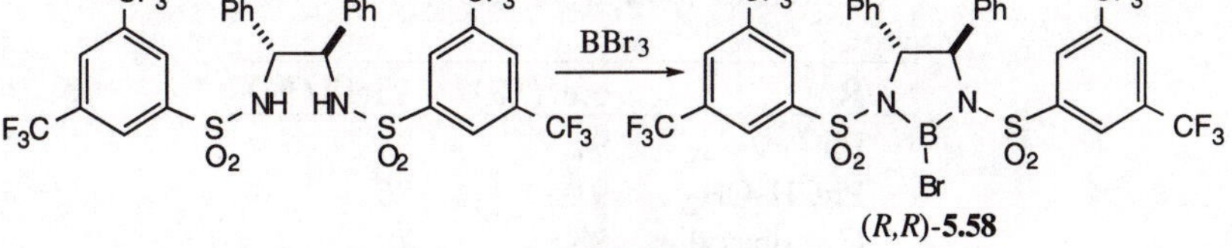

(*R*,*R*)-**5.58**

Fig. 5.35

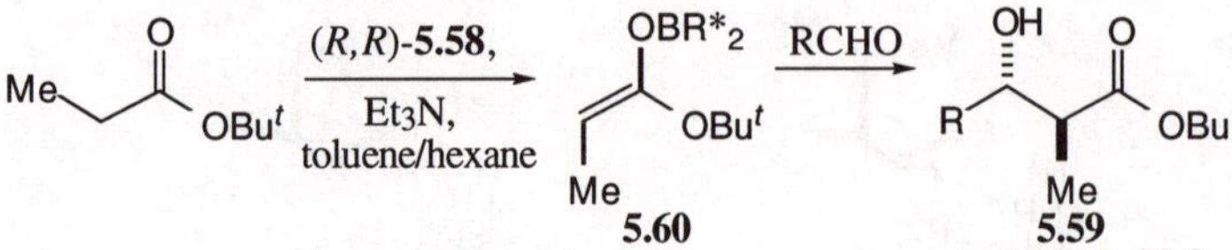

5.60 **5.59** *anti:syn* $\geq$ 98:2

Fig. 5.36

Table 5.7 Stereoselectivity in aldol reactions of **5.60**

R	**e.e. (%)**	**Yield (%)**
Ph	94	93
(*E*)-PhCH=CH	98	81
Cyclohexyl	82	75

ester is used in toluene, and with triethylamine as base, then *anti*-aldol products are obtained in high enantiomeric excess (Table 5.7). This is consistent with the intermediacy of the (*E*)-enolate **5.60** (Fig. 5.36).

The alternative *syn*-aldol product **5.61** is produced when the thiophenyl ester is enolized under the conditions shown in Fig. 5.37. This product arises from the reaction of the (*Z*)-enolate **5.62**, and as with the *tert*-butyl esters the enantioselectivity and diastereoselectivity are excellent (Fig. 5.37 and Table 5.8). Transition state models have been advanced to account for these reactions, and rely on arguments based on chair-like transition states similar to those discussed previously for the aldol reactions of boron enolates.[28]

This type of asymmetric aldol reaction is not limited to enolates carrying simple alkyl substituents. If *tert*-butyl bromoacetate is used, the *anti*-aldol product **5.63** is obtained with high stereoselectivity (Fig. 5.38).[29] Such

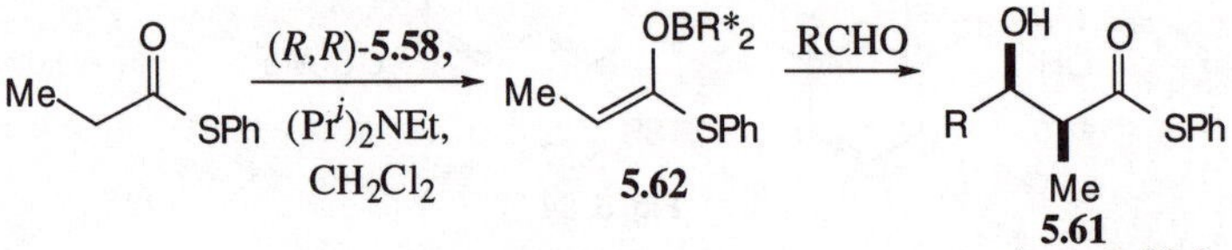

5.62 **5.61** *anti:syn* $\geq$ 98:2

Fig. 5.37

Table 5.8 Stereoselectivity in aldol reactions of **5.62**

R	**e.e. (%)**	**Yield (%)**
Ph	97	93
$PhCH_2CH_2$	91	86
Cyclohexyl	83	79

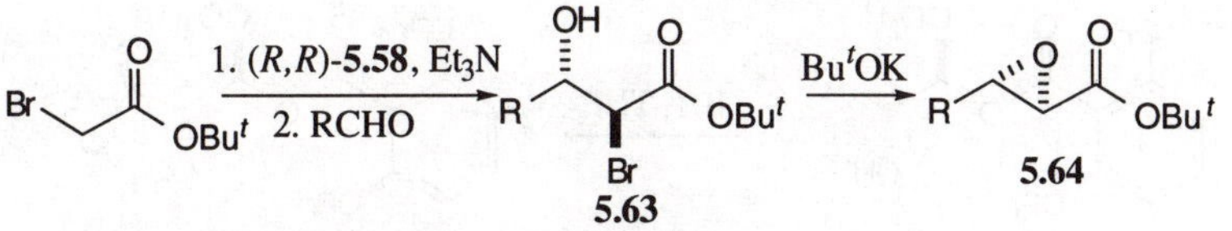

R = Ph, cyclohexyl, Ph(CH$_2$)$_2$, PhCH=CH
anti:syn ≥ 98:2 e.e. 91–98%

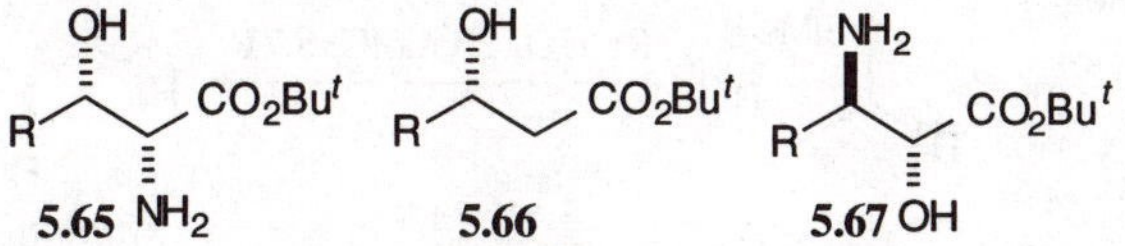

Fig. 5.38

products are valuable intermediates and can easily be transformed into the corresponding epoxide **5.64**, and other systems of interest such as **5.65**, **5.66**, and **5.67**.

Asymmetric aldol reactions are also successful with chiral reagents not based on boron. Complexes derived from tin(II) triflate, tri-*n*-butylfluorostannane, and a chiral diamine such as **5.68** will function as highly effective chiral mediators for aldol reactions with silyl enol ethers of thioesters (**5.69**, Fig. 5.39).[30] It is known that transmetallation of the enol ether is not involved, but detailed transition state models have yet to be proposed.

In aldol reactions, as with most types of reactions in asymmetric synthesis, asymmetric catalysis is the most attractive approach. Several such methods have been developed which clearly have the potential for widespread applications in organic synthesis, and selected examples are discussed in the final part of this chapter.

The complex formed on reaction of the tartaric acid derivative **5.70** with borane, formulated as **5.71**, functions as an asymmetric catalyst for the aldol reaction between silyl enol ethers and aldehydes (Fig. 5.40 and Table 5.9).[31] An interesting and practically useful aspect of this reaction is that the relative and absolute stereochemistry of the products are independent of the geometry of the silyl enol ether. This has been attributed to the fact that an open transition state is involved, and the dominant interaction is between R and R^2 (Fig. 5.40).

A related type of catalyst based on an even more readily available enantiomerically pure starting material, the natural amino acid tyrosine, has also

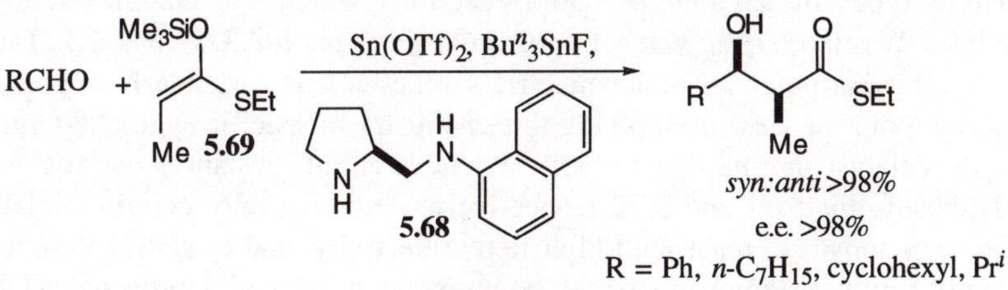

syn:anti >98%
e.e. >98%
R = Ph, *n*-C$_7$H$_{15}$, cyclohexyl, Pr^i

Fig. 5.39

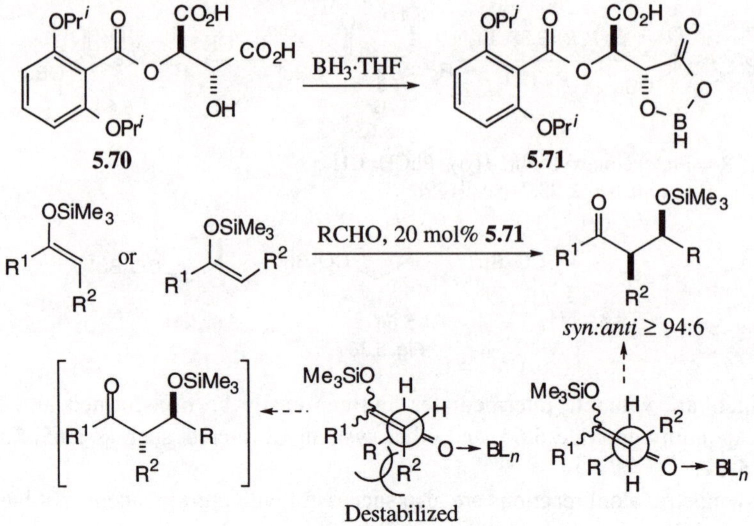

Fig. 5.40

Table 5.9 Stereoselectivity in aldol reactions catalysed by **5.71**

R^1	R^2	R	e.e. (%)	Yield (%)
Bu^n	H	Ph	85	81
Bu^n	H	Bu^n	80	70
Et	Me^a	Ph	96	96
Et	Me^b	Ph	94	97
$-(CH_2)_4-$		Ph	>95	57

[a] (E)-Silyl enol ether.
[b] (Z)-Silyl enol ether.

been used for the aldol reaction of silyl enol ethers derived from methyl ketones. The catalyst **5.72** (Fig. 5.41) is easily prepared as shown, and has been used to catalyse aldol reactions of simple silyl enol ethers, and of **5.73** (Table 5.10).[32] Products from the latter reaction are easily converted into dihydropyranones such as **5.74** which have considerable potential as synthetic intermediates.

These types of asymmetric aldol reactions which use enantioselective catalysis at relatively low levels (20 mole per cent for **5.71** and **5.72**) are clearly of great potential in asymmetric synthesis. It is particularly important from the point of view of applying this chemistry to specific synthetic targets that the catalysts are based on readily available chiral substances, tartaric acid and tyrosine for **5.71** and **5.72** respectively. Although only certain enolates have been shown to react with high stereoselectivity, and a relatively narrow range of product stereoisomers can be prepared at present, approaches which

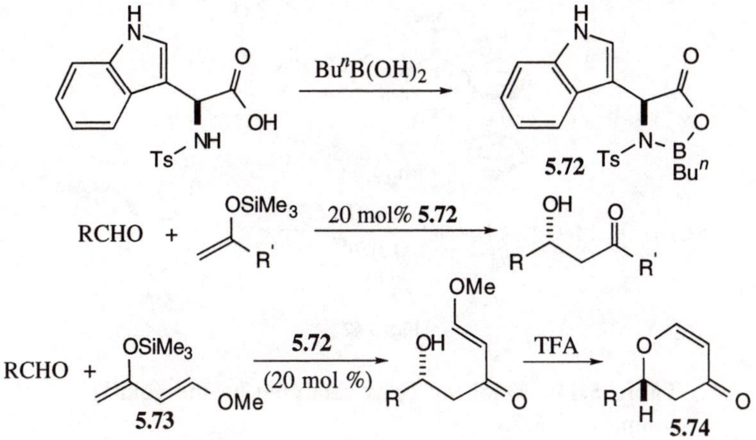

Fig. 5.41

Table 5.10 Aldol reactions catalysed by **5.72**

R	R'	e.e. (%)	Yield (%)
Ph	Ph	89	82
Cyclohexyl	Ph	93	67
Pr^n	Ph	89	94
2-Furyl	Ph	92	100
Ph	Bu^n	90	100
Ph	MeOCH=CH[a]	82	100
2-Furyl	MeOCH=CH[a]	67	83
Cyclohexyl	MeOCH=CH[a]	76	80
$PhCH_2CH_2$	MeOCH=CH[a]	69	57

[a]Silyl enol ether **5.73**.

involve asymmetric catalysis are certain to achieve great importance as the methodology develops. An asymmetric aldol process which takes place with very high enantioselectivity using very low levels of an enantioselective catalyst is provided by the following example.

Remarkable levels of stereoselectivity have been observed in an aldol reaction catalysed by gold(I) complexes of ligand **5.75** and related ligands (Fig. 5.42). This type of complex catalyses the addition of methyl isocyanoacetate **5.76** to aldehydes, which produces the *trans*-oxazolines **5.77** in extremely high enantiomeric excess (Table 5.11).[33] The relatively high cost of the ligand and metal salt are offset by the catalyst functioning at a level of 1 mole per cent, and the reactions taking place at room temperature. The reaction products **5.77** are useful intermediates for the synthesis of β-hydroxy-α-amino acids.

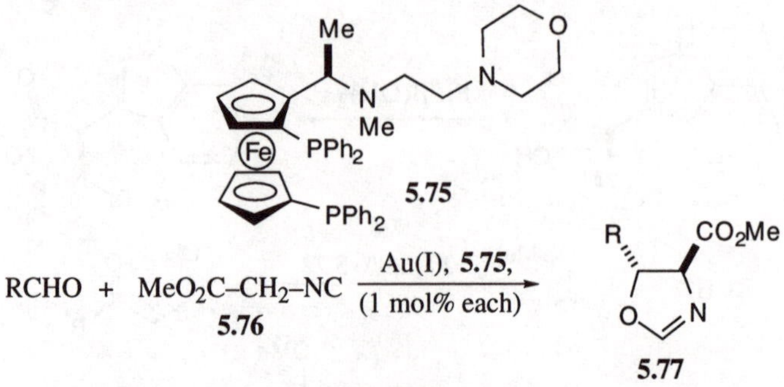

Fig. 5.42

Table 5.11 Aldol reactions catalysed by the gold(I) complex of **5.75**

R	*trans:cis*	e.e. (%)	Yield (%)
Ph	95:5	95	93
o-MeOC$_6$H$_4$	92:8	92	98
p-ClC$_6$H$_4$	94:6	94	97
Pr^n	87:17	92	85
Pr^i	99:1	92	100
Bu^t	100:0	97	94

Experiments involving changes in ligand structure, and careful n.m.r. measurements have led to the proposal that this reaction takes place through a three-coordinate gold complex in which the terminal amino group plays an important role.[34] A representation of the proposed transition state is shown in Fig. 5.43.

Throughout this chapter, emphasis has been laid on control of both relative and absolute stereochemistry in aldol reactions. In some cases it has proved possible to prepare two or three out of the four possible stereoisomeric aldol products. In principle, given control of both the stereochemistry and the facial selectivity of the enolate, then all four possible product stereoisomers could be produced. As a final example in this chapter, a case in which this has been

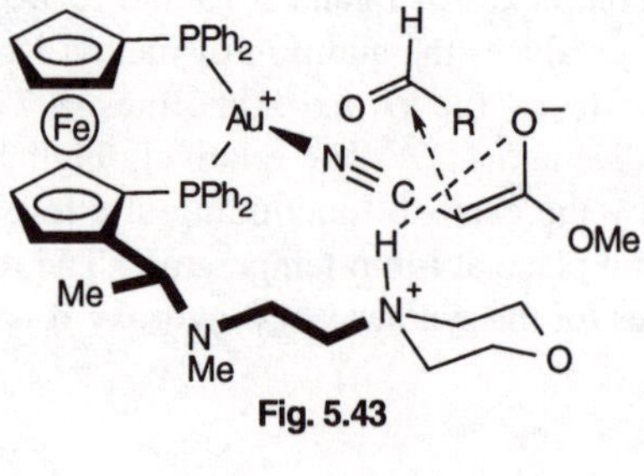

Fig. 5.43

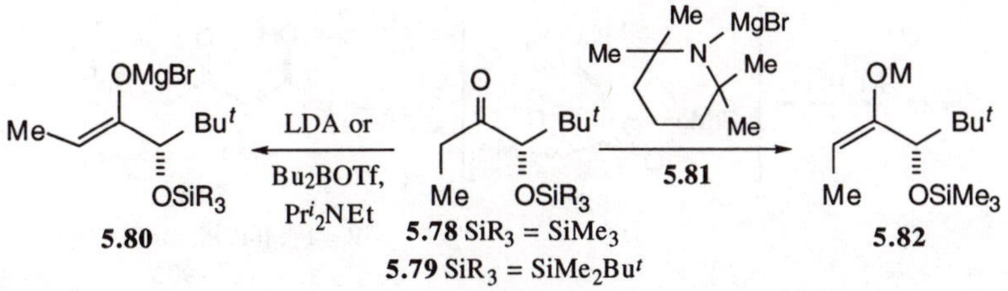

Fig. 5.44

achieved using aldol reactions of the chiral ketones **5.78** and **5.79** (Fig. 5.44) will be discussed.[35]

Enolization of ketone **5.78** with either LDA or dibutylboron triflate produces the (Z)-enolate **5.80**, whereas using **5.81** as base produces the (E)-enolate **5.82** (from either **5.78** or **5.79**). Reaction of the (Z)-boron enolate with aldehydes produces the *syn*-aldol products **5.83** (Fig. 5.45). The facial selectivity of this aldol reaction is consistent with transition state **5.84**, in which dipole–dipole interactions are minimized (cf. Fig. 5.11).

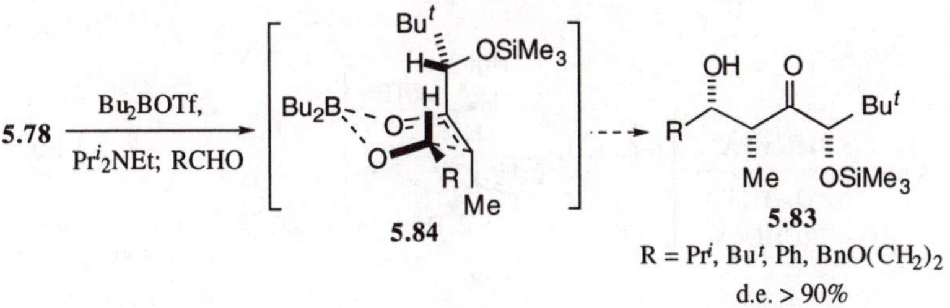

Fig. 5.45

As would be expected for aldol reactions which take place through Zimmerman–Traxler transition states, reaction of the (Z)-lithium enolate also gives a *syn*-aldol product. In this case reaction takes place from the opposite face of the lithium enolate compared with the (Z)-boron enolate to give **5.85**. This is consistent with a transition state **5.86** in which the oxygen of the OSiMe$_3$ group is chelated to the lithium (Fig. 5.46). Such chelation is not possible when boron is involved in a six-membered transition state because the four-coordinate boron cannot accept another ligand (cf. Fig. 5.11 and Fig. 5.23).

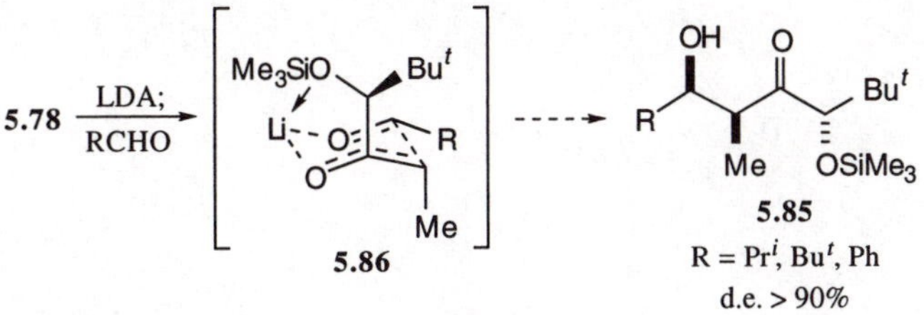

Fig. 5.46

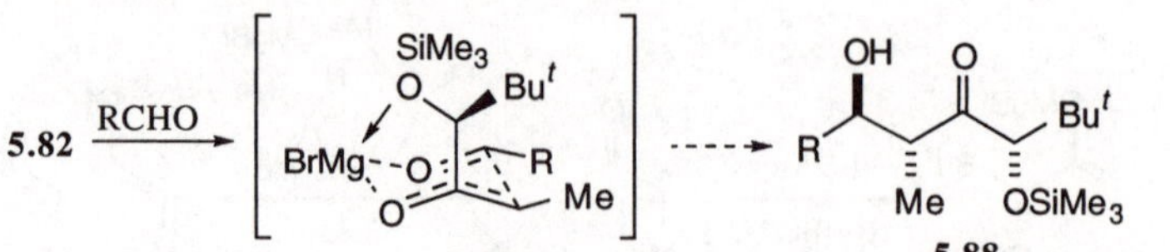

5.87

5.88

R = Pr^i, Bu^t, Ph, BnO(CH_2)$_2$

d.e. > 90%

Fig. 5.47

A similar transition state **5.87** with the oxygen of the $OSiMe_3$ group chelated to the counter ion is thought to be responsible for the formation of the *anti* aldol product **5.88** on reaction of the (*E*)-magnesium enolate **5.82** (Fig. 5.47).

The final possible product isomer **5.89** can be obtained by reaction of the (*E*)-enolate without chelation of the oxygen substituent to the counter ion. This can be achieved by formation of the (*E*)-magnesium enolate followed by transmetallation to the corresponding triisopropoxytitanium enolate (Fig. 5.48), and reaction is thought to take place through **5.90**.

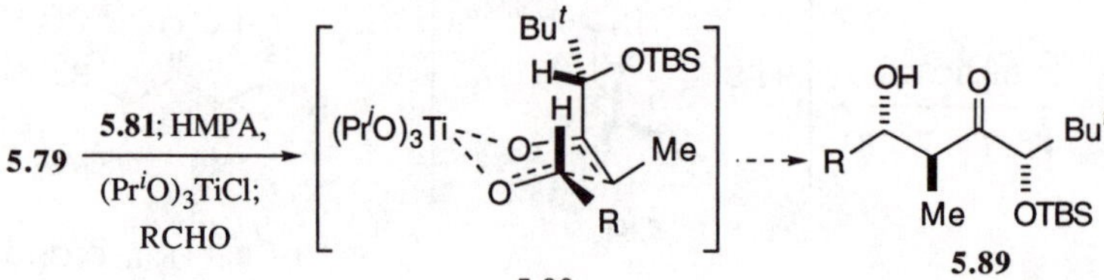

5.90

5.89

R = Me, Pr^i, Bu^t

d.e. > 90%

Fig. 5.48

The aldol reactions discussed in this chapter show that high enantioselectivity can be achieved in a number of different ways. Chiral auxiliary technology is well developed and is often efficient and predictable, and particularly exciting is the emergence of methodology using asymmetric catalysis with catalysts which are relatively easy to obtain. The powerful methods available for asymmetric aldol reactions provide invaluable reactions for use in asymmetric synthesis. Equally efficient and powerful methods for enantioselective synthesis are available for additions to C–C double bonds, which are the topic of the next chapter.

References

1. For reviews, see Heathcock, C. H. (1983), in *Asymmetric Synthesis*, (ed. J. D. Morrison), Vol 3, pp. 111–212, Academic Press, New York; Heathcock, C. H. (1990), *Aldrichimica Acta*, **23**, 99.
2. Ireland, R.E., Wipf, P., and Armstrong III, J. D. (1991), *J. Org. Chem.*, **56**, 650.
3. Evans, D. A. (1983), in *Asymmetric Synthesis*, (ed. J. D. Morrison), Vol. 3, pp. 9–11, Academic Press, New York.
4. Ref. 1, pp. 154–160; Li, Y., Paddon–Row, M. N., and Houk, K. N. (1988), *J. Amer. Chem. Soc.*, **110**, 3684.
5. Masamune, S. and Choy, W. (1982), *Aldrichimica Acta*, **15**, 47.
6. Masamune, S., Ali, S. A., Snitman, D. L., and Garvey, D. S. (1980), *Angew. Chem. Int. Ed. Engl.*, **19**, 557.
7. Masmune, S., Choy, W., Kerdesky, F. A. J., and Imperiali, B. (1981), *J. Amer. Chem. Soc.*, **103**, 1566.
8. Masamune, S., Hirama, M., Mori, S., Ali, S. A., and Garvey, D. S. (1981), *J. Amer. Chem. Soc.*, **103**, 1568.
9. Evans, D. A., Takacs, J. M., McGee, L. R., Mathre, D. J., and Bartroli, J. (1981), *Pure and Appl. Chem.*, **53**, 1109; Evans, D. A. (1982), *Aldrichimica Acta*, **53**, 23; Ref. 1, pp. 184–188.
10. Gage, J. and Evans, D. A. (1989), *Organic Syntheses*, **68**, 83, and references cited therein.
11. Evans, D. A., Bartrolli, J., and Shih, T. L. (1981), *J. Amer. Chem. Soc.*, **103**, 2127.
12. Evans, D. A., Bender, S. L., and Morris, J. (1988), *J. Amer. Chem. Soc.*, **110**, 2506.
13. Evans, D. A., Clark, J. S., Metternich, R., Novack, V. J., and Sheppard, G. S. (1990), *J. Amer. Chem. Soc.*, **112**, 866.
14. Evans, D. A., Ng, H. P., Clark, J. S., and Rieger, D. L. (1992), *Tetrahedron*, **48**, 2127.
15. Walker, M. A. and Heathcock, C. H. (1991), *J. Org. Chem.*, **56**, 5747.
16. Oppolzer, W., Chapuis, C., and Bernardinelli, G. (1984), *Helv. Chim. Acta*, **67**, 1397; Davis, F. A., Towson, J. C., Weismiller, M. C., Lal, S., and Carroll, P. J. (1988), *J. Amer. Chem. Soc.*, **110**, 8477.
17. Oppolzer, W., Blagg, J., Rodriguez, I., and Walther, E. (1990), *J. Amer. Chem. Soc.*, **112**, 2767.
18. Oppolzer, W., Starkemann, C., Rodriguez, I., and Bernardinelli, G. (1991), *Tetrahedron Lett.*, **32**, 61.
19. Davies, S. G., Dodor, I. M., and Warner, P. (1984), *J. Chem. Soc., Chem. Commun.*, 956.
20. Liebeskind, L. S., Welker, M. E., and Fengl, R. W. (1986), *J. Amer. Chem. Soc.*, **108**, 6328.
21. Davies, S. G., Dodor–Hedgecock, I. M., and Warner, P. (1985), *Tetrahedron Lett.*, **26**, 2125.
22. Meyers, A. I. and Yamamoyo, Y. (1984), *Tetrahedron*, **40**, 2309; Meyers, A. I. and Yamamoyo, Y. (1981), *J. Amer. Chem. Soc.*, **103**, 4278.
23. Paterson, I., Goodman, J. M., Lister, M. A., Schumann, R. C., McClure, C. K., and Norcross, R. D. (1990), *Tetrahedron*, **46**, 4663.
24. Bernardi, A., Capelli, A. M., Comotti, A., Gennari, C., Gardner, M., Goodman, J. M., and Paterson, I. (1991), *Tetrahedron*, **47**, 3471.
25. Paterson, I. and Smith, J. D. (1991), *Tetrahedron Lett.*, **34**, 5351.

26. Blanchette, M. A., Malamas, M. S., Nantz, M. H., Roberts, J. C., Somfai, P., Whritenour, D. C., Masamune, S., Kageyama, M., and Tamura, T. (1989), *J. Org. Chem.*, **54**, 2817; Masamune, S., Sato, T., Kim, B. M., and Wollmann, T. A. (1986), *J. Amer. Chem. Soc.*, **108**, 8279; Reetz, M. T., Kunisch, F., and Heitmann, P. (1986), *Tetrahedron Lett.*, **27**, 4721.
27. Corey, E. J. and Kim, S. S. (1990), *J. Amer. Chem. Soc.*, **112**, 4976.
28. Corey, E. J. and Lee, D. H. (1993), *Tetrahedron Lett.*, **34**, 1737.
29. Corey, E. J. and Choi, S. (1991), *Tetrahedron Lett.*, **32**, 2857.
30. Kobayashi, S., Uchiro, H., Fujishita, Y., Shiina, I., and Mukaiyama, T. (1991), *J. Amer. Chem. Soc.*, **113**, 4247.
31. Furuta, K., Maruyama, T., and Yamamoto, H. (1991), *J. Amer. Chem. Soc.*, **113**, 1041.
32. Corey, E. J., Cywin, C.L., and Roper, T.D. (1992), *Tetrahedron Lett.*, **33**, 6907.
33. Hayashi, T., Sawamura, M., and Ito, Y. (1992), *Tetrahedron*, **48**, 1999.
34. Sawamura, M., Ito, Y., and Hayashi, T. (1990), *Tetrahedron Lett.*, **31**, 2723.
35. Van Draanen, N. A., Arseniyadis, S., Crimmins, M. T., and Heathcock, C. H. (1991), *J. Org. Chem.*, **56**, 2499.

6 Additions to C–C double bonds

Additions to C–C double bonds represent an important general area for asymmetric synthesis, not least because several chiral centres can be established in a single reaction. One of the most thoroughly studied class of reactions of this type is the Diels–Alder reaction and related 4+2 cycloadditions.[1]

In its most general form the Diels–Alder reaction involves three important components, the dienophile, the diene, and the catalyst. The last is not always necessary, but it is often difficult to achieve high levels of stereoselectivity without the help of a Lewis acid catalyst. In addition to increasing the rate of reaction, the catalyst usually improves the regioselectivity and stereoselectivity of the cycloaddition.

A detailed discussion of the mechanism of Diels–Alder reactions, and related theory, is not within the scope of this volume. Nevertheless, a much simplified brief outline follows, in order to aid understanding of the models that have been proposed to account for the stereochemical outcome of the reactions which will be discussed.

The Diels–Alder reaction is generally considered to be a concerted cycloaddition in which both new σ bonds are formed more or less simultaneously. The most common type of Diels–Alder reaction involves a relatively electron-rich diene and an electron-deficient dienophile. Such reactions are often rationalized by consideration of the highest occupied molecular orbital (HOMO) of the diene and the lowest unoccupied orbital of the dienophile (LUMO). This is shown schematically in Fig. 6.1.[2]

One important consequence of the mechanism is that the stereochemistry at the termini of the diene is conserved in the adduct (Fig. 6.1). This means that the relative stereochemistry at these two centres in the adduct can be controlled predictably by choice of diene geometry.

In many Diels–Alder reactions chiral centres are likely to originate from the termini of the diene and the dienophile. A typical general case is shown in Fig. 6.2, in which the diene **6.1** carries an electron-donating group R. The dienophile **6.2** is electron deficient by virtue of the electron-withdrawing substituent Z. Assuming that neither reactant is chiral, there are now two diastereoisomeric transition states which lead to (racemic) diastereoisomeric

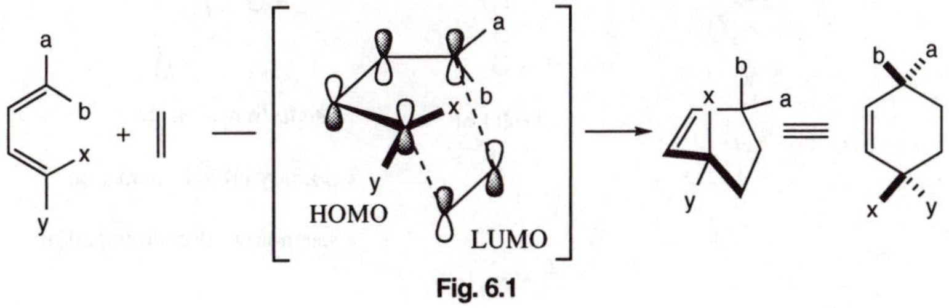

Fig. 6.1

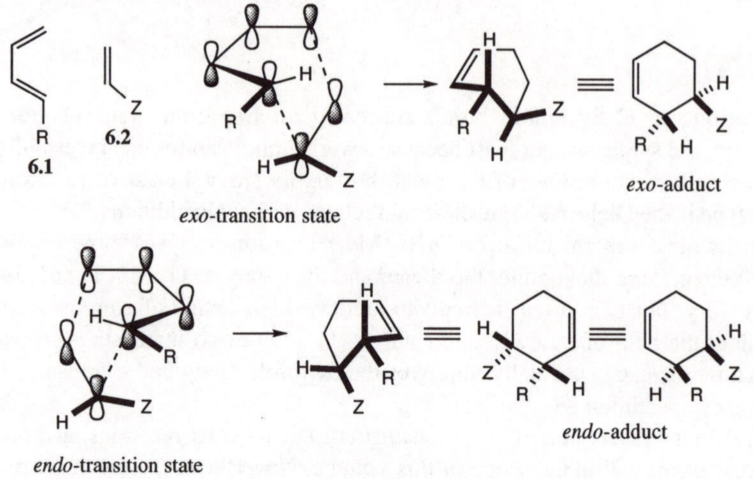

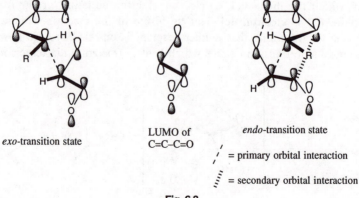

Fig. 6.2

products, the *endo*- and *exo*-adducts. The outcome of reaction from either face of the diene and from the same face of the dienophile is illustrated in Fig. 6.2.

The *exo*-transition state and the *exo*-adduct usually involve less steric interactions than the corresponding *endo*-transition state and adduct, but in most Diels–Alder cycloadditions under kinetic control it is the *endo*-adduct which predominates. The *endo*-adduct is usually thermodynamically less stable than the *exo*-adduct, but it is formed more rapidly. This is often accounted for by invoking 'secondary orbital interactions' (Fig. 6.3).

Primary orbital interactions result in bonding between the atoms involved, whereas secondary orbital interactions do not. In the case of the Diels–Alder reaction shown in Fig. 6.2, the *endo*-transition state is thought to be stabilized (compared to the *exo*-transition state) by the secondary orbital interaction shown in Fig. 6.3. The overall effect is to make the less stable *endo*-adduct the major

exo-transition state   LUMO of C=C–C=O   *endo*-transition state

/ ′ = primary orbital interaction

/ ′ = secondary orbital interaction

Fig. 6.3

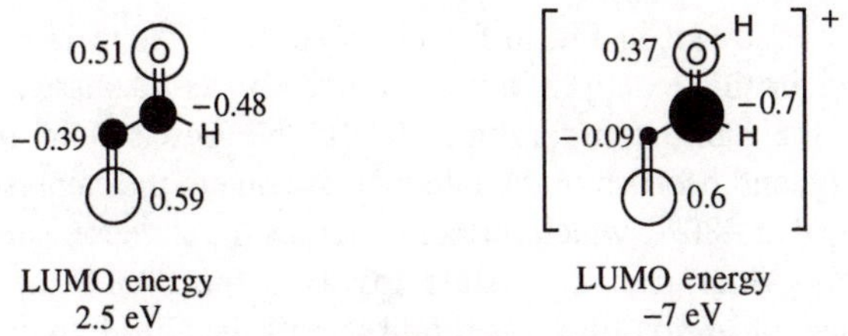

LUMO energy
2.5 eV

LUMO energy
−7 eV

Fig. 6.4

product under conditions of kinetic control, when the reaction is effectively irreversible. These are just the conditions which are encountered in asymmetric synthesis, depending as it does on differences in the relative energies of transition states.

As stated earlier in this chapter, Diels–Alder reactions are strongly affected by Lewis acid catalysts. The catalyst is thought to act by coordination to the dienophile, and this has several important effects on the LUMO of the dienophile. These are illustrated (Fig. 6.4) by reference to a typical dienophile (acrolein) and its protonated form which serves as a model for the dienophile–Lewis acid complex. The energy of the LUMO is lowered (2.5 eV to -7.0 eV), leading to a lowering of the energy gap between the HOMO of the diene and the LUMO of the dienophile and in consequence an increase in rate of reaction.[3] The polarization of the dienophile is also increased on coordination. This is represented by the orbital coefficients, which may loosely be considered as equivalent to the relative size and phase of the p orbitals which make up the LUMO. Often a larger difference in the coefficients of the C–C double bond will increase the regioselectivity, and the increase of the coefficient on the carbonyl carbon will increase the secondary orbital interaction (Fig. 6.3) in the transition state and lead to an increase in *endo*-selectivity.

A typical example of the effect of Lewis acid catalysis on a simple Diels–

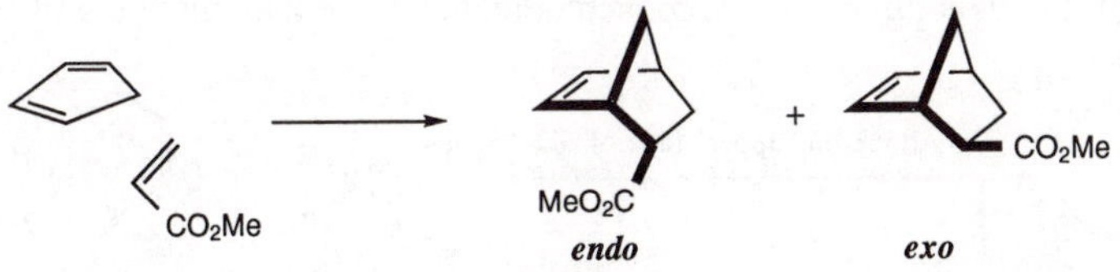

Fig. 6.5

Table 6.1   Effect of Lewis acid on a Diels–Alder addition

Catalyst	Temp. (°C)	*endo* (%)	*exo* (%)
None	0	88	12
AlCl₃	0	96	4
AlCl₃	-80	99	1

Alder reaction is shown in Fig. 6.5 and Table 6.1.[4] There is no question of regioselectivity in this example, but the other effects discussed above are in evidence. At the same temperature (0°C), the Lewis acid increases the stereoselectivity, and because of the rate increase due to the catalyst the reaction can be performed at −80°C which further increases the stereoselectivity.

As the Diels–Alder transition state involves the diene, dienophile, and catalyst, if one or more of these components is chiral then asymmetric synthesis becomes possible. A consideration of the general example, the cycloaddition of **6.1** and **6.2** (Fig. 6.2), reveals that there are four possible stereoisomeric products in a typical reaction, as shown in Fig. 6.6. If either **6.1** or **6.2** are chiral, then adducts **6.3**, **6.4**, **6.5**, and **6.6** are diastereoisomeric. Adducts **6.3** and **6.4** correspond to reaction from both faces of the diene on to the 'upper' face of the dienophile, and **6.5** and **6.6** arise from addition to the 'lower' face of the dienophile.

The absolute configuration at C-1 depends on which face of the dienophile **6.2** is attacked, and that at C-2 on which face of the diene **6.1** reacts. This situation is analogous to that already encountered in the analysis of asymmetric crotyl additions (Chapter 3) and aldol reactions (Chapter 5), and is typical of the situation whenever a reaction takes place in which both reactants are prochiral.

For efficient asymmetric Diels–Alder reactions one of the four possible adducts should predominate and high facial selectivity in one of the reactants is not a guarantee of high stereoselectivity. For example, a chiral dienophile **6.2** might react exclusively on the 'upper' face but still give a 1:1 mixture of *exo*- and *endo*-adducts **6.3** and **6.4**. Nevertheless, as discussed above, the *endo*-adduct usually predominates under the conditions of most enantioselective reactions and in practice it is often possible to achieve high levels of stereoselectivity.

The use of a dienophile attached to a chiral auxiliary has been the most intensively studied type of asymmetric Diels–Alder reaction. The chiral auxiliary is usually attached via a carbonyl group conjugated with the double bond of the dienophile. This approach demands more than blocking of one face

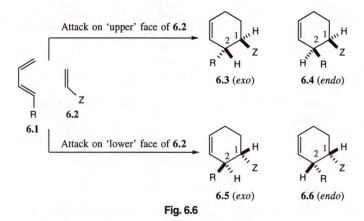

Fig. 6.6

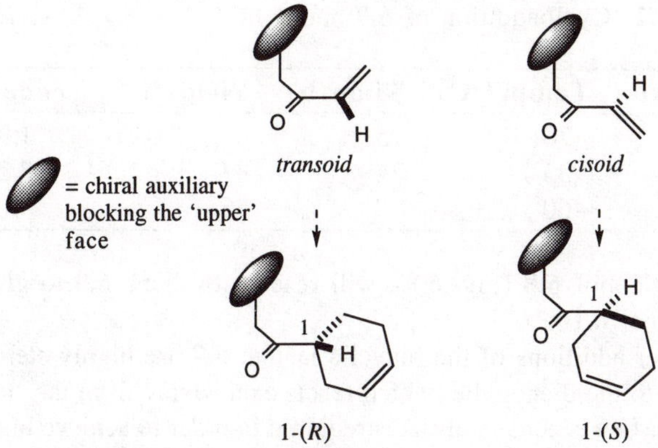

Fig. 6.7

of the dienophile. The orientation of the C–C double bond with respect to the chiral auxiliary must also be controlled, as illustrated in Fig. 6.7. Reaction is directed to the 'lower' face of the dienophile by the chiral auxiliary, but the absolute configuration at C-1 depends on whether reaction takes place from the *transoid*- or *cisoid*-conformation of the dienophile. Which of these is preferred can depend on the structural features of the dienophile and chiral auxiliary, and on the presence or absence of a catalyst.

There are few asymmetric Diels–Alder reactions controlled by chiral auxiliaries attached to the dienophile which give high levels of stereoselectivity in the absence of a catalyst. However, if the dienophile is sufficiently activated, then the reaction can occur under conditions which are mild enough to provide high diastereoisomeric excesses.

The bicyclic lactam **6.7**, prepared from the readily available enantiomerically

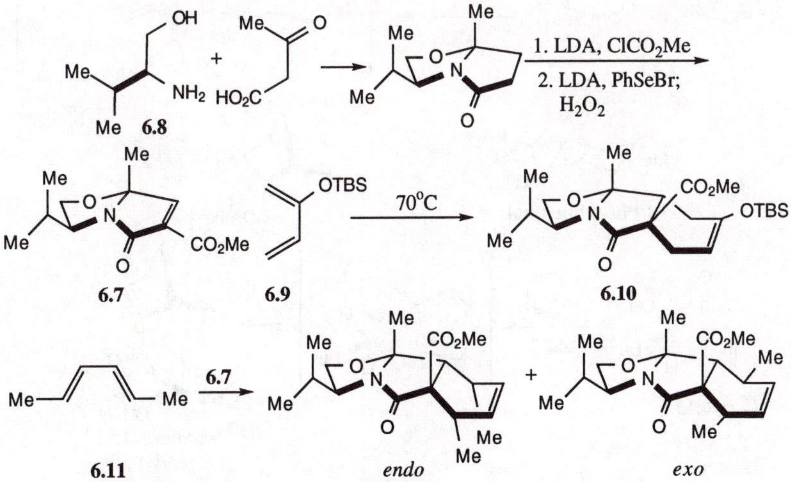

Fig. 6.8

Table 6.2 Cycloaddition of **6.7** and **6.11**

Catalyst	Temp. (°C)	Time (h)	Yield (%)	*endo:exo*
None	70	15	87	1.3:1
ZnCl$_2$	25	24	87	2.2:1
SnCl$_4$	−60	3	55	10:1

pure aminoalcohol **6.8** (Fig. 6.8), will react with diene **6.9** to give a single diastereoisomer **6.10**.[5]

Diels–Alder additions of the bicyclic lactam **6.7** are highly stereoselective with respect to the dienophile, which reacts exclusively from the 'lower' face. Nevertheless, Lewis acid catalysis is required in order to achieve high levels of regioselectivity and *endo*-cycloadditions; a typical example is shown in Fig. 6.8 and Table 6.2.

Many other chiral auxiliaries have been used in conjunction with Lewis acid catalysts, and selected examples will now be discussed. Efficient Lewis acid catalysis is possible using substrates which allow only single point complexation as for **6.12**, derived from 8-phenylmenthol[6] (Fig. 6.9), or with systems designed for chelation such as **6.13**[7] (Fig. 6.9). The high diastereoselectivity of its cycloadditions is consistent with the expected chelation, and an *endo*-reaction from a *cisoid*-conformation. Although the chiral auxiliary must be destroyed in its removal (similar to the use of this type of ketol in aldol reactions discussed in Chapter 5) this approach has been used to prepare intermediates for several natural product syntheses. The diastereoselectivity of Diels–Alder reactions of **6.13** and related ketols can be

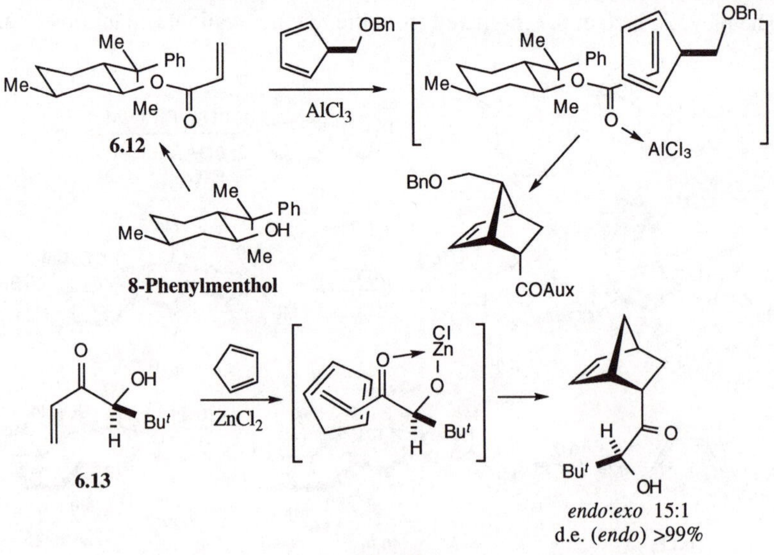

Fig. 6.9

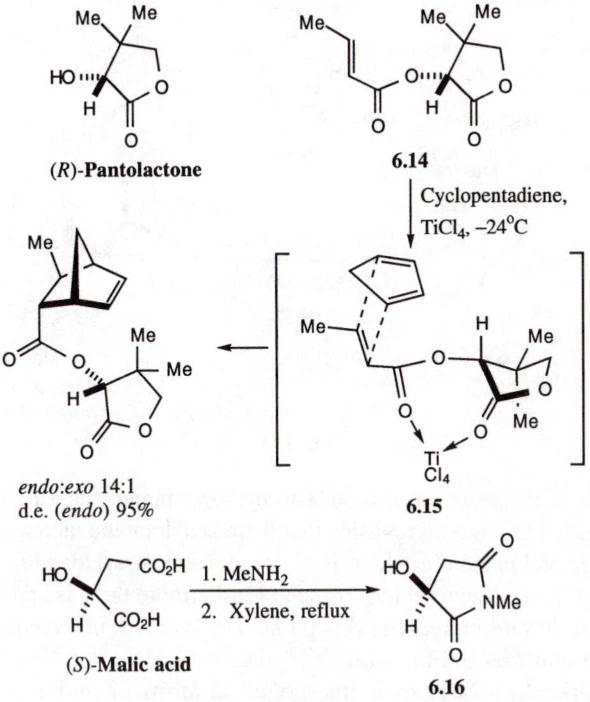

(R)-Pantolactone

6.14

Cyclopentadiene, TiCl₄, –24°C

endo:exo 14:1
d.e. (endo) 95%

6.15

(S)-Malic acid

1. MeNH₂
2. Xylene, reflux

6.16

Fig. 6.10

further enhanced by the use of chiral dienes, which allows for the principle of 'double asymmetric synthesis'[8] (Chapter 2). Two closely related systems have been developed which also rely on chelation of the Lewis acid to provide the necessary conformational rigidity. α,β-Unsaturated esters of (R)-pantolactone such as **6.14** (Fig. 6.10) undergo highly stereoselective *endo*-cycloaddition which is accounted for in terms of the chelated transition state **6.15** (Fig. 6.10).[9] (R)-Pantolactone is commercially available, but the other enantiomer is not, and to broaden the scope of this type of system the related (S)-lactam **6.16** was introduced (Fig. 6.10). This is readily prepared from (S)-malic acid and cycloadditions of α,β-unsaturated esters of this chiral auxiliary are also highly stereoselective. As a consequence of the opposite C-3 absolute configuration and an analogous transition state, the adducts correspond to addition from the opposite face of the dienophile (Fig. 6.10).

In the presence of dimethylaluminium chloride (DMAC), α,β-unsaturated *N*-acyl derivatives of the readily available chiral oxazolidinones **6.17, 6.18**, and **6.19** (Fig. 6.11) undergo *endo*-cycloaddition with high diastereoselectivity.[10] As above, and as previously discussed in relation to diastereoselective alkylation (Chapter 4) of *N*-acyl derivatives of these oxazolidinones, chelation is important in understanding these reactions. This Diels–Alder reaction has been studied particularly carefully, and the results of these studies will be considered briefly.

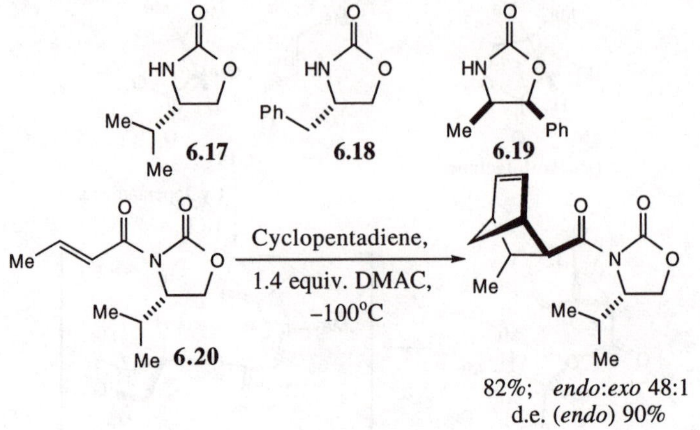

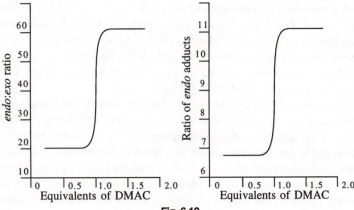

Fig. 6.11

This study of the reaction of **6.20** with cyclopentadiene as a function of the amount of added Lewis acid revealed that there is a dramatic increase in both the *endo:exo* ratio and in the diastereoselectivity of the *endo*-cycloaddition when one equivalent (or more) Lewis acid is present. Furthermore there is approximately a one hundredfold rate increase at this point. The increase in stereoselectivity is illustrated graphically in Fig. 6.12.

This remarkable behaviour is interpreted in terms of initial complexation followed by reversible ionization of the complex promoted by reaction with more Lewis acid. This is illustrated in Fig. 6.13. The positively charged intermediate **6.21** formed on ionization would be expected to be much less conformationally mobile than **6.22**, and much more reactive and stereoselective in Diels–Alder reactions. The major adducts from this and related acyloxazolidinones correspond to reaction via a transition state resembling **6.23**. Using **6.18** (Fig. 6.11) as the chiral auxiliary results in even higher

Fig. 6.12

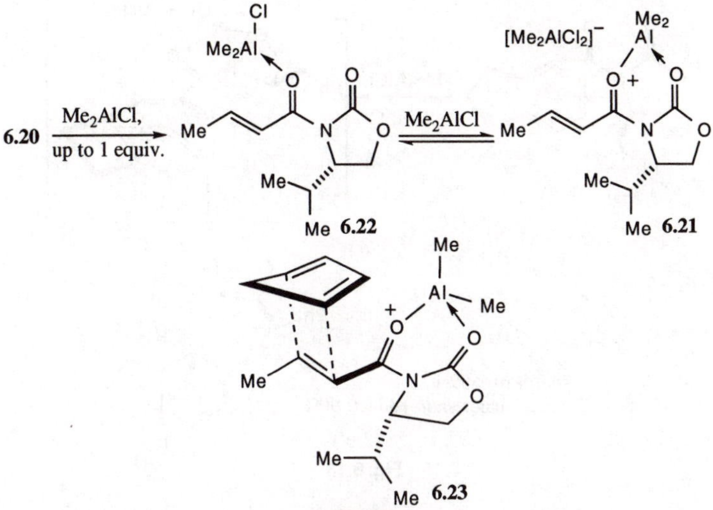

Fig. 6.13

levels of stereoselectivity (in the same sense as for **6.17**), thought to be a result of an 'enhanced steric effect' promoted by electronic interactions of the aromatic ring. High stereoselectivity in the opposite sense is observed when **6.19** is used as the chiral auxiliary. Analogous intermediates and transition states are involved but the opposite face of the dienophile is blocked by the oxazolidinone methyl group.

The adducts derived from these cycloadditions are easily purified and the chiral auxiliary can usually be removed cleanly. In a simple example of this methodology, α-terpineol **6.24** was prepared in very high (>98 per cent) enantiomeric excess as shown (Fig. 6.14).[10]

Oxazolidinone chiral auxiliaries can also control the stereochemical outcome

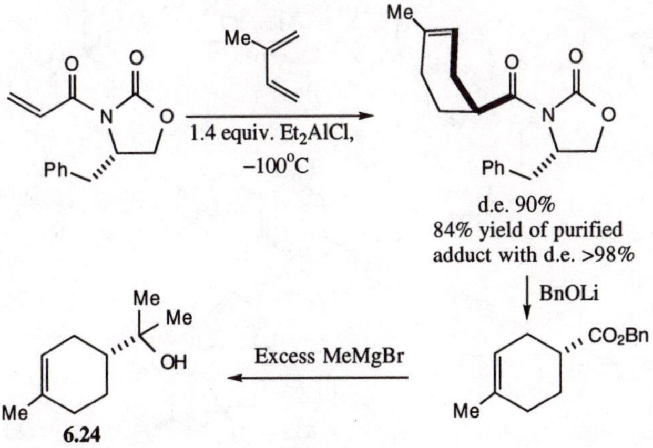

d.e. 90%
84% yield of purified
adduct with d.e. >98%

Fig. 6.14

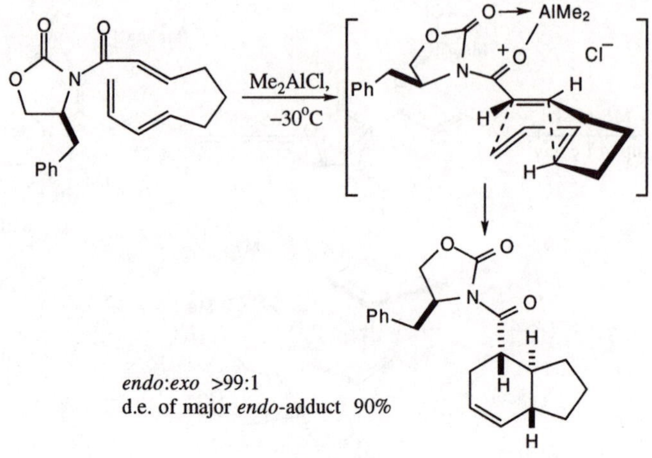

endo:*exo* >99:1
d.e. of major *endo*-adduct 90%

Fig. 6.15

of intramolecular Diels–Alder reactions. The reaction is highly *endo*-selective, and high diastereoisomeric excesses are obtained for the *endo*-adducts, as shown for a simple case in Fig. 6.15.

This intramolecular version of the Diels–Alder reaction was used as a key feature of a total synthesis of the aglycone **6.25** of the macrolide lepicidin.[11] This key cyclization is shown in Fig. 6.16, and illustrates the potential of such highly controlled and predictable cycloadditions in synthesis. In this example, the presence of other chiral centres and functionality does not interfere with the cycloaddition, which establishes two new rings and four new chiral centres in a

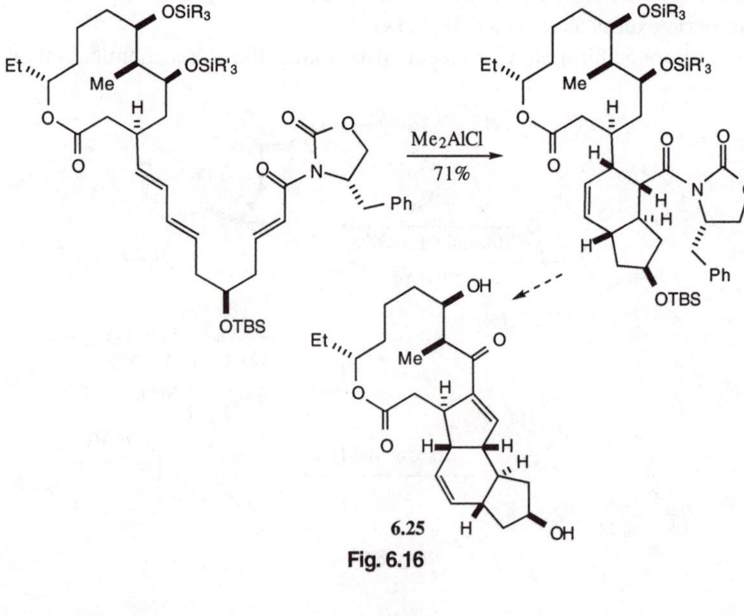

6.25

Fig. 6.16

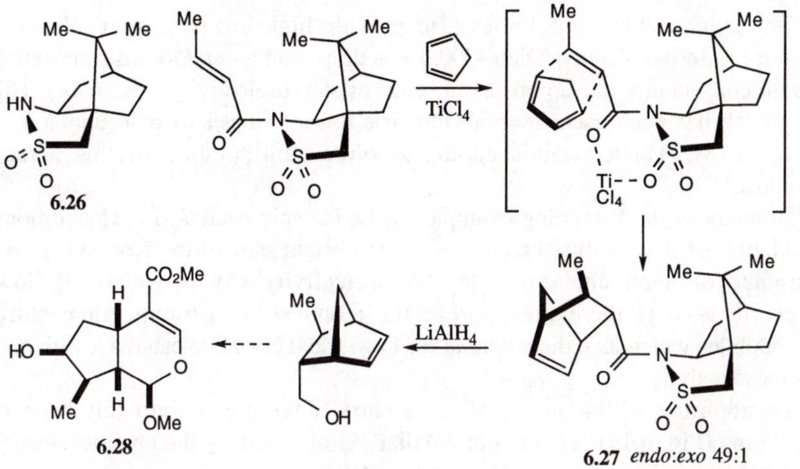

6.26

6.28

6.27 *endo:exo* 49:1
d.e. (*endo*) ≥97%

Fig. 6.17

single reaction.

α,β-Unsaturated derivatives of the sultam chiral auxiliary **6.26** (Fig. 6.17) and its enantiomer, both readily available (Chapter 4), will undergo Lewis acid catalysed Diels–Alder reactions with high stereoselectivity. A chelated transition state (Fig. 6.17) involving *endo*-addition is proposed to account for the observed stereoselectivity.[12] Adduct **6.27** has been used in an asymmetric total synthesis of 1-*O*-methyl loganin aglucone **6.28**.[13]

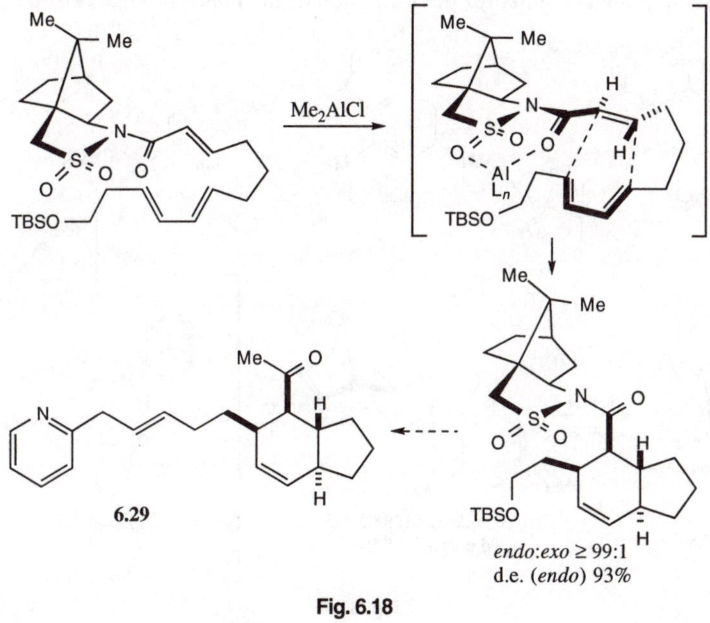

6.29

endo:exo ≥ 99:1
d.e. (*endo*) 93%

Fig. 6.18

The sultam chiral auxiliaries also provide high levels of stereochemical control in intramolecular Diels–Alder reactions, and an *endo*-transition state is again consistent with the observed sense of stereoselectivity (Fig. 6.18). This approach has been used in an asymmetric total synthesis of pulo'upone **6.29** (Fig. 6.18), which established the absolute configuration of this natural product.[14]

In many of the foregoing examples of Lewis acid catalysed cycloadditions, chelation of the catalyst has played an important part in the reaction. In the absence of such chelation the stereoselectivity can be relatively low. Nevertheless, if there are appropriate geometric constraints, then chiral dienophiles which lack the potential for Lewis acid chelation can react with high stereoselectivity.

Dienophiles which use **6.30** as a chiral auxiliary fit into this class of reactions (Fig. 6.19). This chiral auxiliary is obtained by the enantioselective reduction of the corresponding ketone **6.31** using **6.32** as an asymmetric catalyst (this type of reaction is discussed more fully in Chapter 7). Diels–Alder reaction of **6.33** with cyclopentadiene in the presence of dimethylaluminium chloride affords the *endo*-adduct in very high diastereoisomeric excess (Fig. 6.19), and the auxiliary is easily removed by alkaline hydrolysis.[15]

The trifluoromethyl group of **6.33** is locked between the *ortho*-methyl groups on the aromatic ring, and the plane of the aromatic ring is therefore approximately at right angles with the CF_3–C bond. The Lewis acid coordinates as shown, resulting in the dienophile C–C double bond adopting the *transoid*-conformation. The usual *endo*-approach of the diene from the less hindered face of the dienophile accounts for the formation of the major product. Furthermore,

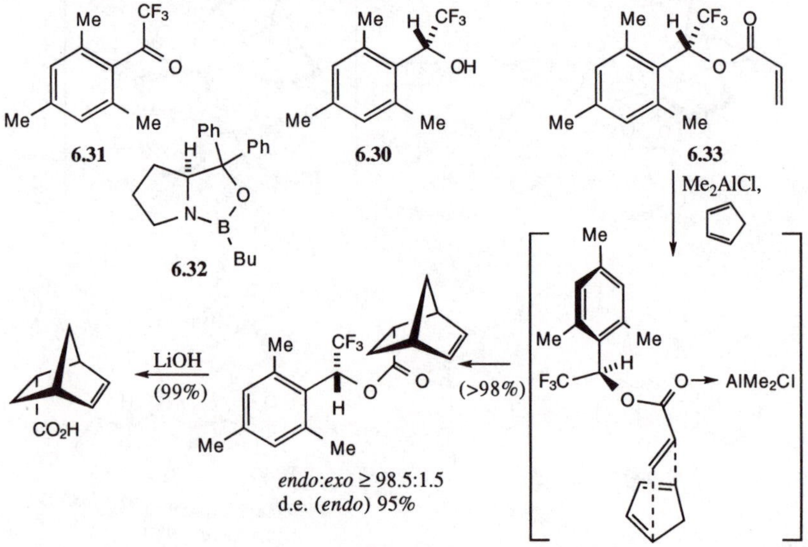

endo:exo ≥ 98.5:1.5
d.e. (*endo*) 95%

Fig. 6.19

as both enantiomers of this chiral auxiliary are readily available, either enantiomer of the final product may be obtained.

In principle, asymmetric Diels–Alder reactions could be carried out by using a chiral diene rather than the dienophile. In practice, this is a much less successful approach. In a normal Diels–Alder reaction the dienophile is electron deficient, often by virtue of a carbonyl group, providing for convenient linkage to, and removal of, the various chiral auxiliaries. Most chiral dienes which have been used consist of a chiral group linked to the diene through oxygen. This poses several potential problems, not the least of which is the synthesis of the required dienes. Moreover, such dienes are very electron rich and tend to be rather unstable towards acids, including Lewis acids commonly used as catalysts in Diels–Alder reactions. Finally, careful design is necessary in order for the chiral group to be removed easily, as simple ethers are often very difficult to cleave under mild conditions.

In spite of these difficulties, several classes of such dienes have been used in asymmetric Diels–Alder reactions, and some examples are presented in Fig. 6.20.[16]

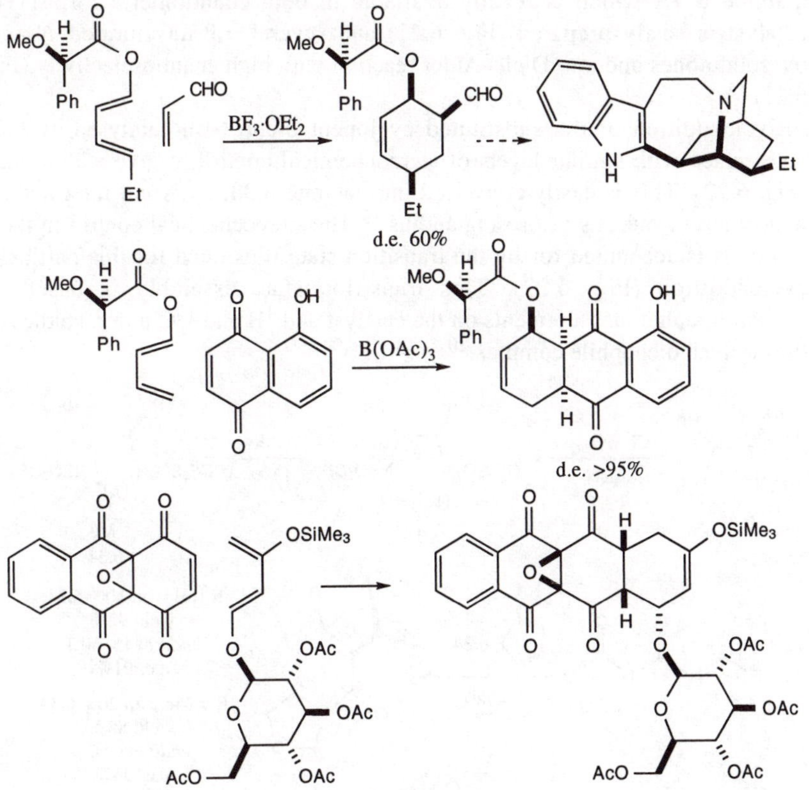

Fig. 6.20

In principle, an even more attractive general approach to asymmetric Diels–Alder reactions than using chiral auxiliaries is to use a chiral Lewis acid as catalyst. Neither dienophile nor diene require any 'extra' chiral groups attaching and there are no extra steps to cleave chiral auxiliaries. Such Lewis acid mediated cycloadditions are subject to the requirements that the enantiomerically pure catalyst is readily available, and that the stereoselectivity is high. The latter is important as addition from either face of the dienophile gives rise to enantiomers rather than diastereoisomers as when a chiral auxiliary is used. In spite of these difficulties, various efficient catalyst systems have been developed, especially involving aluminium, titanium, and boron Lewis acids, and this approach is rapidly becoming a feasible alternative to methods involving chiral auxiliaries.

Chiral Lewis acidic aluminium complexes have provided some outstanding results in this area. Aluminium complexes derived from menthol give moderate to good enantioselectivity, but the enantioselectivity is highly dependent on the diene used.[17] Very high enantioselectivity can be achieved by using the Lewis acid **6.34** (Fig. 6.21). This reagent is one of a family of useful chiral controllers for a number of important reactions (Chapters 3 and 5) based on the diamine **6.35**, which is readily available in both enantiomeric forms. The catalyst is easily prepared (Fig. 6.21) and several α,β-unsaturated *N*-acyl oxazolidinones undergo Diels–Alder reaction with high enantioselectivity (Fig. 6.21).[18]

Cycloaddition of the substituted cyclopentadiene **6.36** catalysed by **6.34** takes place with similar levels of stereochemical control to give adduct **6.37** (Fig. 6.22). This is easily converted into lactone **6.38**, a pivotal intermediate in a general synthesis of prostaglandins.[19] The stereochemical control in these reactions is accounted for by the transition state illustrated for this particular cycloaddition (Fig. 6.22). This transition state assembly is based on crystallographic measurements on the catalyst and ^{1}H and ^{13}C n.m.r. studies on the catalyst–dienophile complex.[20]

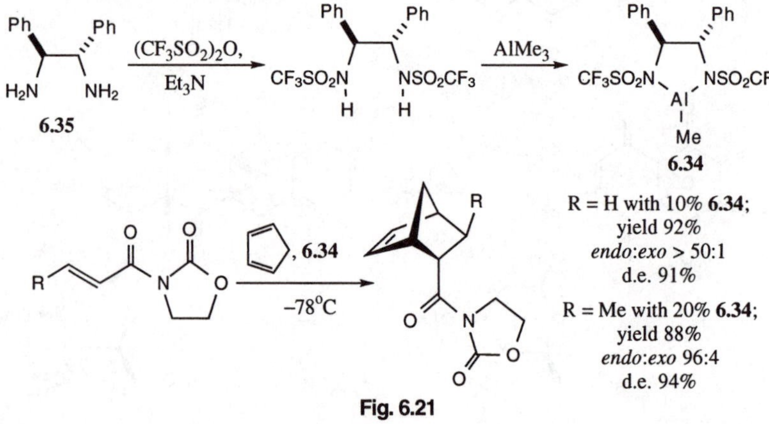
Ph
Ph
H2N
NH2
6.35
(CF3SO2)2O,
Et3N
CF3SO2N
NSO2CF3
H
H
AlMe3
Al
Me
6.34
R
O
O
N
O
, 6.34
−78°C
R = H with 10% 6.34;
yield 92%
endo:exo > 50:1
d.e. 91%
R = Me with 20% 6.34;
yield 88%
endo:exo 96:4
d.e. 94%

Fig. 6.21

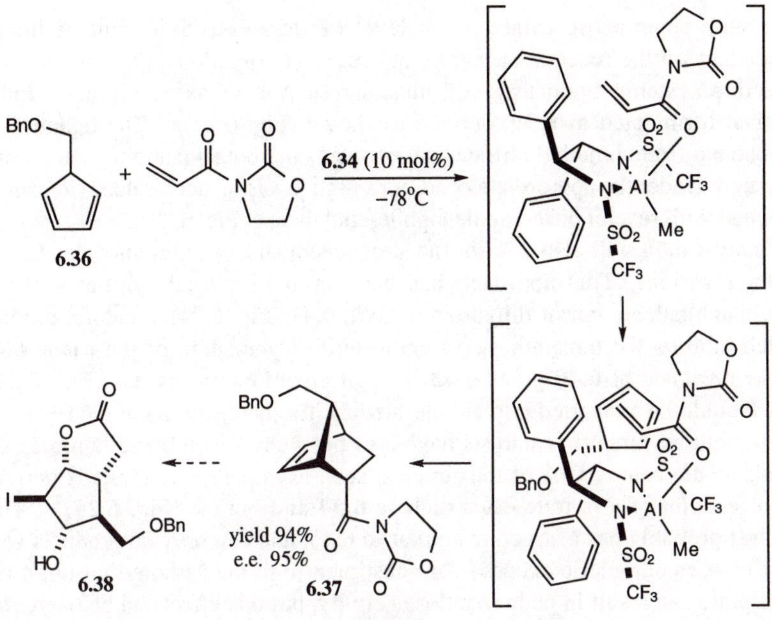

Fig. 6.22

Various chiral titanium(IV) complexes have been used as asymmetric catalysts for Diels–Alder reactions. The most successful catalysts use chiral chelating diols derived from either enantiomer of tartaric acid.[21] As is often the

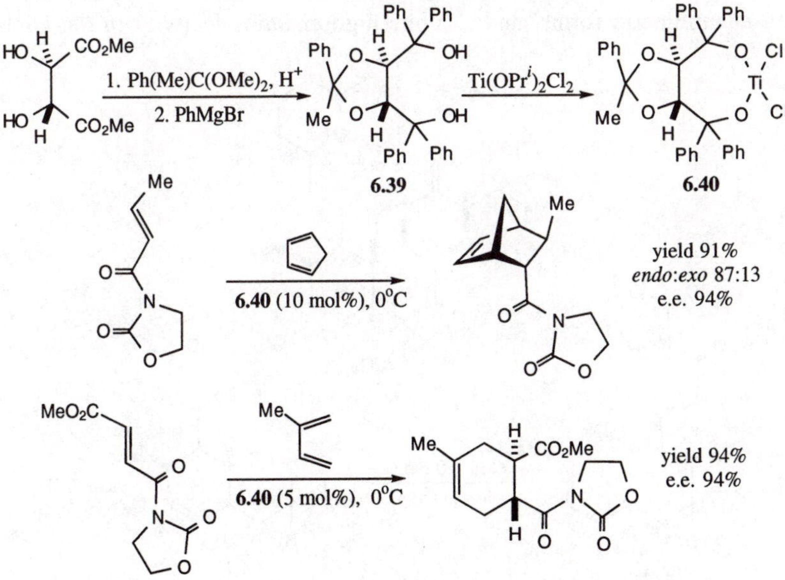

Fig. 6.23

case with asymmetric catalysis the level of stereoselectivity can be highly dependent on the reactant structure and reaction conditions. One of the most effective systems again uses α,β-unsaturated *N*-acyl oxazolidinones and a catalyst formulated as **6.39**, derived as shown (Fig. 6.23).[22] The ligand **6.40** is obtained from dimethyl tartrate as illustrated, and both enantiomers are easily prepared. Under the appropriate conditions high levels of stereoselectivity can be obtained with several different dienophiles and dienes (Fig. 6.23).

Catalyst **6.40** will also control the stereochemistry in intramolecular Diels–Alder reactions. This approach has been used in a total synthesis of the hydronaphthalene unit of dihydromevinolin **6.41** (Fig. 6.24), a natural product which inhibits the biosynthesis of cholesterol. Cyclization of the triene **6.42** in the presence of **6.40** gives **6.43** in high enantiomeric excess (Fig. 6.24), which could be converted into an intermediate for the synthesis of **6.41**.[23]

Efficient asymmetric catalysts have also been developed using boron as the Lewis acidic centre. One of the simplest such asymmetric catalysts is derived from reaction of a tartrate ester such as **6.44** and borane (Fig. 6.25).[24] This catalyst probably has a structure similar to **6.45**, and can provide products with very high enantiomeric excesses. A substituent β to the carbonyl group of the dienophile can result in poor enantioselectivity, but this effect can be overcome by the presence of another substituent α to the carbonyl group (Fig. 6.25). It is interesting to note that the *exo*-diastereoisomer is favoured in this last class of reactions. As with the titanium(IV) catalyst **6.40** (Fig. 6.23), this type of boron derivative will also catalyse enantioselective intramolecular Diels–Alder reactions.

The simple chiral dichloroborane catalyst **6.46** is relatively easy to obtain in both enantiomeric forms, and exhibits high enantioselectivity in the Diels–

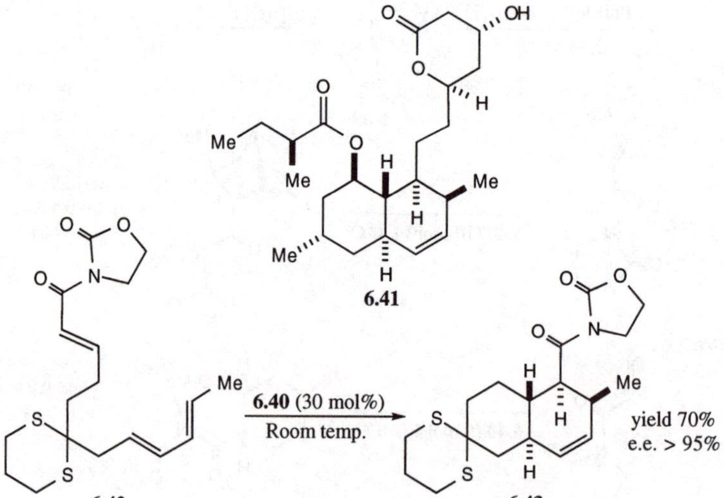

Fig. 6.24

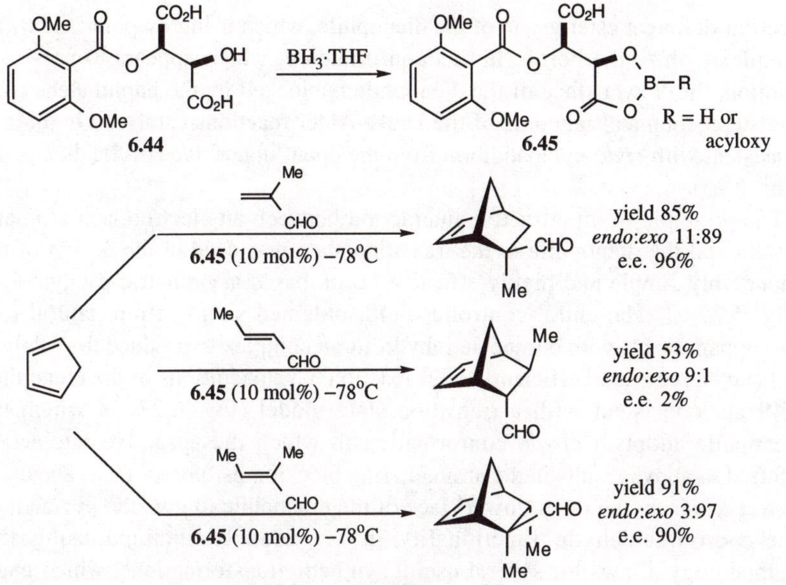

Fig. 6.25

Alder reaction (Fig. 6.26).[25] The proposed transition state (Fig. 6.26) is supported by an X-ray crystal structure of the complex between **6.46** and a dienophile (methyl (*E*)-2-butenoate).

The conformation of this complex is thought to be stabilized by electrostatic and dipole–dipole interactions between the electron-rich aromatic system and the

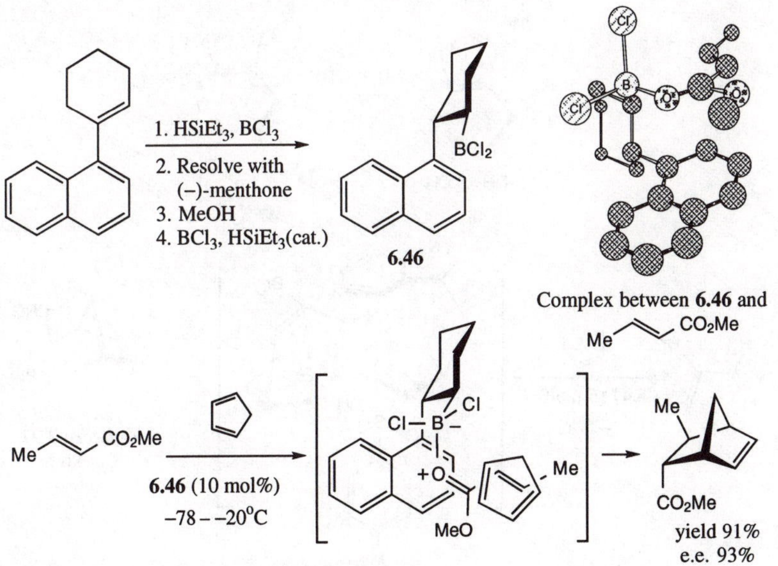

Complex between **6.46** and

Fig. 6.26

electron-deficient ester group of the dienophile, which is highly polarized by its complexation to the boron. In this conformation, which appears to persist in solution, the 'lower' face of the dienophile is blocked by the naphthalene ring. The stereochemical outcome of the Diels–Alder reactions catalysed by **6.46** is consistent with *endo*-cycloaddition from the open 'upper' face of the dienophile (Fig. 6.26).

The principle of an attractive interaction between an electron-rich aromatic system and the dienophile in the transition state was used in the design of the remarkably simple and highly effective boron-based asymmetric catalyst **6.47** (Fig. 6.27).[26] The chiral controller **6.48**, obtained simply from natural (*S*)-tryptophan, reacts with borane–tetrahydrofuran complex to produce the catalyst, and can be recovered efficiently after reaction. Cycloadditions of the dienophile **6.49** are consistent with a transition state model (Fig. 6.27) in which the dienophile adopts a *cisoid*-conformation in which the attractive interaction referred to above results in the aromatic ring blocking its 'upper' face. The diene then reacts from the open 'lower' face of the dienophile to give the *exo*-adduct. The α-bromoaldehyde functionality in the adducts obtained using this methodology allows for several useful synthetic transformations, which make these adducts versatile synthetic intermediates.

Compared to the Diels–Alder reaction, less progress has been made in the development of enantioselective versions of other types of additions to C–C double bonds. This is not too surprising given the mechanistic characteristics of

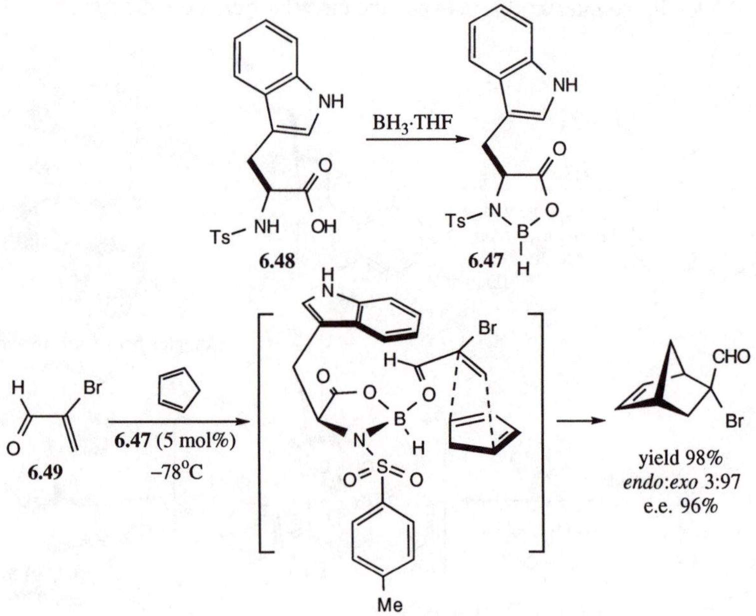

Fig. 6.27

the Diels–Alder reaction, and its great importance in synthesis. The following section illustrates selected examples of some asymmetric cycloadditions which could be of general application.

Additions of 1,3-dipoles to C–C double bonds are mechanistically similar to the Diels–Alder reaction, being concerted cycloadditions involving six π electrons. They differ in that a five-membered ring is formed as the 1,3-dipole supplies only three atoms to the new ring, and in that at least one of the atoms derived from the 1,3-dipole is a heteroatom. Such 1,3-dipolar cycloadditions are often highly regioselective and stereoselective. However, when the product possesses two chiral centres, one from each component (1,3-dipole and dipolarophile), neither the level nor the sense of diastereoselectivity is easy to predict. In effect, compared to the Diels–Alder reaction the '*endo*-rule' is often much less important, and steric effects can dominate and so favour the sterically less hindered *exo* product.

1,3-Dipolar cycloadditions of nitrile oxides are not complicated by this as no new chiral centres originate from the 1,3-dipole, and highly diastereoselective cycloadditions have been observed using sulphonamide-based chiral auxiliaries. The sultam **6.26** (Fig. 6.17) has been used as a chiral auxiliary to control the stereoselectivity of nitrile oxide cycloadditions, and although the level of

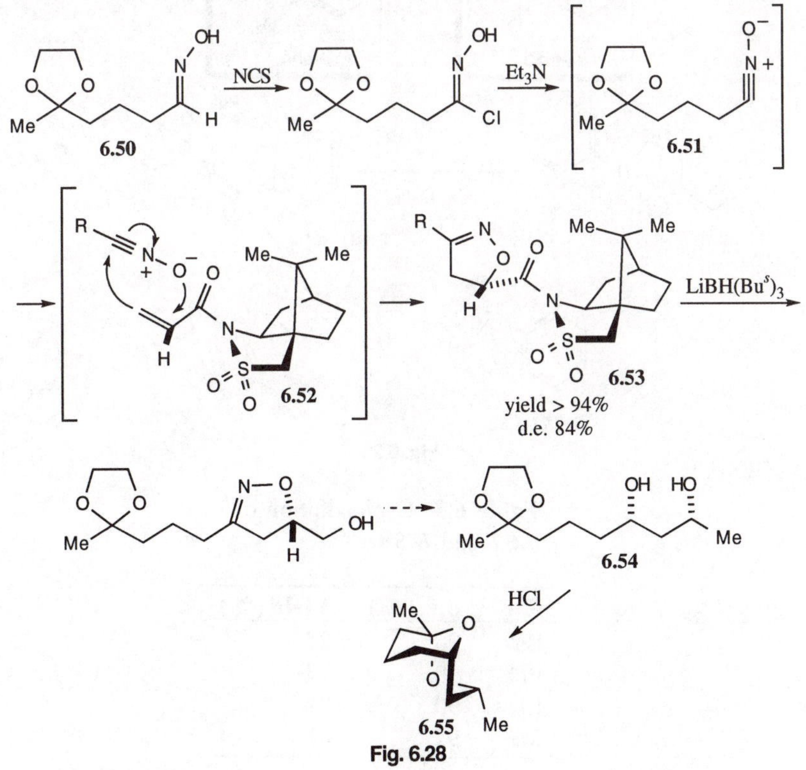

Fig. 6.28

stereoselectivity varies considerably with the structure of the dienophile, this approach has been used in the synthesis of some simple natural products (Fig. 6.28).[27]

Several general methods are available for the formation of nitrile oxides, which usually must be generated *in situ*, and one such method is illustrated in Fig. 6.28. Treatment of an aldoxime **6.50** with *N*-chlorosuccinimide forms the chloro-oxime, which readily eliminates to the nitrile oxide **6.51** on treatment with weak base. Generation of **6.51** in the presence of the dipolarophile **6.52** results in efficient diastereoselective 1,3-dipolar cycloaddition to give **6.53**.

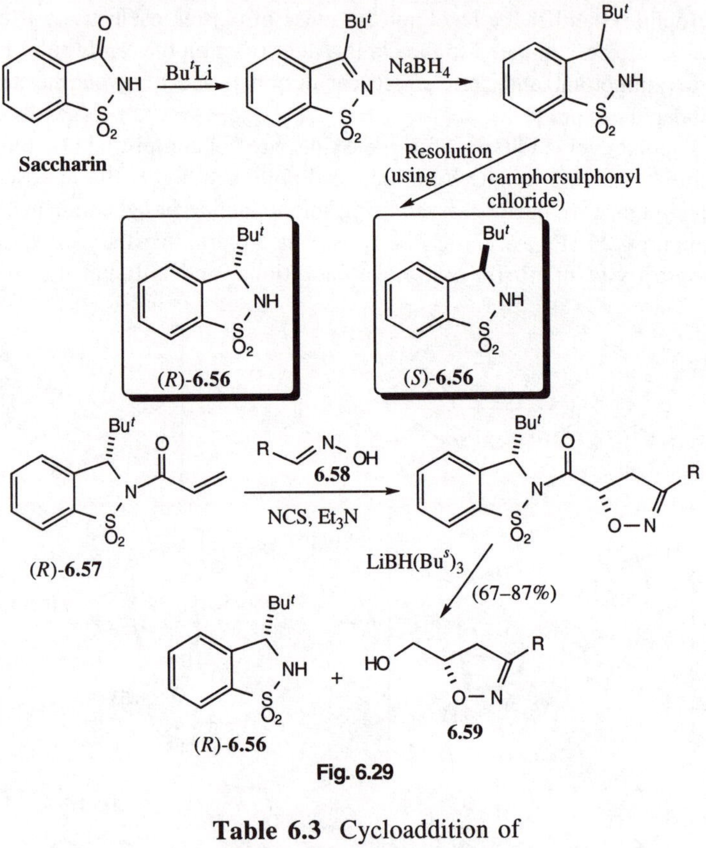

Fig. 6.29

Table 6.3 Cycloaddition of **6.57** and **6.58**

R	d.e. (%)	Yield (%)
Bu^t	96	87
Ph	90	77
Et	90	81
Me	92	81

Reductive removal of the chiral auxiliary and further manipulations gave the 1,3-diol **6.54**, which was converted into the natural product **6.55** on treatment with acid (Fig. 6.28). The stereoselectivity observed in these cycloadditions is consistent with a transition state model similar to that shown in Fig. 6.28, in which the 1,3-dipole approaches the *cisoid*-conformation of the dipolarophile from the 'upper' face.

The stereoselectivity of nitrile oxide cycloadditions has been improved through the development of the chiral auxiliary **6.56** (Fig. 6.29).[28] The racemate is readily available from saccharin and can be resolved easily. Both enantiomers are therefore available and, as can be seen from Fig. 6.29 and Table 6.3, the cycloaddition of dipolarophiles derived from **6.56**, such as **6.57**, is highly efficient. The adducts from nitrile oxides generated from oximes **6.58** are easy to crystallize to provide pure diastereoisomers, and the chiral auxiliary can be removed by reduction (and recovered if desired) to provide enantiomerically pure isoxazolines **6.59**.

Methods for the asymmetric synthesis of cyclobutane derivatives by stereoselective 2+2 cycloaddition to C–C double bonds are relatively scarce, but it is possible to achieve high levels of stereoselectivity with the appropriate chiral auxiliary. One such method uses enantiomerically pure bicyclic lactams such as **6.60** (Fig. 6.30), analogous to **6.7** (Fig. 6.8) and prepared by the same route, which undergo highly diastereoselective photochemical 2+2 cycloadditions with ethylene (Fig. 6.30).[29] It is interesting to note that this cycloaddition takes place from the 'upper' face of the lactam, in contrast to the Diels–Alder reactions of **6.7** (Fig. 6.8) in which the diene approaches from the opposite face. This photochemical approach has been used in a synthesis of grandisol **6.61**, the sex pheromone of the boll weevil, as outlined in Fig. 6.30.

It is also possible to carry out non-photochemical 2+2 cycloadditions to C–C double bonds by using a ketene (or related compound) attached to a chiral auxiliary as one of the reactants. An example of this approach is shown in Fig. 6.31, in which the chiral auxiliary **6.62** is easily prepared from natural proline. The reactant in this case is not a ketene but a keteneiminium salt **6.63** derived as shown (Fig. 6.31).[30]

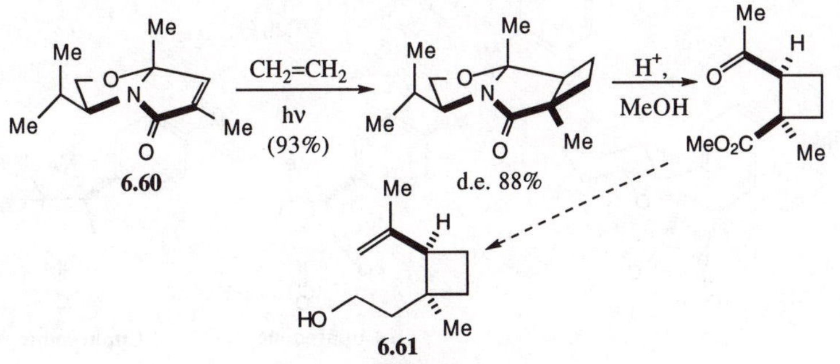

Fig. 6.30

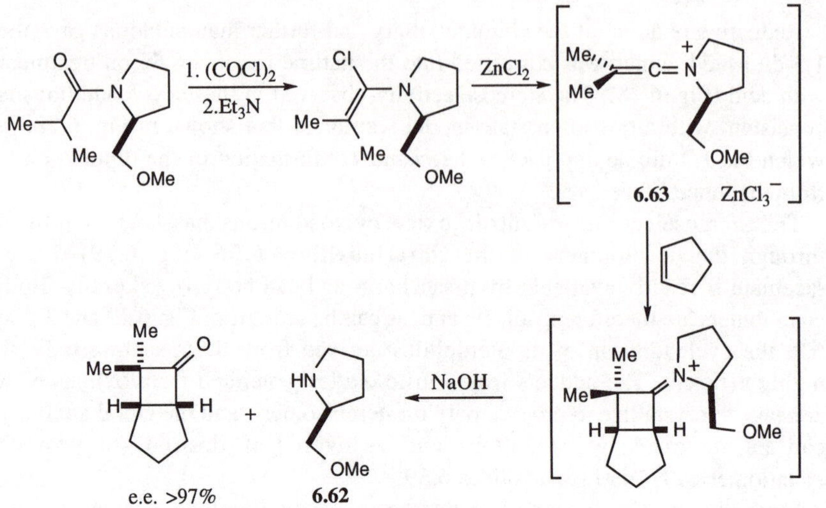

Fig. 6.31

The chiral auxiliary can be attached to the other reactant in this type of ketene cycloaddition. (1S,2R)-2-Phenylcyclohexanol (obtained by a resolution procedure) serves as an effective chiral auxiliary for the cycloaddition of dichloroketene to vinyl ethers (Fig. 6.32). The cycloaddition shown is highly diastereoselective, and was used in enantioselective synthesis of the natural products α- and β-cuparenone (Fig. 6.32).[31] This cycloaddition is thought to take place through the conformation of the vinyl ether shown, where the phenyl group effectively blocks attack of the ketene from the 'lower' face of the vinyl ether double bond.

Much effort has gone into the development of asymmetric synthesis of

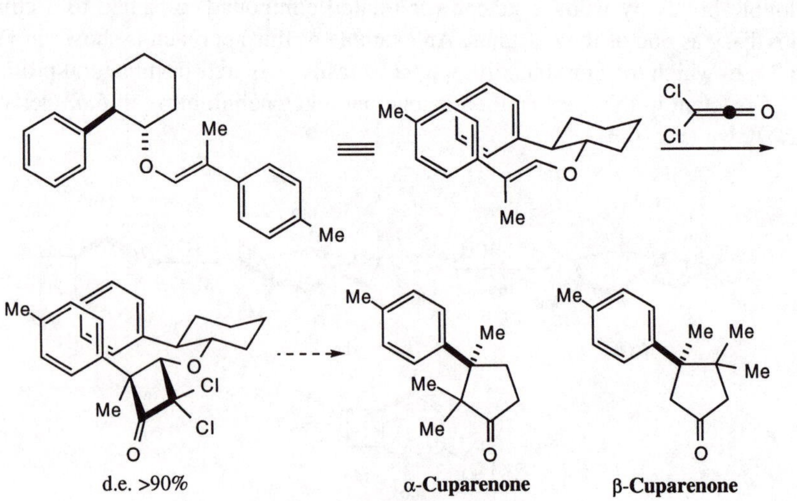

Fig. 6.32

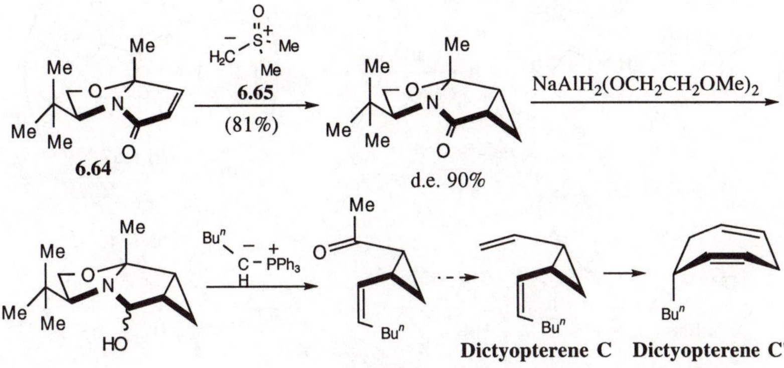

Fig. 6.33

cyclopropanes by '1+2' addition to C–C double bonds. The stimulus for this work can be found in the presence of cyclopropane units in natural products, and their analogues, which possess useful biological activity.

Chiral auxiliaries can be used in asymmetric versions this type of reaction, as exemplified in Fig. 6.33. Bicyclic lactams such as **6.64** (related to **6.7** and **6.60** in Figs 6.8 and 6.30) undergo cyclopropanation with sulphonium ylids such as **6.65**, and this approach has been used in an asymmetric synthesis of dictyopterene C and C' (Fig. 6.33).[32] These are components of certain brown seaweeds and possess remarkable biological activity related to the sexual reproduction of these seaweeds.

The reaction of alkenes with esters of diazoacetic acid **6.66** (Fig. 6.34) to give cyclopropanes is catalysed by certain metal complexes, and can be carried out to provide products of high enantiomeric excess. As with other reactions in which chiral centres originate from both reactants, the possibility exists for the formation of two 'classes' of product. In the reaction illustrated in Fig. 6.34 these have been classified as *cis*- and *trans*-products.[33] Analogous situations have been discussed for aldol (*syn*- and *anti*-products, Chapter 5, Fig. 5.2) and Diels–Alder (*endo*- and *exo*-adducts, Fig. 6.2) reactions.

Much of the stimulus for work in this area has been provided by the 'pyrethroid' insecticides, important and effective compounds which are esters of

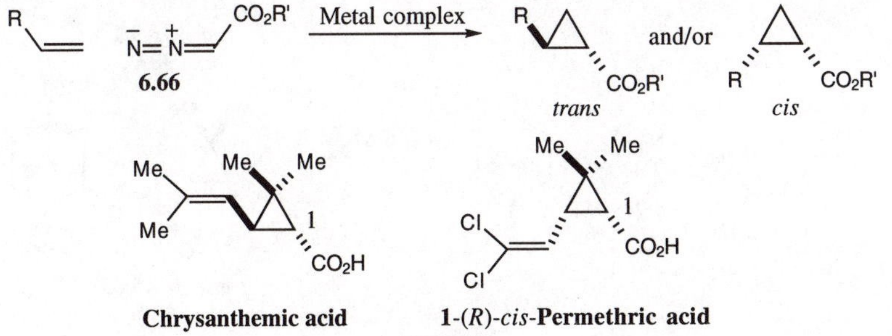

Chrysanthemic acid   **1-(*R*)-*cis*-Permethric acid**
Fig. 6.34

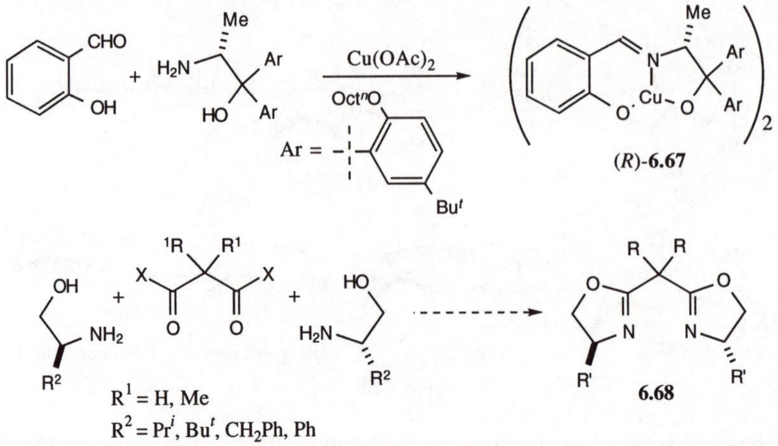

Fig. 6.35

various cyclopropane carboxylic acids, such as chrysanthemic acid and permethric acid (Fig. 6.34).

The reaction between diazoacetates and alkenes is catalytic in the metal complex, and the most effective methods for asymmetric synthesis use chiral metal complexes as enantioselective catalysts. The metal involved is usually copper, and two classes of catalyst merit particular attention, **6.67** and copper complexes of **6.68** (Fig. 6.35). These types of catalyst are relatively easy to prepare, as outlined in Fig 6.35, and give excellent results at low catalyst:substrate ratios (1:100 to 1:1000).[34] The aminoalcohols on which these catalysts are based are available either from natural amino acids or by resolution.

Catalyst **6.67** is effective in promoting the enantioselective cyclopropanation of diene **6.69** using the *l*-menthyl ester of diazoacetic acid. The chrysanthemic

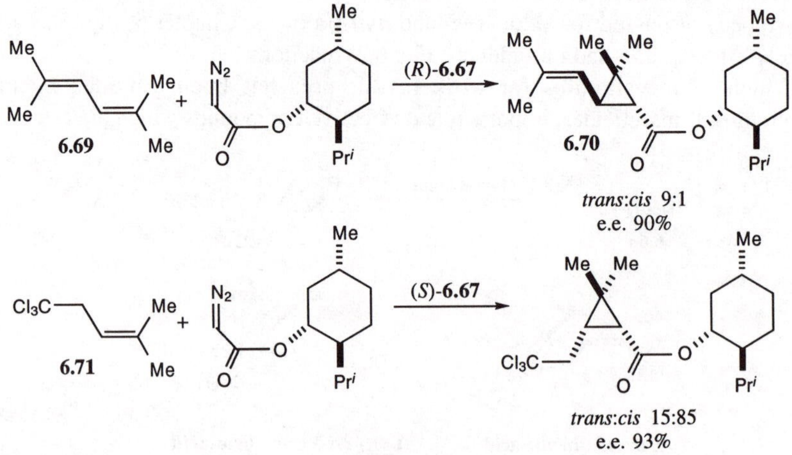

Fig. 6.36

ester **6.70** is obtained with high enantioselectivity (Fig. 6.36). When alkene **6.71** is used in place of diene **6.69** the product distribution changes dramatically. The *cis*-isomer now predominantes, and in order to obtain the 1(*R*)-enantiomer, a precursor to permethrinic acid, the *opposite* enantiomer of the catalyst is required.

Complexes derived from enantiomerically pure bis(oxazoline) ligands of general structure **6.68** (Fig. 6.35) and various copper salts can be used as highly effective asymmetric cyclopropanation catalysts. The catalytically active species is the copper(I) complex, which can be formed either directly from the appropriate copper(I) salt (e.g. copper(I) triflate), or indirectly by *in situ* reduction of the copper(II) complex. Representative ligands **6.72**, **6.73**, and examples of their reactions are shown in Fig. 6.37 and Table 6.4.[35,36] The enantioselectivity and diastereoselectivity (*cis:trans* ratio) of the process are sensitive to the structure of the ligand, diazoacetate, and olefin. With monosubstituted, *trans*-disubstituted, and terminal disubstituted olefins, high

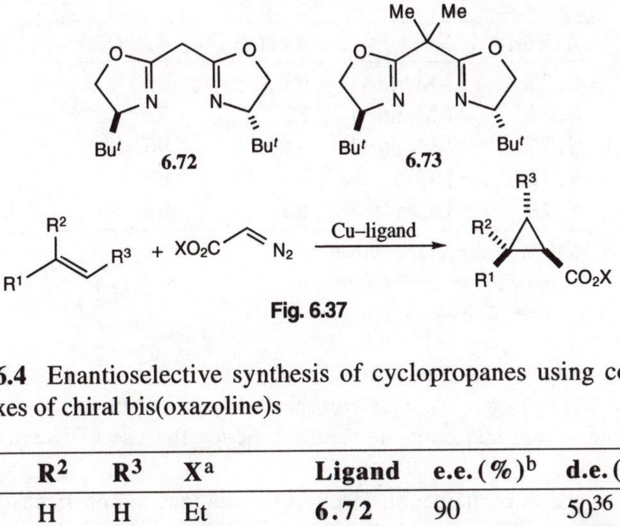

Fig. 6.37

Table 6.4 Enantioselective synthesis of cyclopropanes using copper complexes of chiral bis(oxazoline)s

R^1	R^2	R^3	X^a	Ligand	e.e. (%)[b]	d.e. (%)[c]
Ph	H	H	Et	**6.72**	90	50[36]
Ph	H	H	Bu^t	**6.72**	98	68[36]
Ph	H	H	BHT	**6.73**	99	88[35]
CH$_2$Ph	H	H	BHT	**6.73**	99	86[35]
Me	Me	H	Et	**6.73**[d]	99	—[35]
Ph	Ph	H	Et	**6.73**[d]	99	—[35]
Pr^n	H	Pr^n	*l*-Menthyl	**6.72**	88	—[36]

[a]BHT = 2,6-di-*tert*-butyl-4-methylphenol.

[b]e.e. of the *trans*-cyclopropane.

[c]d.e. = (% *trans*) - (% *cis*).

[d]The enantiomer of **6.73** was used to produce cyclopropanes enantiomeric with those shown in Fig. 6.37.

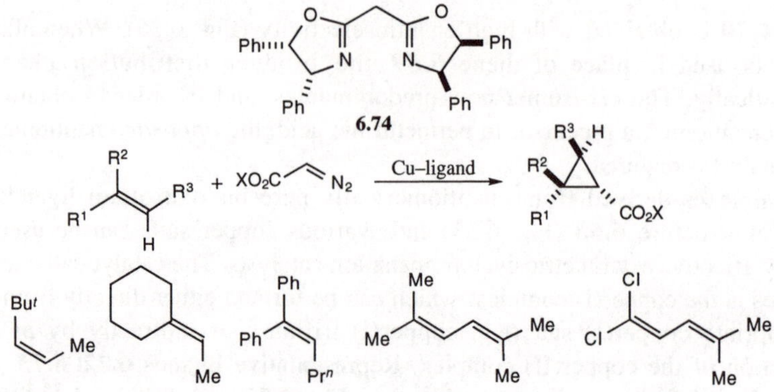

Fig. 6.38

Table 6.5   Enantioselective synthesis of cyclopropanes using copper complexes of **6.74**

Alkene	X[a]	e.e. ($\%$)[b]	d.e. ($\%$)[c]
6.75	*l*-Menthyl	95	76
6.76	*l*-Menthyl	82	72
6.77	*l*-Menthyl	84	96
6.78	DCM	94	90
6.79	DCM	92	98

[a]DCM = dicyclohexylmethyl.
[b]e.e. of the *trans*-cyclopropane.
[c]d.e. = (% *trans*) − (% *cis*).

enantioselectivity is observed when the R group in the bis(oxazoline) ligand **6.68** (Fig. 6.35) is large. The diastereoselectivity increases as the diazoacetate ester becomes more sterically demanding, hence the use of esters of highly hindered alcohols.

In order to obtain high enantioselectivity in the reaction of trisubstituted and unsymmetrical *cis*-disubstituted alkenes, copper(I) complexes of ligand **6.74** are required (Fig. 6.38 and Table 6.5).[37]

Almost all of the preceding additions have been cycloadditions in which the addend was a neutral species. Alkenes with appropriate functionalization will undergo addition reactions with either electrophiles or nucleophiles. These two addition processes encompass a large number of reactions, and in order to illustrate some of these, asymmetric versions of one important example of each will be discussed.

The addition of a nucleophile, in most useful cases, requires the alkene to be electron deficient by virtue of an electron-withdrawing group (EWG). The general reaction, often referred to as conjugate addition, is shown in Fig. 6.39.

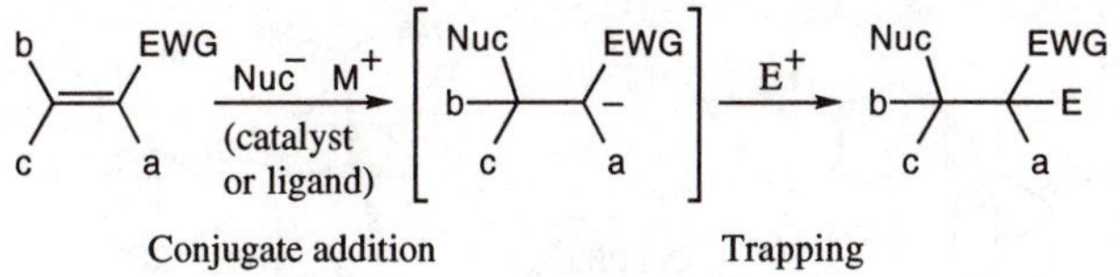

Conjugate addition   Trapping

Fig. 6.39

Asymmetric conjugate addition reactions can be carried out using chiral auxiliaries in either the nucleophile or the acceptor, chiral ligands (if a metal is involved), or a chiral catalyst. In this type of process two new chiral centres can be created along with the sequential formation of two new bonds. The two new bonds are made to two different reagent types, nucleophile and electrophile, which significantly increases the scope and versatility of this method.

The electron-withdrawing group (EWG, Fig. 6.39) is often a carbonyl substituent, and acceptors with a various chiral auxiliaries attached via this carbonyl group have been used in asymmetric conjugate additions. As is usually the case, the level and sense of stereoselectivity depends on the conformation of the reacting system, the structure and stereochemistry of the chiral auxiliary, and the mechanism of the reaction. The mechanism of conjugate additions is not well understood, and the details certainly vary with reagent, conditions, and substrate. This contributes to the variability in stereoselectivity which is typical in this type of reaction. Nevertheless, it is possible to achieve high levels of diastereoselectivity using even simple chiral auxiliaries.

Unsaturated esters of 8-phenylmenthol (Fig. 6.40) undergo stereoselective conjugate addition, high diastereoisomeric excesses being obtained with *trans*-enoates (Fig. 6.40).[38] As in the Diels–Alder reactions controlled by this chiral auxiliary (Fig. 6.9), the aromatic ring is thought to be important in shielding one face of the double bond, and the reaction is proposed to take place via the conformation shown in Fig. 6.40.

Several chiral auxiliaries derived from camphor have been developed which provide high levels of stereochemical control in conjugate additions; two examples are shown in Fig. 6.41.[39] In these reactions, and in that shown in Fig. 6.40, the stereochemistry of the major product is consistent with attack on the less hindered face of the enoate, in a *transoid*-conformation with the carbonyl *syn*-planar with the alkoxy C–H bond. These chiral auxiliaries can be removed by hydrolysis to provide carboxylic acids in high enantiomeric excess. Both enantiomers are readily available, and the directing effect of the chiral auxiliaries

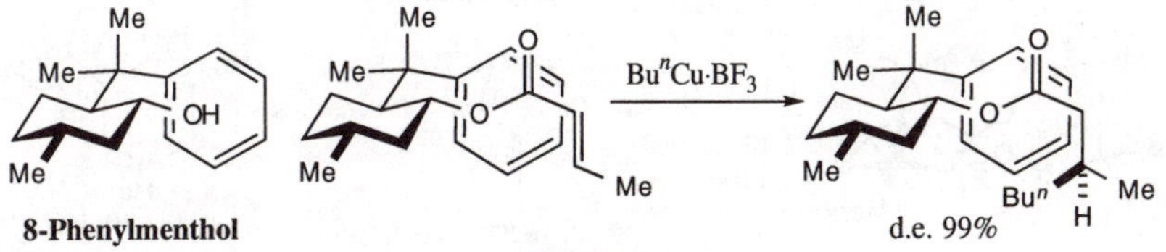

8-Phenylmenthol   d.e. 99%

Fig. 6.40

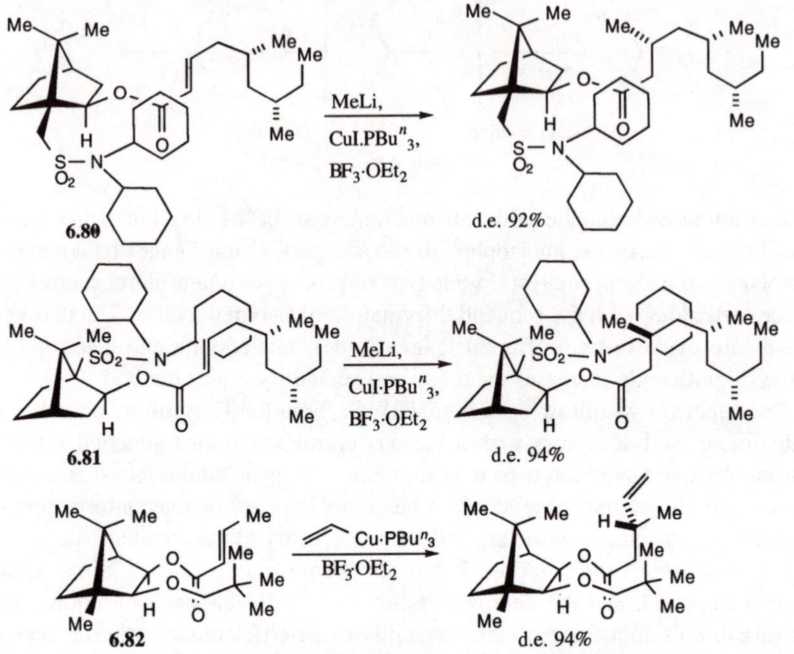

Fig. 6.41

usually overrides any 'intrinsic' facial bias due to the presence of other chiral centres. For example, additions to **6.80** and **6.81**, which differ only in the enantiomer of the chiral auxiliary used, take place with effectively the same level of stereoselectivity but in the opposite sense.

The attributes of these chiral auxiliaries make this approach a powerful and

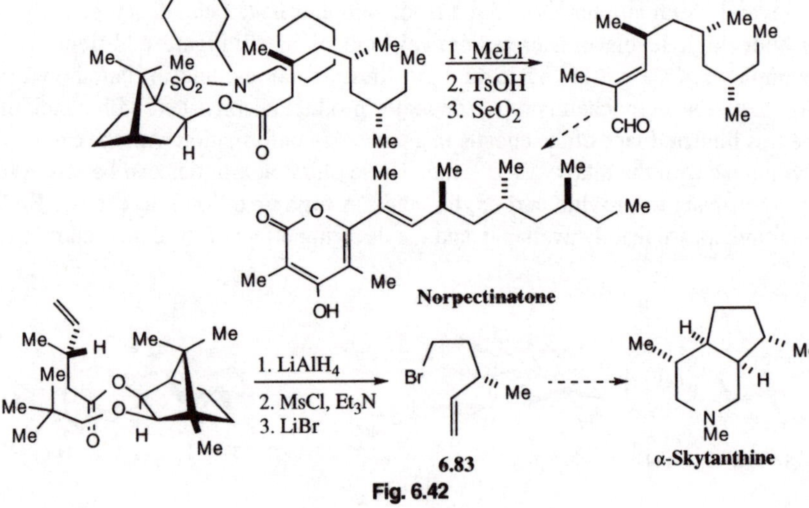

Fig. 6.42

general tool for asymmetric synthesis, which has been used in the synthesis of several natural products. Conjugate addition to the enantiomer of **6.82** (Fig. 6.41) was used in the preparation of the bromide **6.83** (Fig. 6.42),[40] a key intermediate in a synthesis of the alkaloid α-skytanthine. The conjugate addition product of **6.81** (Fig. 6.41) was used in a total synthesis of the putative structure of the metabolite norpectinatone (Fig. 6.42).

Derivatives of the sultam chiral auxiliary **6.26** (Fig. 6.17 and Fig. 6.28) and its enantiomer **6.84** (Fig. 6.43) also undergo highly diastereoselective conjugate additions (Fig. 6.43 and Table 6.6), and this approach has been used in the synthesis of β-silylcarboxylates and derivatives thereof.[41] The stereochemistry of the addition is consistent with chelation of the aluminium followed by attack of the organocopper reagent from the less hindered face, with the enoate in a *cisoid*-conformation **6.85** (Fig. 6.43).

The chiral auxiliary can be cleaved by methanolysis to provide the β-silylcarboxylates, which are valuable and versatile intermediates; for example, such esters can be converted into '*anti*-aldol' products by sequential alkylation *anti*- to the silicon, and conversion of the silyl substituent into a hydroxyl

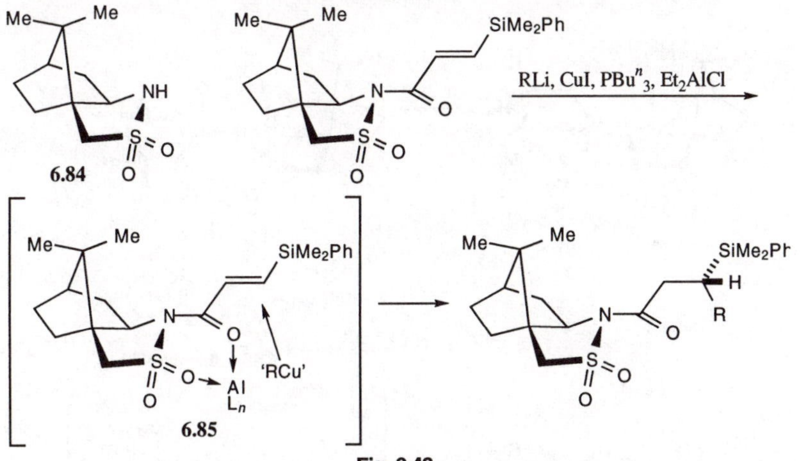

Fig. 6.43

Table 6.6 Conjugate additions to **6.84**

R	d.e. (%)
$CH_2=CH-$	90
(Z)-MeCH=CH–	96
(E)-MeCH=CH–	96
Et	86
Pr^i	86

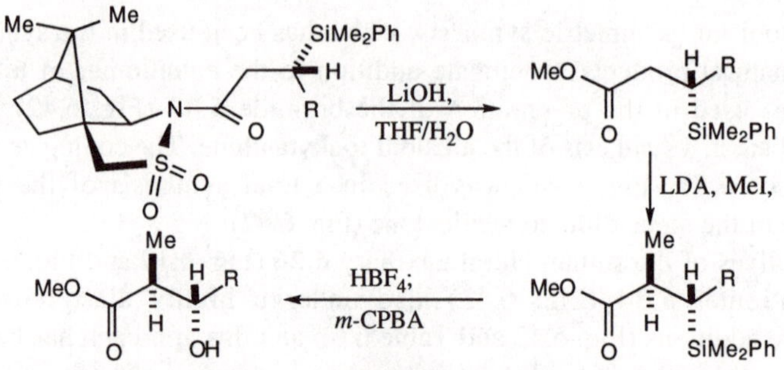

Fig. 6.44

group with retention of configuration (Fig. 6.44).[41]

These sultam chiral auxiliaries perform equally well with enoates which contain simple carbon substituents in place of the silyl group. Moreover, this auxiliary can be used to control both conjugate addition and *in situ* trapping (see Fig. 6.39). The stereochemical outcome of this conjugate addition–trapping approach depends on the substitution pattern of the C–C double bond, particularly on the presence or absence of a substituent α to the carbonyl group.

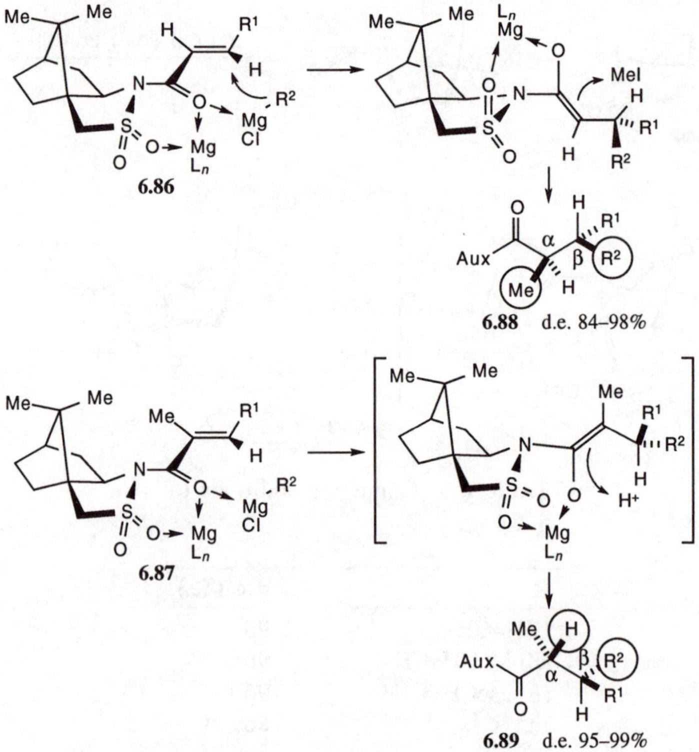

Fig. 6.45

Conjugate addition of Grignard reagents (Fig. 6.45) is thought to proceed through a chelated transition state **6.86** analogous to that proposed in Fig. 6.43.[42] Strong coordination to the magnesium ion is the dominating factor, and the facial selectivity of the conjugate addition step is independent of substitution α to the carbonyl group. This is not the case for the alkylation step, and although the same enolate geometry (Z-) results from the conjugate addition, its *conformation* depends on α substitution. This difference in conformation is thought to be responsible for the difference in stereochemistry in the α-substitution step. This is illustrated by the selected examples shown in Fig. 6.45, where the encircled groups in **6.88** and **6.89** are the nucleophile and electrophile.

Conjugate addition of hydride followed by trapping of the resulting enolate is also a highly stereoselective process (Fig. 6.46). Taken together, the examples of conjugate addition–trapping discussed here represent a flexible approach to the asymmetric synthesis of α,β-disubstituted carbonyl compounds. This versatility is exemplified by Fig. 6.46, which shows three different routes to the same product, all of which would provide very high levels of stereoselectivity.[43]

Asymmetric conjugate addition–trapping is also possible using the iron-centred chiral auxiliary **6.90** (Fig. 6.47), which is also effective in controlling enolate alkylation (Chapter 4) and aldol reactions (Chapter 5). Although the example shown in Fig. 6.47 involves addition of a lithium amide,[44] simple alkyllithium reagents will also add from the same face of the enoate.[45] In both cases the resulting enolate reacts with electrophiles from the same face as the initial addition. These reactions are consistent with the models discussed previously (Figs 4.23 and 4.25) in which the 'back' face of the enolate is

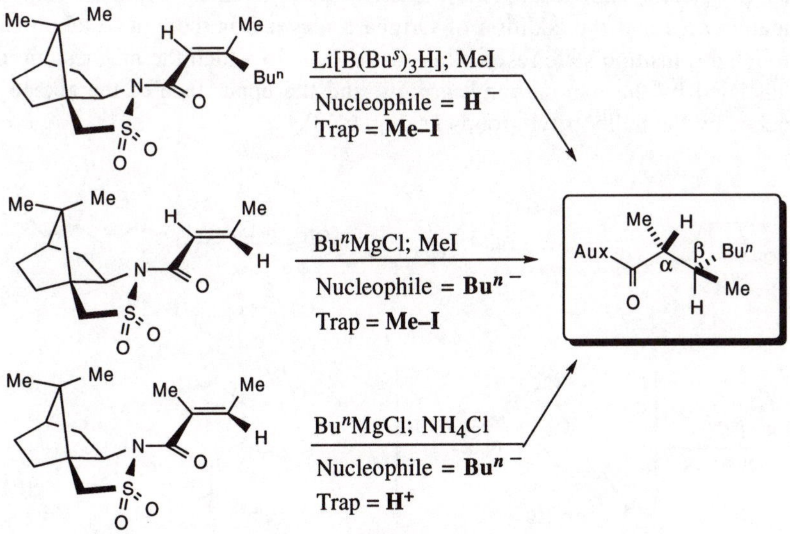

Fig. 6.46

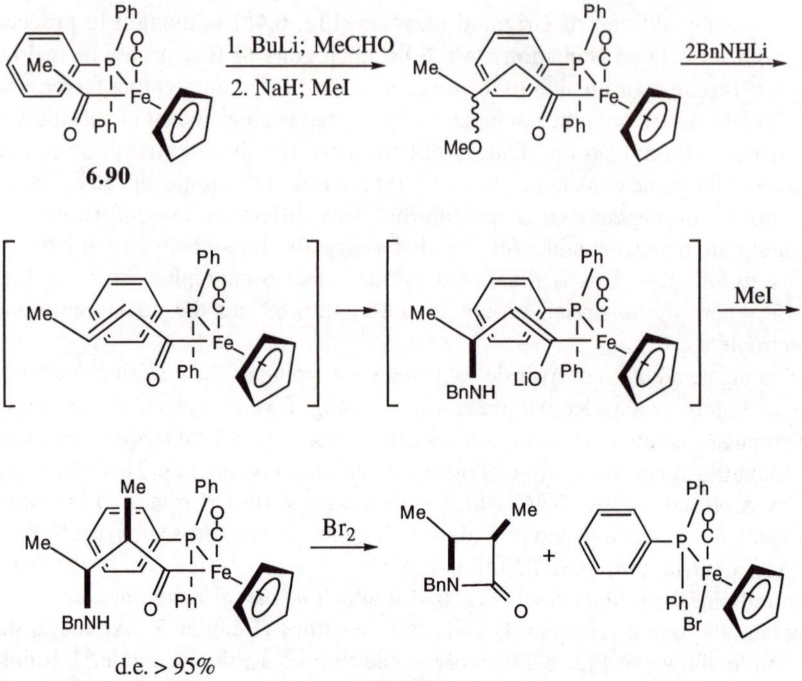

Fig. 6.47

blocked by one of the phenyl rings of the phosphine ligand.

In much of this chapter, chelation effects have been used to account for the observed stereoselectivity, and this is also the case for addition to derivatives of **6.91** (Fig. 6.48, Table 6.7). This chiral auxiliary is readily available from L-glutamic acid and the addition of Grignard reagents is thought to take place through a transition state resembling that shown, in which the magnesium ion is chelated by the two carbonyl groups, and the upper face of the alkene is blocked by the bulky trityl group (Tr, Fig. 6.48).[46]

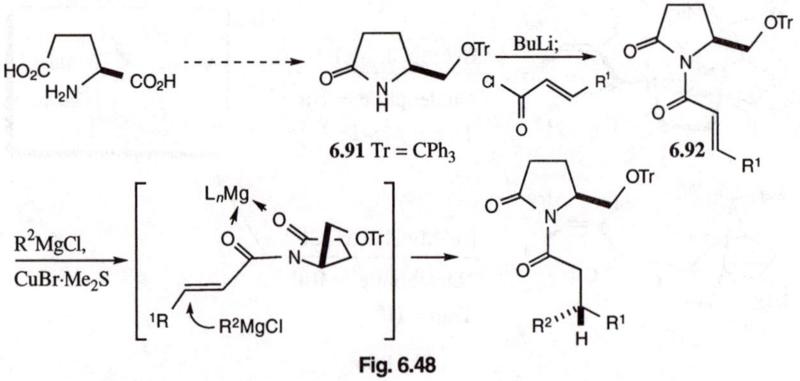

Fig. 6.48

Table 6.7 Conjugate additions to **6.92**

R^1	R^2	d.e. (%)
Me	*p*-Tol	91
Me	*c*-Hex	87
Me	Bu^n	91
Me	CH$_2$=CH–	88
Bu^n	Et	81
Bu^n	CH$_2$=CH–	84

The idea of chelation of the metal in the intermediates (and presumably the transition state) can be taken further by incorporating into the auxiliary a group which will react with the organometallic reagent and thereby produce even stronger binding to the metal. An appropriately positioned hydroxyl group has been used in several cases to provide high levels of diastereoselectivity.

α,β-Unsaturated amides derived from the readily available aminoalcohols

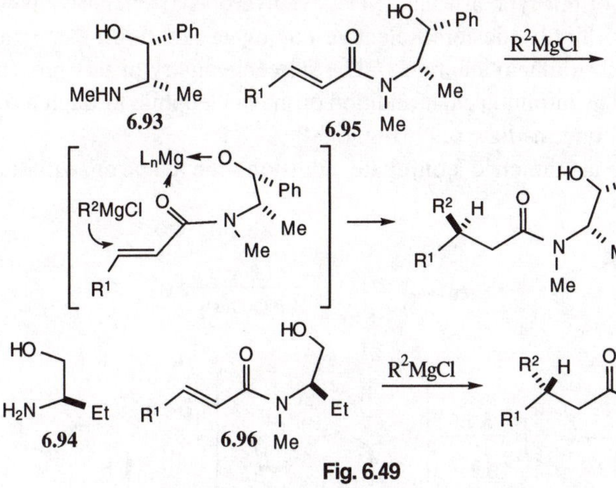

Fig. 6.49

Table 6.8 Conjugate additions to **6.95** and **6.96**

Acceptor	R^1	R^2	d.e. (%)
6.95	Me	Bu^n	85
6.95	Ph	Et	98
6.95	Et	Ph	93
6.96	Ph	Et	92
6.96	Ph	Bu^n	99

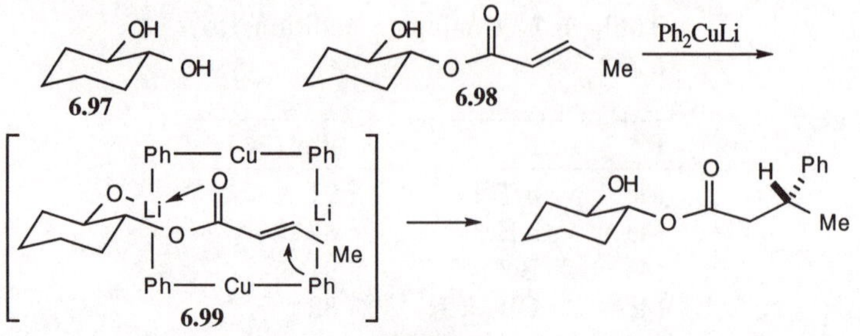

Fig. 6.50

6.93[47] (ephedrine) and **6.94**[48] (resolved) have been used in this way (Fig. 6.49, Table 6.8). In both cases reaction is proposed to take place from the less hindered face of the intermediate produced by reaction of the hydroxyl group with the Grignard reagent and coordination to the carbonyl group to the magnesium ion as shown for addition to **6.95**.

A similar approach is possible by using α,β-unsaturated esters of diols in which only one of the alcohol functions is esterified. Diol **6.97** (Fig. 6.50) is easily obtained by enzymatic hydrolysis of the racemic diacetate (see Chapter 9 for a discussion of this type of enantioselective hydrolysis) and derivatives such as **6.98** undergo highly diastereoselective conjugate addition of cuprates, as shown for lithium diphenylcuprate.[49] The stereochemistry of this process has been interpreted as intramolecular addition of the nucleophile through a square-planar 'dimeric' intermediate **6.99** (Fig. 6.50).

For efficient asymmetric conjugate addition, the electron-withdrawing

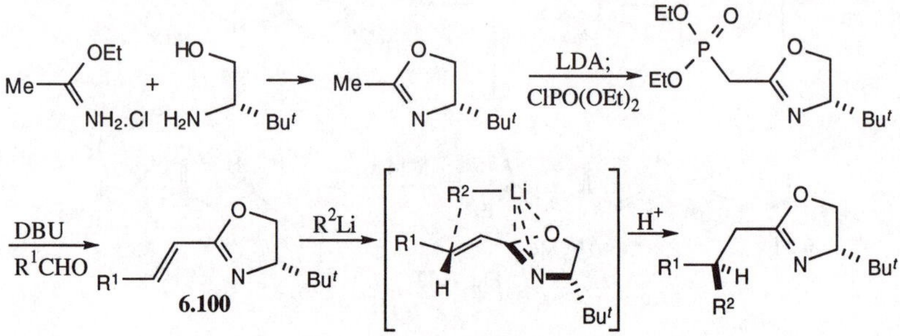

Fig. 6.51

Table 6.9 Conjugate additions to **6.100**

R^1	R^2	d.e. (%)
Bu^n	Ph	97
c-Hex	Ph	96
Ph	Bu^n	96

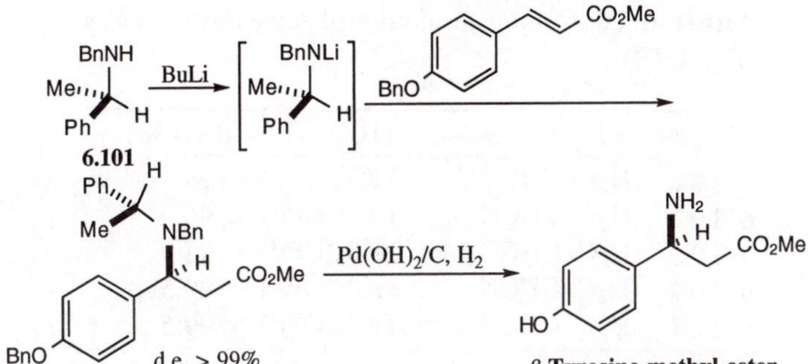

Fig. 6.52

substituent need not be a carbonyl group, and several methods have been developed which use imine (C=N) functionality. One such method involves the use of chiral oxazolines as the electron-withdrawing group, analogous to their use in alkylation (Chapter 4) and other reactions (Figs 6.8, 6.30, and 6.33).[50] The best diastereoselectivity is found when the face-blocking group is *tert*-butyl. The appropriate unsaturated oxazolines **6.100** are easy to prepare (Fig. 6.51) and undergo highly stereoselective conjugate addition on treatment with organolithium derivatives (Fig. 6.51, Table 6.9).

In all the preceding examples of asymmetric conjugate addition, it is the acceptor which is chiral. A chiral nucleophile should also allow for asymmetric conjugate addition, although there are far fewer examples of this approach which take place with high stereoselectivity.

Lithium amides derived from simple enantiomerically pure secondary amines such as **6.101** will add to some α,β-unsaturated esters with high diastereoselectivity, as exemplified by the synthesis of β-tyrosine methyl ester (Fig. 6.52).[51]

Conjugate addition of chiral enolates takes place with high diastereoselectivity in some cases. Titanium enolates derived from acylated oxazolidinones such as **6.102**, **6.103**, and **6.104** will add to α,β-unsaturated ketones, esters, and nitriles with very high diastereoselectivity (Fig. 6.53 and Table 6.10).[52] The

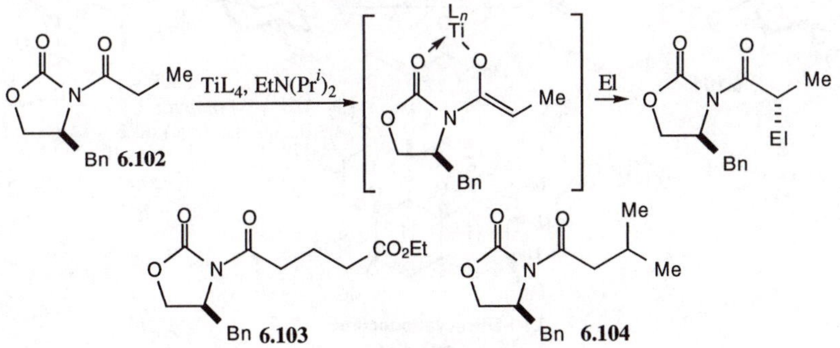

Fig. 6.53

Table 6.10 Conjugate addition of acyloxazolidinones
(Fig. 6.53)

Nuc	El	TiL_4	d.e. (%)
6.102	$H_2C=CHCOEt$	$TiCl_4$	>98
6.102	$H_2C=CHCN$	$TiCl_3(OPr^i)$	96
6.102	$H_2C=CHCO_2Me$	$TiCl_3(OPr^i)$	98
6.103	$H_2C=CHCN$	$TiCl_3(OPr^i)$	>99.5
6.104	$H_2C=CHCN$	$TiCl_3(OPr^i)$	>99.5

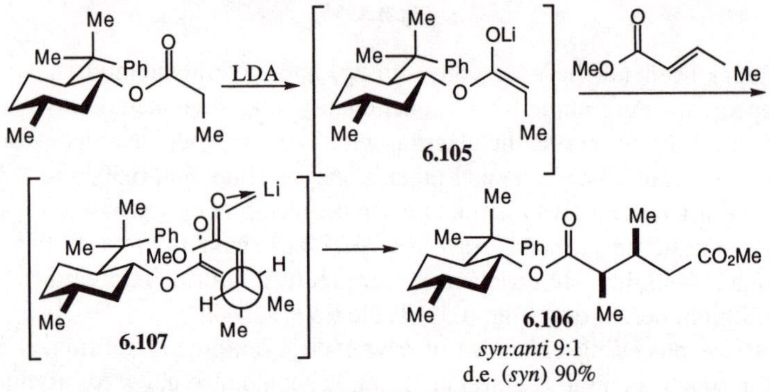

Fig. 6.54

stereochemistry of these reactions is consistent with addition of the electrophile
to the less hindered face of the chelated (Z)-enolate as shown in Fig. 6.53.

8-Phenylmenthol (Fig. 6.9 and Fig. 6.40) has also been used to control the
stereochemistry of conjugate additions of enolates.[53] The enolate **6.105** (Fig.
6.54) reacts to give mainly the *syn*-product **6.106**. A transition state model
which is consistent with this and other reactions of this type of system involves

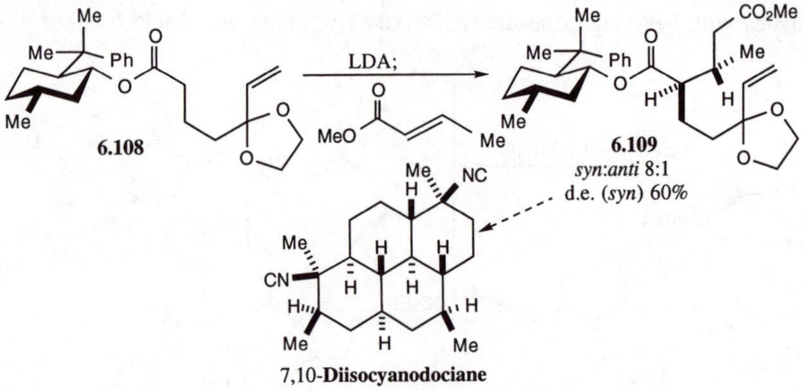

7,10-**Diisocyanodociane**

Fig. 6.55

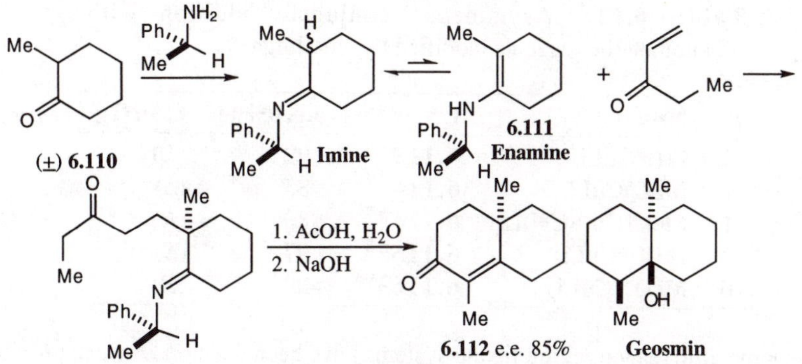

(±) **6.110** **6.111** **6.112** e.e. 85% **Geosmin**

Fig. 6.56

coordination of the enoate to the lithium atom of the enolate and reaction via the conformation represented by **6.107**.

This method for asymmetric conjugate addition has been used at an early stage in a total synthesis of the marine natural product 7,10-diisocyanodociane (Fig. 6.55).[54] The lithium enolate derived from **6.108** provides the conjugate addition product **6.109** in high enough stereochemical purity to allow for completion of the total synthesis, which proved the absolute configuration to be as shown.

Certain imines derived from cyclic ketones and chiral primary amines will add to electron deficient alkenes with high stereoselectivity, the reaction taking place via formation of a small amount of the corresponding enamine **6.111** (Fig. 6.56).[55] In effect, the conversion of **6.110** to **6.112** is an asymmetric version of the Robinson annulation sequence, an important general method for the synthesis of bicyclic systems which has been widely used in natural product synthesis. Enamine **6.111** was used in the total synthesis of geosmin, a metabolite of many actinomycetes.

Enantioselective nucleophilic species can also be generated by reaction of a 'normal' organometallic reagent such as a dialkylzinc or an alkylcopper derivative with a chiral ligand. In spite of the relative lack of knowledge about the transition state and intermediates involved in such conjugate additions,

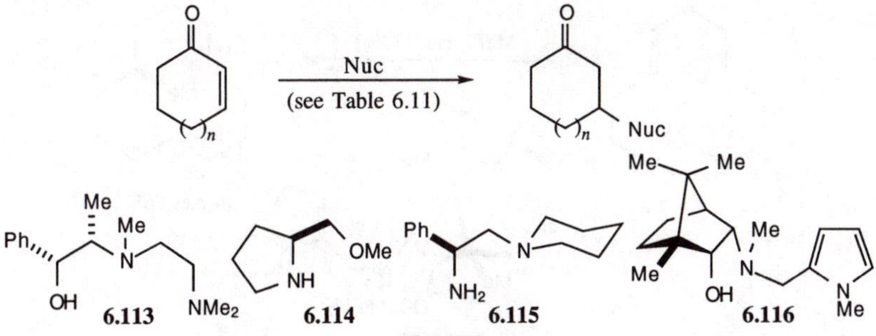

6.113 **6.114** **6.115** **6.116**

Fig. 6.57

Table 6.11   Asymmetric conjugate addition with organometallic reagents modified by chiral ligands

n	Nuc	L*	e.e. (%)	Config.
1	EtL*CuLi	6.113[56]	92	(R)
1	MeL*CuLi	6.114[57]	83	(S)
1	MeL*Cu(SCN)Li$_2$	6.114[57]	83	(R)
2	Bu^{n}L*CuLi	6.115[58]	97	(S)
10	Me$_2$L*CuLi$_2$	6.116[59]	>99	(R)

organometallic reagent/chiral ligand systems have been developed which provide good levels of stereoselectivity. A selection of some of these is given in Fig. 6.57 and Table 6.11.

In some cases it has proved possible to develop conditions for asymmetric conjugate addition which use only a catalytic amount of the chiral ligand, as shown in Fig. 6.58.[60,61]

The most important electrophilic addition to a double bond which can be carried out with high enantioselectivity is hydroboration.[62] The relative importance of this reaction is due to the ready availability of reagents for enantioselective hydroboration, the stereoselectivity of their reactions, and the versatility of the boranes which are produced.

Some important aspects and characteristics of hydroboration are illustrated by the preparation of diisopinocampheylborane **6.117**, usually abbreviated to Ipc$_2$BH (Fig. 6.59). The boron becomes attached to the more nucleophilic end

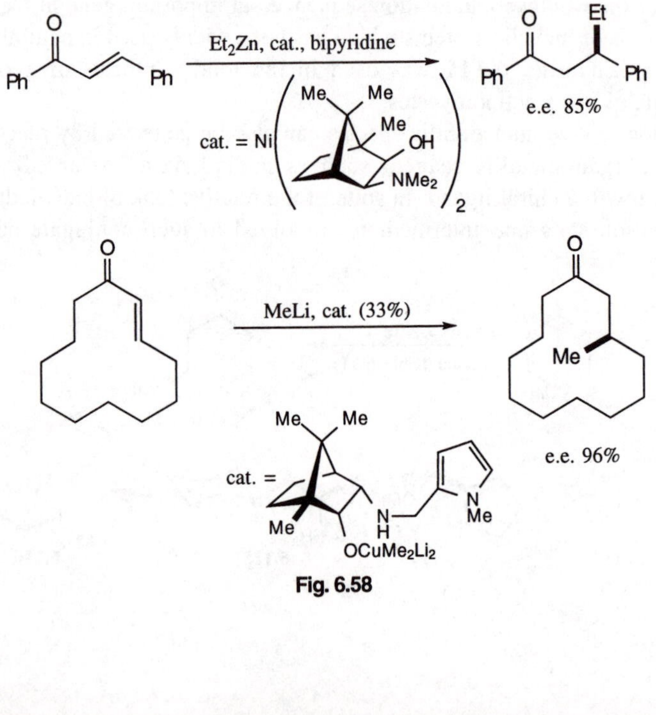

Fig. 6.58

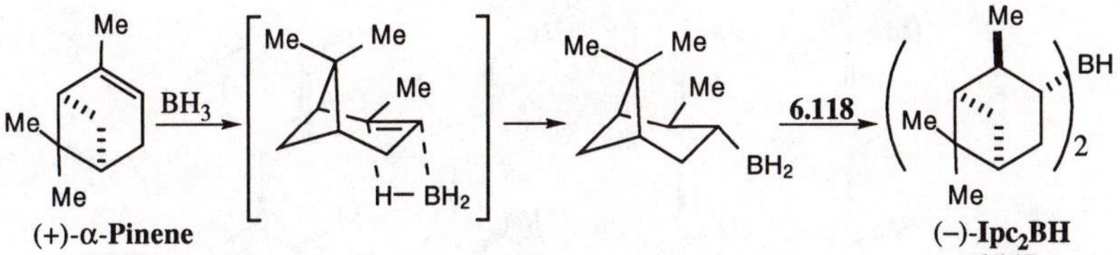

(+)-α-Pinene
6.118

(−)-Ipc₂BH
6.117

Fig. 6.59

of the double bond, attacks from the less hindered face, and more than one of the B–H bonds may react. In general, the reaction is sensitive to steric factors, which is why the hydroboration of pinene **6.118** stops after addition of two moles of the alkene.

Although Ipc_2BH is too hindered to add to a third mole of pinene, it will hydroborate less hindered alkenes with high enantioselectivity, particularly *cis*-alkenes. The trialkylboranes which are produced will undergo numerous subsequent reactions, usually with retention of configuration, the most common being oxidation to the corresponding alcohol(s).

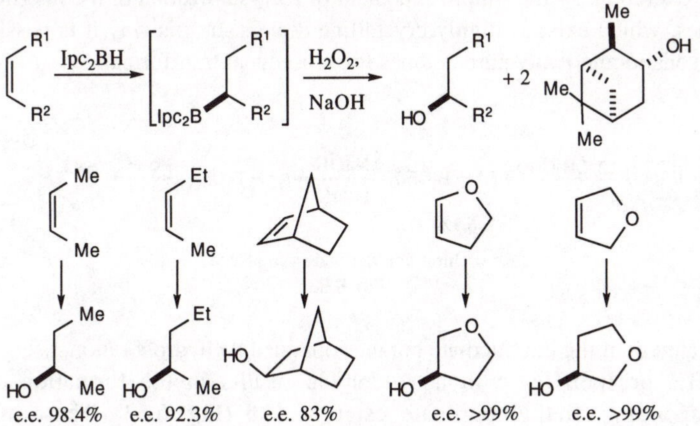

Fig. 6.60

This general approach to the enantioselective synthesis of secondary alcohols is illustrated in Fig. 6.60, along with a selection of examples.[63] The hindered nature of Ipc_2BH results in unsatisfactory reaction with *trans*-alkenes and

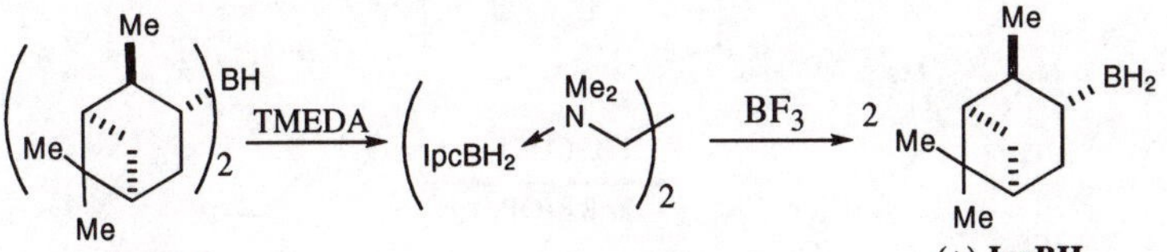

(−)-Ipc₂BH
6.117

(+)-IpcBH₂
6.119

Fig. 6.61

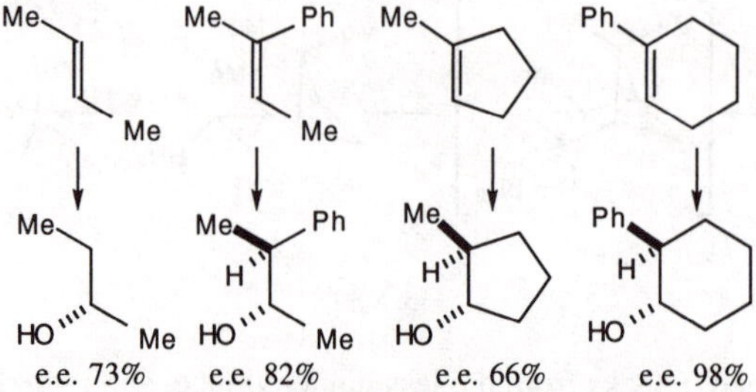

e.e. 73%   e.e. 82%   e.e. 66%   e.e. 98%

Fig. 6.62

trisubstituted alkenes. Monoisopinocampheylborane, IpcBH$_2$ **6.119**, is much less hindered, and is easily prepared from Ipc$_2$BH as shown (Fig. 6.61).[64]

This reagent reacts with *trans*-alkenes and with trisubstituted alkenes with good to high enantioselectivity to give intermediate dialkylboranes which can be oxidized to give the corresponding alcohols (Fig. 6.62).[65]

Although the enantiomeric excesses in these reactions are somewhat lower than those obtained in the hydroboration of *cis*-alkenes with Ipc$_2$BH, this is easily overcome by the simple expedient of recrystallization of the intermediate boranes, which exist as highly crystalline dimers. In this way it is possible to obtain enantiomerically pure boranes for subsequent transformations.

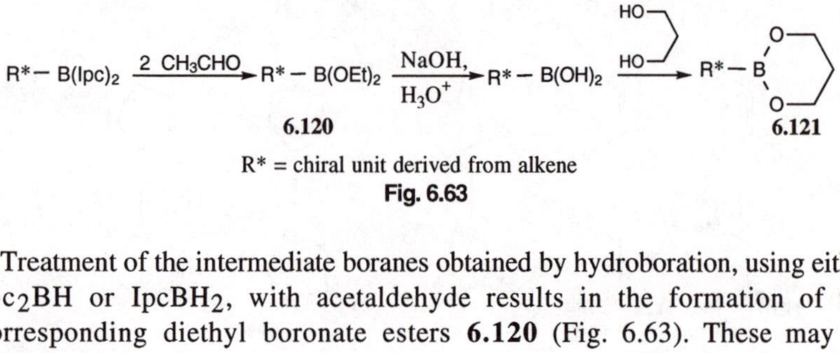

R* = chiral unit derived from alkene

Fig. 6.63

Treatment of the intermediate boranes obtained by hydroboration, using either Ipc$_2$BH or IpcBH$_2$, with acetaldehyde results in the formation of the corresponding diethyl boronate esters **6.120** (Fig. 6.63). These may be

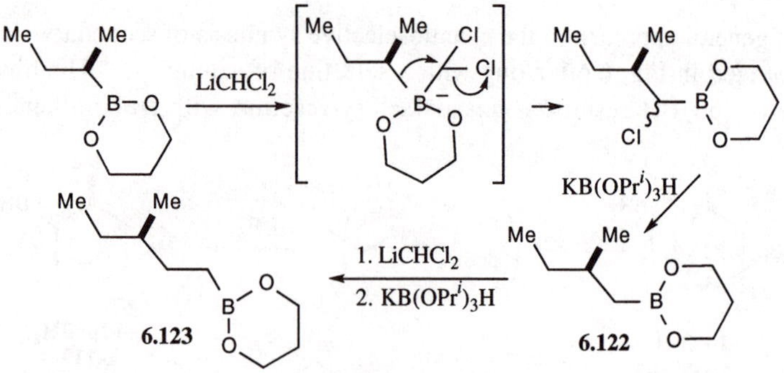

Fig. 6.64

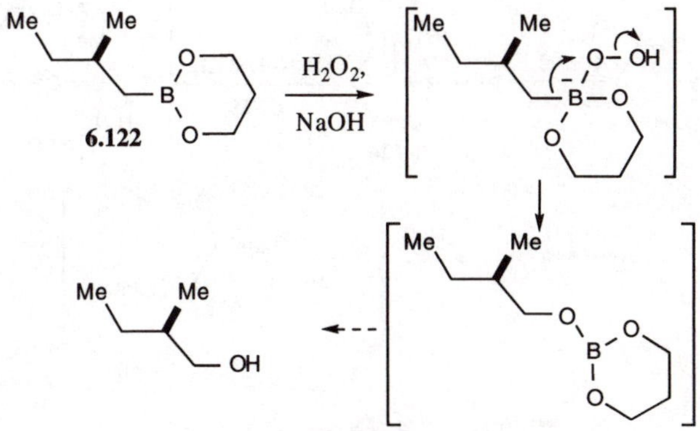

Fig. 6.65

converted into other boronate esters such as **6.121** by hydrolysis and re-esterification, shown schematically in Fig. 6.63.[66,67] An additional advantage of this procedure is that the α-pinene which is liberated can be isolated easily and recycled.

Enantiomerically pure boronate esters such as **6.121** have great potential for asymmetric synthesis, as they undergo a variety of valuable synthetic transformations. Of particular interest is the general procedure available for homologation, as exemplified in Fig. 6.64, a sequence which can be repeated.[68,69]

Boronate esters are easily converted into the corresponding enantiomerically pure alcohols on oxidation with basic hydrogen peroxide, as shown in Fig. 6.65 for the homologated ester **6.122**. This is a general reaction and the migration from boron to oxygen takes place with retention of configuration where applicable.[69]

The homologous aldehyde can be obtained from a boronate ester by the sequence outlined in Fig. 6.66. Once again, clean retention of configuration and high maintenance of chirality is observed.[67]

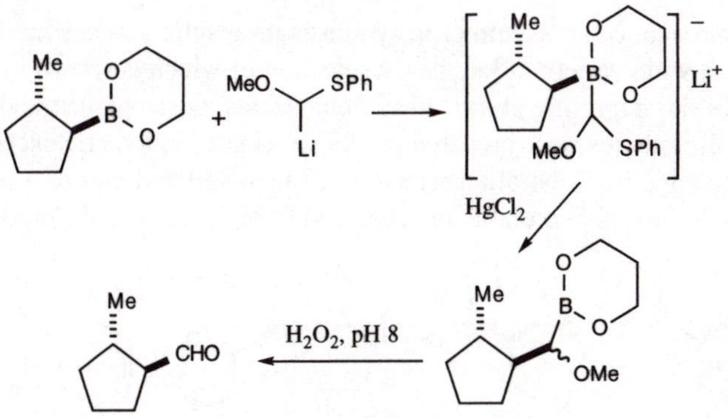

Fig. 6.66

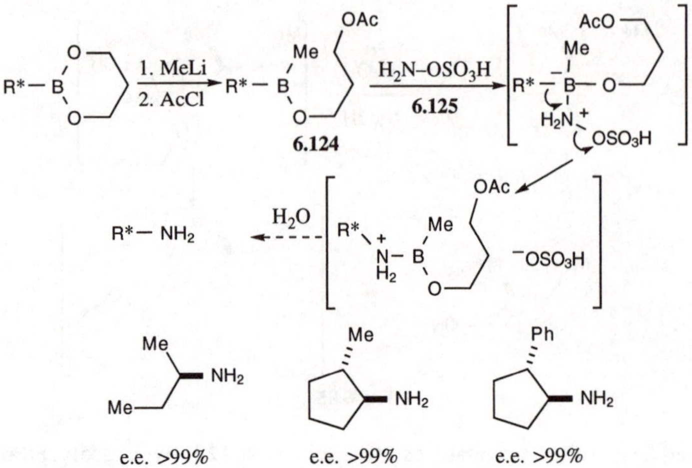

Fig. 6.67

The conversion of enantiomerically pure boronate esters into the corresponding primary amine, analogous to oxidation to the alcohol (Fig. 6.65), can be achieved by treatment of intermediate alkylmethylborinic esters such as **6.124** with hydroxylamine *O*-sulphonic acid **6.125** (Fig. 6.67). This provides a general approach to the enantioselective synthesis of primary amines.[70]

Dialkyl borinates analogous to **6.124** can also be used in a general asymmetric synthesis of ketones. This approach relies on the reaction of such intermediates with dichloromethyl methyl ether under basic conditions, and the general scheme is illustrated in Fig. 6.68.[71]

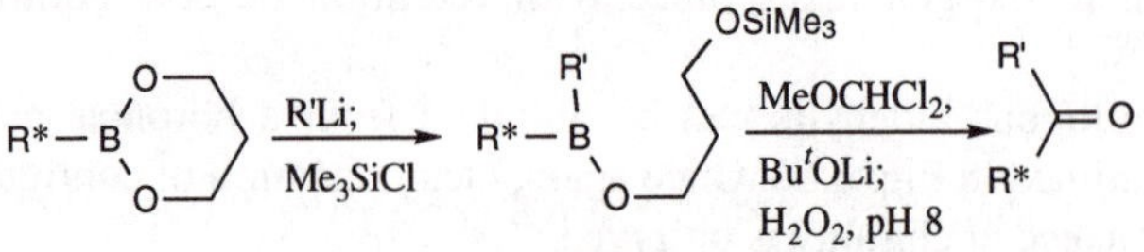

Fig. 6.68

This approach to the asymmetric synthesis of acyclic ketones has been used to prepare a wide variety of ketones, a selection of which is shown in Fig. 6.69. In all cases the migrating chiral substituent retains its stereochemical integrity, and all of the ketones were prepared in >99 per cent enantiomeric excess.[71]

The preceding hydroboration reactions all involved addition of a borane to a mono-alkene. It is also possible to achieve very high regio- and stereoselectivity

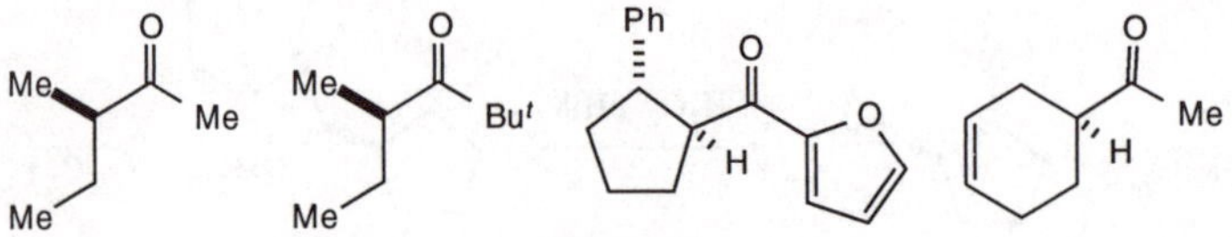

Fig. 6.69

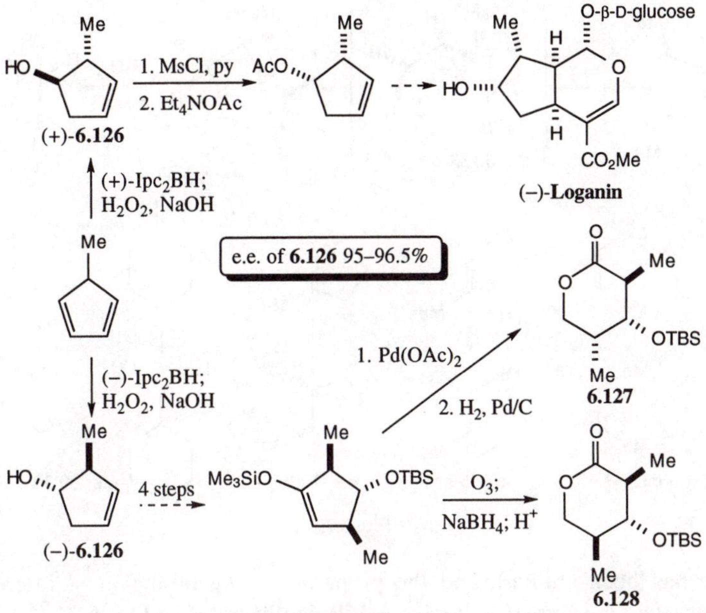

Fig. 6.70

in the hydroboration of some cyclic dienes, as in the enantioselective hydroboration of 5-methylcyclopentadiene (Fig. 6.70) which provides **6.126** in high enantiomeric purity. Both enantiomers of this alcohol have been used in natural product synthesis as outlined in Fig. 6.70. The natural enantiomer of loganin, an important precursor of many plant-derived natural products, was prepared from (+)-**6.126**.[72] The enantiomeric alcohol (−)-**6.126** was used to prepare lactones **6.127** and **6.128**, which were used in a total synthesis of an intermediate which contained all ten chiral centres of erythronolide.[73]

Asymmetric hydroboration with boranes derived from sources other than α-

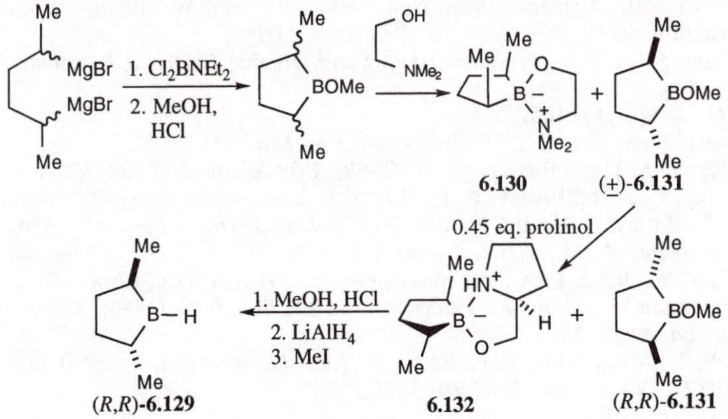

Fig. 6.71

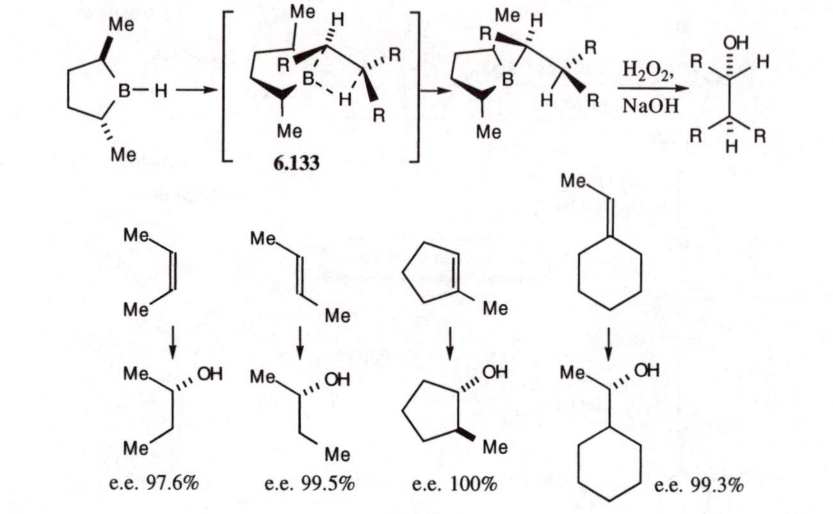

Fig. 6.72

pinene has been studied, and the preparation one of the most interesting reagents, the C_2 symmetric borane **6.129**, is illustrated in Fig. 6.71.[74]

Although the preparation and resolution of this borane limits its application somewhat, it reacts with a range of olefin types with very high enantioselectivity, the sense of which is consistent with a simple transition state model **6.133** (Fig. 6.72).

References

1. Krohn., K. (1991), in *Organic Synthesis Highlights*, (eds J. Mulzer, H.-J. Altenbach, M. Braun, K. Krohn, and H.-H. Reissig), pp. 54–65, VCH, New York; Paquette, L. A., in *Asymmetric Synthesis*, (ed. J. D. Morrison), Vol. 3, pp. 455–501, Academic Press, New York; Carruthers, W. (1990), *Cycloaddition Reactions in Organic Synthesis*, Pergamon, Oxford.
2. Fleming, I. (1976), *Frontier Orbitals and Organic Chemical Reactions*, Wiley, London.
3. Ref. 2, pp. 161–165.
4. Sauer, J. and Kredel, J. (1966), *Tetrahedron Lett.*, 731.
5. Meyers, A. I. and Busacca, C. A. (1989), *Tetrahedron Lett.*, **30**, 6973.
6. Corey, E. J. and Ensley, H. E. (1975), *J. Amer. Chem. Soc.*, **97**, 6908; Corey, E. J., Ensley, H. E., and Parnell, C. A. (1978), *J. Org. Chem.*, **43**, 1610.
7. Masamune, S., Murakami, S., and Tobita, H. (1983), *J. Org. Chem.*, **48**, 4441; Choy, W., Reed, L. A., and Masamune, S., (1983), *J. Org. Chem.*, **48**, 1137.
8. Masamune, S., Choy, W., Petersen, J. S., and Sita, L. S. (1985), *Angew. Chem. Int. Ed. Engl.*, **24**, 1.
9. Poll, T., Abdel Hady, A. F., Karge, R., Linz, G., Weetman, J., and Helmchen, G. (1989), *Tetrahedron Lett.*, **30**, 5595.

10. Evans, D. A., Chapman, K. T., and Bisaha, J. (1988), *J. Amer. Chem. Soc.*, **110**, 1238.
11. Evans, D. A. and Black, W. C., (1993), *J. Amer. Chem. Soc.*, **115**, 4497.
12. Oppolzer, W., Chapuis, C., and Bernardinelli, G. (1984), *Helv. Chim. Acta*, **67**, 1397.
13. Vandewalle, M., Van der Eycken, J., Oppolzer, W., and Vullioud, C. (1986), *Tetrahedron*, **42**, 4035.
14. Oppolzer, W., Dupuis, D., Poli, G., Raynham, T.M., and Bernardinelli, G. (1988), *Tetrahedron Lett.*, **29**, 5885.
15. Corey, E. J., Cheng, X.-M., and Cimprich, K. A. (1991), *Tetrahedron Lett.*, **32**, 6839.
16. Trost, B. M., Godelski, S. A., and Genêt, J. P. (1978), *J. Amer. Chem. Soc.*, **100**, 3930; Trost, B. M., O'Krongly, D., and Belletire, J. L. (1980), *J. Amer. Chem. Soc.*, **102**, 7595; Gupta, R. C., Harland, P. A., and Stoodley, R. J. (1983), *J. Chem. Soc.*, *Chem. Commun.*, 754.
17. Hashimoto, S.-I., Komeshima, N., and Koga, K. (1979), *J. Chem. Soc.*, *Chem. Commun.*, 437; Takemura, H., Komeshima, N., Takahashi, I., Ikota, N., Tomioka, K., and Koga, K. (1987), *Tetrahedron Lett.*, **28**, 5687.
18. Corey, E. J., Imwinkelried, R., Pikul, S., and Xiang, Y. B. (1989), *J. Amer. Chem. Soc.*, **111**, 5493.
19. Corey, E. J., Imai, N., and Pikul, S. (1991), *Tetrahedron Lett.*, **32**, 7517.
20. Corey, E. J. and Sarshar, S. (1992), *J. Amer. Chem. Soc.*, **114**, 7938.
21. Narasaka, K., Iwasawa, N., Inoue, M., Yamada, T., Nakashima, M., and Sugimori, J. (1989), *J. Amer. Chem. Soc.*, **111**, 5340.
22. Corey, E. J. and Matsumura, Y. (1991), *Tetrahedron Lett.*, **32**, 6289.
23. Narasaka, K., Saitou, M., and Iwasawa, N. (1991), *Tetrahedron Asymmetry*, **2**, 1305.
24. Furuta, K., Shimizu, S., Miwa, Y., and Yamamoto, H. (1989), *J. Org. Chem.*, **54**, 1483.
25. Hawkins, J. M. and Loren, S. (1991), *J. Amer. Chem. Soc.*, **113**, 7794.
26. Corey, E. J. and Loh, T.-P. (1991), *J. Amer. Chem. Soc.*, **113**, 8966.
27. Curran, D. P. and Heffner, T. A. (1990), *J. Org. Chem.*, **55**, 4585.
28. Oppolzer, W., Kingma, A. J., and Pillai, S. K. (1991), *Tetrahedron Lett.*, **32**, 4893.
29. Meyers, A. I. and Fleming, S.A. (1986), *J. Amer. Chem. Soc.*, **108**, 306.
30. Houge, C., Frisque–Hesbain, A. M., Mockel, A., Ghosez, L., Declercq, J. P., Germain, G., and Van Meerssche, M. (1982), *J. Amer. Chem. Soc.*, **104**, 2920.
31. Greene, A. E. and Charbonnier, F. (1985), *Tetrahedron Lett.*, **26**, 5525.
32. Romo, D., Romine, J. L., Midura, W., and Meyers, A. I. (1990), *Tetrahedron*, **46**, 4951.
33. Aratani, T., Yoneyoshi, Y., and Nagase, T. (1975), *Tetrahedron Lett.*, 1707.
34. Aratani, T., Yoneyoshi, Y., and Nagase, T. (1982), *Tetrahedron Lett.*, **23**, 685.
35. Evans, D. A., Woerpel, K. A., Hinman, M. M., and Faul, M. M. (1991), *J. Amer. Chem. Soc.*, **113**, 726.
36. Lowenthal, R. E., Abiko, A., and Masamune, S. (1990), *Tetrahedron Lett.*, **31**, 6005.
37. Lowenthal, R. E. and Masamune, S. (1991), *Tetrahedron Lett.*, **32**, 7373.
38. Oppolzer, W. and Löher, H. J. (1981), *Helv. Chim. Acta*, **64**, 2808.
39. Oppolzer, W. (1987), *Tetrahedron*, **43**, 1969.

40. Oppolzer, W. and Jacobsen, E. J. (1986), *Tetrahedron Lett.*, **27**, 1141; Oppolzer, W., Moretti, R., and Bernardinelli, G. (1986), *Tetrahedron Lett.*, **27**, 4713.
41. Oppolzer, W., Mills, R. J., Pachinger, W., and Stevenson, T. (1986), *Helv. Chim. Acta*, **69**, 1542.
42. Oppolzer, W., Poli, G., Kingma, A. J., Starkemann, C., and Bernardinelli, G. (1987), *Helv. Chim. Acta*, **70**, 2201.
43. Oppolzer, W. and Poli, G. (1986), *Tetrahedron Lett.*, **27**, 4717.
44. Davies, S. G., Dordor–Hedgecock, I. M., Sutton, K. H., Walker, J. C., Jones, R. H., and Prout, K. (1986), *Tetrahedron*, **42**, 5123.
45. Davies, S. G. and Walker, J. C. (1986), *J. Chem. Soc., Chem. Commun.*, 495.
46. Tomioka, K., Suenaga, T., and Koga, K. (1986), *Tetrahedron Lett.*, **27**, 369.
47. Mukaiyama, T. and Iwasawa, N. (1981), *Chem. Lett.*, 913.
48. Touet, J., Baudouin, S., and Brown, E. (1992), *Tetrahedron Asymmetry*, **3**, 587.
49. Fang, C., Ogawa, T., Suemune, H., and Sakai, K. (1991), *Tetrahedron Asymmetry*, **2**, 389.
50. Meyers, A. I. and Shipman, M. (1991), *J. Org. Chem.*, **56**, 7098.
51. Davies, S. G. and Ichiara, O. (1991), *Tetrahedron Asymmetry*, **2**, 183.
52. Evans, D.A., Bilodeau, M. T., Somers, T. C., Clardy, J., Cherry, D., and Kato, Y. (1991), *J. Org. Chem.*, **56**, 5750.
53. Corey, E. J. and Peterson, R. T. (1985), *Tetrahedron Lett.*, **26**, 5025.
54. Corey, E. J. and Magriotis, P. A. (1987), *J. Amer. Chem. Soc.*, **109**, 287.
55. Revial, G. (1989), *Tetrahedron Lett.*, **30**, 4121.
56. Corey, E. J., Naef, R., and Hannon, F. J. (1986), *J. Amer. Chem. Soc.*, **108**, 7114.
57. Dieter, R. K. and Tolkes, M. (1987), *J. Amer. Chem. Soc.*, **109**, 2040.
58. Rossiter, B. E., Miao, G., Swingle, N. M., Eguchi, M., Hernández, A. E., and Patterson, R. G. (1992), *Tetrahedron Asymmetry*, **3**, 231.
59. Tanaka, K., Matsui, J., Kawabata, Y., Suzuki, H., and Watanabe, A. (1991), *J. Chem. Soc., Chem. Commun.*, 1632.
60. Jansen, J. F. G. A. and Feringa, B. L. (1992), *Tetrahedron Asymmetry*, **3**, 581.
61. Tanaka, K., Matsui, J., Suzuki, H., and Watanabe, A. (1992), *J. Chem. Soc., Perkin Trans 1*, 1193.
62. Brown, H. C. and Jadhav, P. K. (1983), in *Asymmetric Synthesis*, (ed. J. D. Morrison), Vol. 2, pp. 1–43, Academic Press, New York.
63. Brown, H. C. (1988), *Chemtracts*, **1**, 77.
64. Brown, H. C., Schwier, J. R., and Singaram, B. (1978), *J. Org. Chem.*, **43**, 4395.
65. Brown, H. C., Jadhav, P. K., and Mandal, A. K. (1981), *Tetrahedron*, **37**, 3547.
66. Brown, H. C. and Singaram, B. (1984), *J. Amer. Chem. Soc.*, **106**, 1797.
67. Brown, H. C., Imai, T., Desai, M. C., and Singaram, B. (1985), *J. Amer. Chem. Soc.*, **107**, 4980.
68. Brown, H. C., Naik, R. G., Singaram, B., and Pyun, C. (1985), *Organometal.*, **4**, 1925.
69. Brown, H. C., Naik, R. G., Bakshi, R. K., Pyun, C., and Singaram, B. (1985), *J. Org. Chem.*, **50**, 5586.
70. Brown, H. C., Kim, K. W., Singaram, B., and Cole, T. E. (1986), *J. Amer. Chem. Soc.*, **108**, 6761.
71. Brown, H. C., Srebnik, M., Bakshi, R. K., and Cole, T. E. (1987), *J. Amer. Chem. Soc.*, **109**, 5420.

72. Partridge, J. J., Chadha, N. K., and Uskokovic, M. R. (1973), *J. Amer. Chem. Soc.*, **95**, 532.
73. Stork, G., Paterson, I., and Lee, F. K. C. (1982), *J. Amer. Chem. Soc.*, **104**, 4686.
74. Masamune, S., Kim, B. M., Petersen, J. S., Sato, T., and Veenstra, S. J. (1985), *J. Amer. Chem. Soc.*, **107**, 4549.

7 Reduction and oxidation

Asymmetric hydrogenation has been studied intensively for a relatively long time. In its usual form the reaction involves the addition of hydrogen to a double bond in the presence of a transition metal catalyst and a chiral, non-racemic ligand.[1] As with many enantioselective reactions which involve catalysis, a substrate, ligand, metal atom, and stoichiometric reagent are all involved in the transition state of the rate determining step. A typical catalytic cycle for asymmetric hydrogenation is shown schematically in Fig. 7.1.

In this schematic representation (Fig. 7.1) the chiral ligand, usually a chelating diphosphine, remains attached to the metal throughout the cycle. The complex **7.1** (which is shown without its coordinating solvent molecules) binds the substrate **7.2** to form intermediate **7.3** (step 1). Oxidative addition of molecular hydrogen to the metal produces **7.4** (step 2), and is followed by hydrogen atom transfer to the substrate to give **7.5** (step 3). Transfer of the second hydrogen atom and subsequent decomplexation of the fully reduced product **7.6** regenerates the catalyst **7.1** and completes the cycle. Efficient asymmetric synthesis is possible because each new chiral centre is created while the substrate is part of a chiral complex (therefore the transition states leading to the possible stereoisomers will be diastereoisomeric). The cycle illustrated in Fig. 7.1 is not the only possible mechanism, but it probably represents the basis of the mechanism of most homogeneous asymmetric hydrogenations.

In practice, the metal involved is almost always either rhodium(I) or ruthenium(II), and the ligand is usually a chiral, chelating diphosphine. A large number of different diphosphines have been used, a few of which are shown in Fig. 7.2 along with their acronyms.[2]

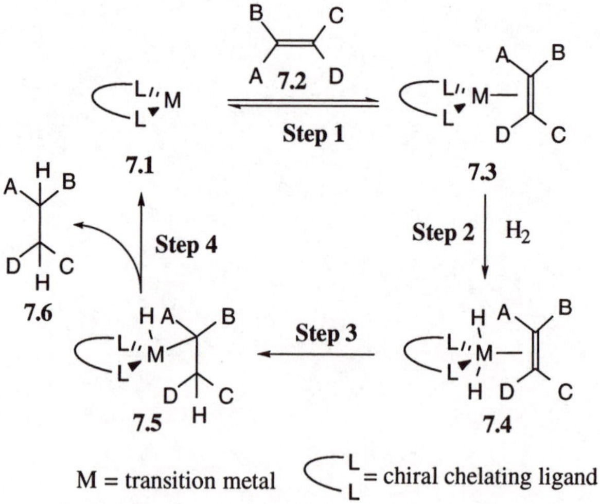

Fig. 7.1

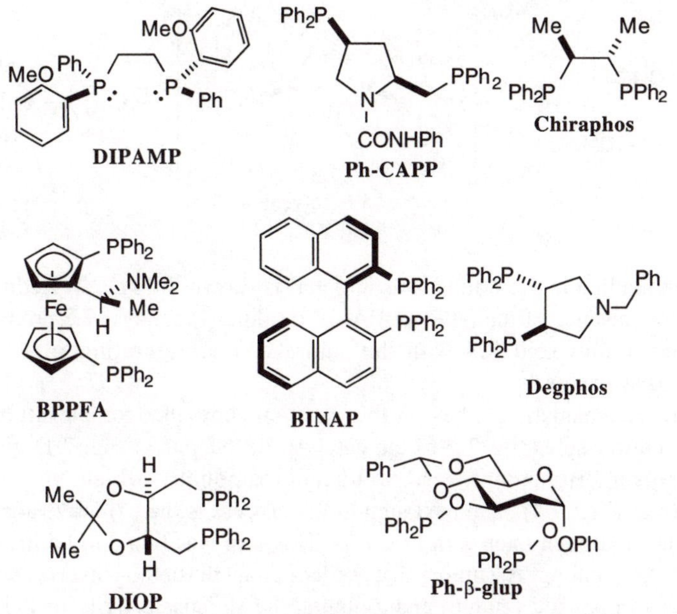

Fig. 7.2

As can be seen from this short selection of ligands (Fig. 7.2), a wide range of structural types of diphosphines can function as part of a catalytic system. All those in Fig. 7.2 have been shown to provide enantiomeric excesses of greater than 90 per cent with the appropriate substrate. These catalysts are relatively easy to prepare, and are obtained in high enantiomeric excess either by resolution of an intermediate (e.g. DIPAMP, BINAP) or by using an enantiomerically pure starting material (e.g. DIOP, Chiraphos, and Degphos, which can all be prepared from tartaric acid).[3] Several are commercially available.

An important aspect of any enantioselective catalyst is the relative amount which is needed. This is often reported as the substrate:catalyst ratio, and for the type of reaction under consideration here this ratio can be very high. An enantiomeric excess of 96.5 per cent has been reported for a hydrogenation using a rhodium(I) complex of Degphos (Fig. 7.2) at a substrate:catalyst ratio of 50 000:1.

Discussion of all the efficient systems which are available for asymmetric hydrogenation is beyond the scope of this chapter. Instead, a typical well-studied example will be considered in detail, followed by examples of unsaturated compounds which are usually good substrates, and certain other selected enantioselective catalyst systems.

Hydrogenation of (Z)-α-acylamidocinnamates in the presence of rhodium complexes of various diphosphines often gives very high (≥95 per cent) enantiomeric excess in high chemical yield. An example of this reaction which

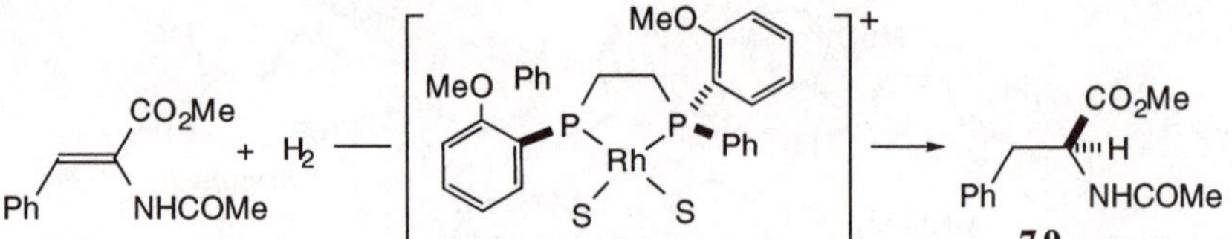

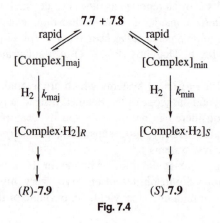

Fig. 7.3

has been studied in great mechanistic detail is shown in Fig. 7.3. Reduction of **7.7** in the presence of the (R,R)-DIPAMP rhodium(I) catalyst **7.8** produces the protected amino acid **7.9** with the natural (S)-configuration in very high enantiomeric excess.[4]

Careful mechanistic studies on this reaction show that complexation of the hydrogenation substrate **7.7** to the catalyst **7.8** (step 1 in Fig. 7.1) gives two diastereoisomeric complexes, in unequal amounts, which are in rapid equilibrium (Fig. 7.4). The next step in the process is the oxidative addition of molecular hydrogen (step 2 in Fig. 7.1). This step was found to be irreversible and rate determining. Assuming that the individual diastereoisomeric complexes give rise to opposite (single) enantiomers, the stereoselectivity of the overall reaction depends strongly on the relative rates of oxidative addition to the two diastereoisomeric complexes (Fig. 7.4).

It was found that the minor complex ([Complex]$_{min}$, Fig. 7.4) reacts with hydrogen much more rapidly than the major complex. The ratio k_{min}:k_{maj} was found to be 573:1, and the equilibrium ratio of [Complex]$_{min}$:[Complex]$_{maj}$ was 1:11. Taken together these values give a product ratio of ~52:1 in favour of the (S)-enantiomer, which corresponds to an enantiomeric excess of ~96 per cent, close to that observed. It has therefore been shown clearly that *the major enantiomer of the product is derived from the minor catalyst–substrate complex.*

The structures of the complexes formed on oxidative addition of hydrogen are thought to resemble **7.10** and **7.11** (Fig. 7.6), and correspond to coordination

$$\textbf{7.7} + \textbf{7.8}$$

rapid ⇄ ⇄ rapid

[Complex]$_{maj}$ [Complex]$_{min}$

H$_2$ | k_{maj} H$_2$ | k_{min}

[Complex·H$_2$]$_R$ [Complex·H$_2$]$_S$

(R)-**7.9** (S)-**7.9**

Fig. 7.4

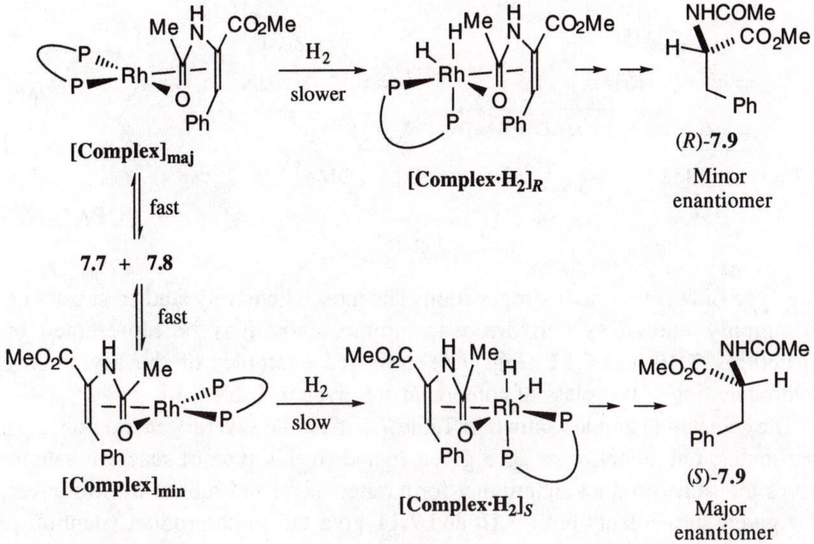

Fig. 7.5

of either face of the prochiral C=C bond. Bonding of the hydrogen atom then takes place on the face of the alkene which is coordinated to the metal.

The intermediates in Fig. 7.5 are all shown with the carbonyl group coordinating to the metal centre throughout the reaction. This is an important structural feature in many substrates which give high enantiomeric excesses in

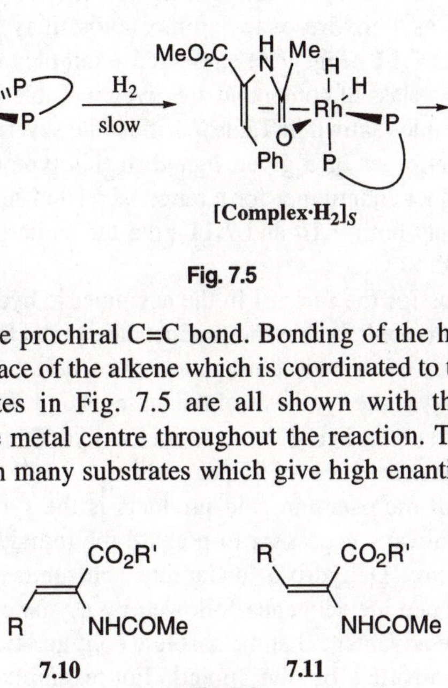

Fig. 7.6

Table 7.1 Asymmetric hydrogenation of **7.10** and **7.11** in the presence of Rh(I)(ligand) complex

Substrate	R	R'	Ligand	e.e. (%)
7.10	H	H	(S,S)-Chiraphos	92(R)
7.10	Pr^i	H	(S,S)-Chiraphos	100(R)
7.10	Ph	H	(R,R)-DIPAMP	96(S)
7.10	$MeOCH_2$	Me	(R,R)-DIPAMP	86(S)
7.11	Pr^n	Me	(R,R)-DIPAMP	95(S)
7.11	Pr^i	Me	(R,R)-DIPAMP	78(S)
7.11	Ph	H	(S)-BINAP	87(R)
7.11	$MeOCH_2$	Me	(R,R)-DIPAMP	94(S)

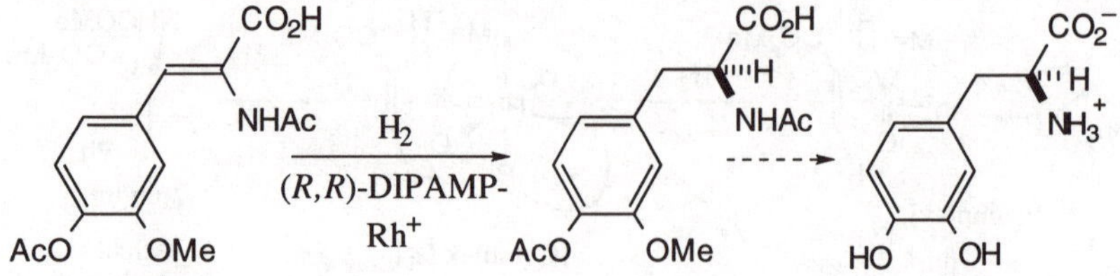

Fig. 7.7

this type of asymmetric hydrogenation. The most extensively studied substrates, commonly known as dehydro-α-acylamino acids, may be represented by structures **7.10** and **7.11** (Fig. 7.6). Selected examples of the asymmetric hydrogenation of this class of compound are given in Table 7.1.[5]

The selected examples shown in Table 7.1 illustrate several general points. In particular, one enantiomer of a given ligand in this type of reaction usually gives the same product enantiomer for a range of related substrates. Moreover, for given substituents both **7.10** and **7.11** give the same product enantiomer with the same ligand.

One of the reasons for the interest in the asymmetric hydrogenation of such dehydro-α-acylamino acids is that the products are protected α-amino acids, a most important class of organic compounds. The first commercial catalytic asymmetric process was the synthesis of the amino acid L-DOPA developed by Monsanto, the crucial step being the asymmetric hydrogenation of **7.12** (Fig. 7.7) with an enantiomeric excess of 95 per cent.[6]

One of the uses of the α-amino acid products is the synthesis of peptides. However, it is not always necessary to prepare the individual α-amino acids followed by coupling. Dehydro-α-acylamino acid units themselves can be incorporated into a peptide sequence followed by asymmetric hydrogenation. This strategy has the advantage that the absolute configuration of the new chiral centres can be controlled by the ligand. For example, the two peptide diastereoisomers **7.13** and **7.14** can be prepared with high stereoselectivity by reduction of a single starting material **7.15** using different ligands (Fig. 7.8, see Fig. 7.2 for ligand structures).[7]

Studies with a range of other substrates and chiral diphosphine ligands have shown that, in general, two structural features are helpful for high enantioselectivity in the asymmetric hydrogenation of alkenes. These features, an electron-withdrawing group (e.g. CO_2R, Ph, CN) on the α-carbon, and a basic carbonyl group positioned β to the olefinic bond, are accommodated by the general structure **7.16** proposed for the intermediate complexes (Fig. 7.9). Accordingly, the substrates **7.17** to **7.20** can be reduced with good enantioselectivities (up to 90 per cent), whereas for **7.21**, which lacks the electron-withdrawing group, the enantiomeric excess is less than 50 per cent.[8]

Tetrasubstituted alkenes such as **7.22** are poor substrates for the usual catalysts; however, rhodium(I) complexes of the ferrocenyldiphosphines **7.23**

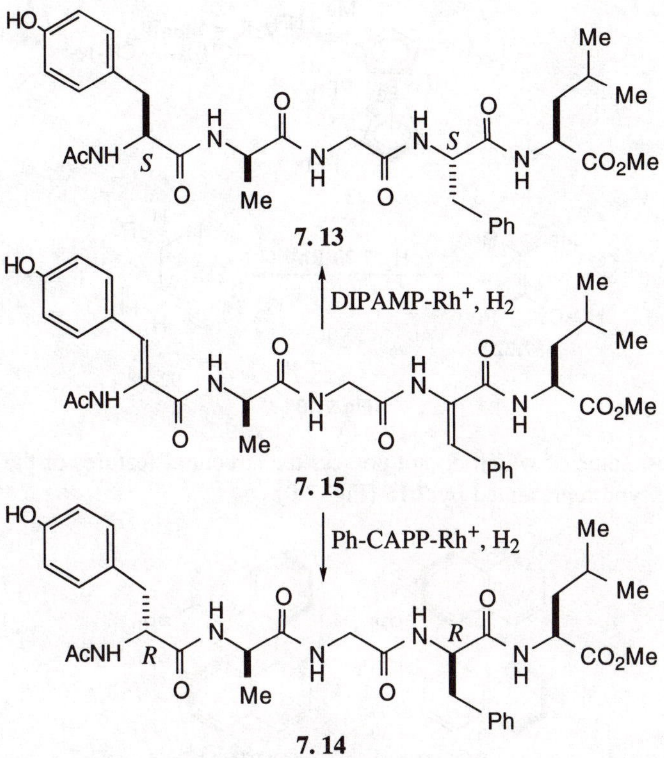

7. 13

DIPAMP-Rh⁺, H₂

7. 15

Ph-CAPP-Rh⁺, H₂

7. 14

Fig. 7.8

will catalyse the *cis*-hydrogenation of **7.22** (R = Me, Et, Ph) in high enantiomeric excess (Fig. 7.10).[9]

All the catalyst systems which have been considered so far have used rhodium(I) as the metal. Ruthenium(II) carboxylate complexes of readily available (*R*)- and (*S*)-BINAP (Fig. 7.11) have been found to be extremely selective and versatile catalysts for the asymmetric hydrogenation of a range of

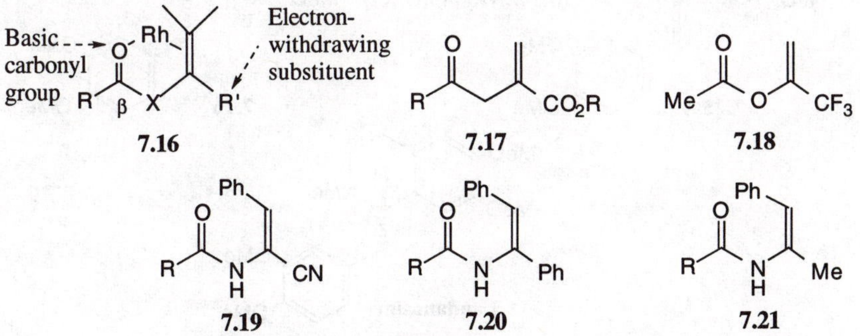

7.16   **7.17**   **7.18**

7.19   **7.20**   **7.21**

Fig. 7.9

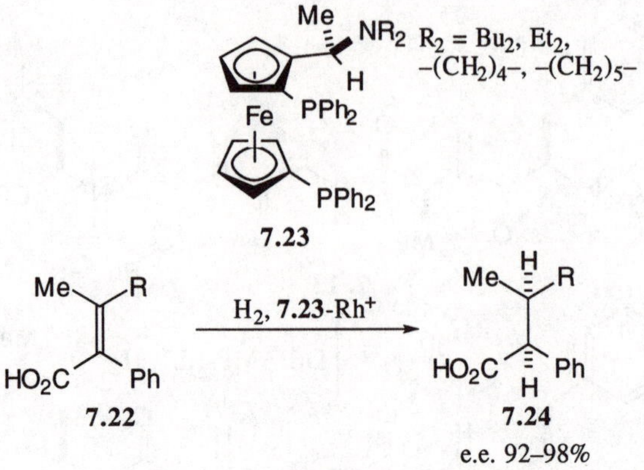

7.23

Fig. 7.10

substrates, some of which do not possess the structural features of the general substrate type represented by **7.16** (Fig. 7.9).

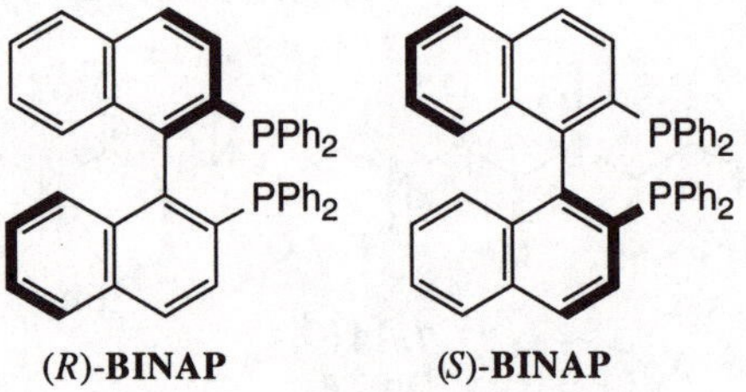

(*R*)-**BINAP**   (*S*)-**BINAP**

Fig. 7.11

Hydrogenation of the enamides **7.25** (for example R = CH$_3$, CF$_3$, Bu^t, H) with [(*R*)-BINAP]Ru(OAc)$_2$ gives **7.26** in >99.5 per cent enantiomeric excess. Only (*Z*)-enamides are reduced, the (*E*)-isomers do not react. These reduced

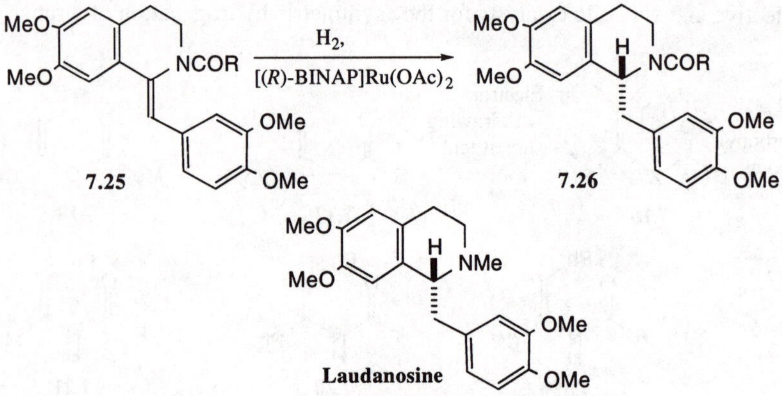

Laudanosine

Fig. 7.12

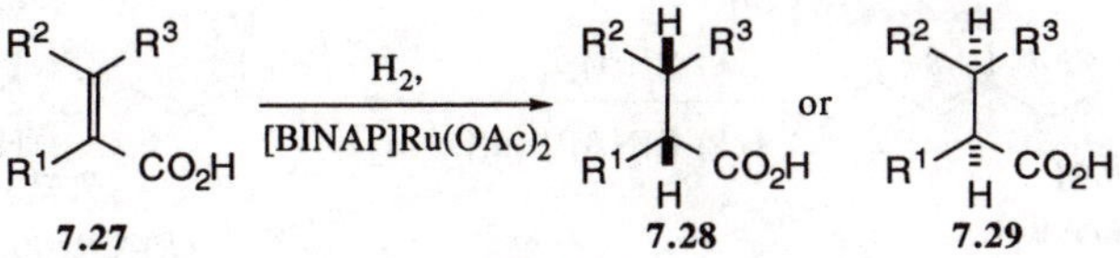

Fig. 7.13

Table 7.2 Asymmetric reduction of **7.27** to **7.28** with $[(R)\text{-BINAP}]Ru(OAc)_2$

R^1	R^2	R^3	e.e. (%)
Me	Me	H	91
H	$Me_2C=C(CH_2)_2$	Me	87
H	$HOCH_2$	Me	93
Me	$AcOCH_2$	H	83

products represent key intermediates for the synthesis of isoquinoline alkaloids such as laudanosine (Fig. 7.12).[10]

The catalysts $[(R)\text{-BINAP}]$- or $[(S)\text{-BINAP}]Ru(OAc)_2$ can be used for efficient asymmetric hydrogenation of α,β-unsaturated carboxylic acids which do not contain an acylamino group. As outlined above, these are usually poor substrates for the rhodium(I) catalysts.[11]

This and other studies suggest that the mechanism of asymmetric hydrogenation with the BINAP–ruthenium(II) complexes is different from that of the chiral rhodium(I) catalysts. It is proposed that the mechanism might involve reaction of the alkene with a ruthenium hydride intermediate, rather than oxidative addition of molecular hydrogen to a catalyst–alkene complex. Whatever the detailed mechanism, the BINAP–ruthenium(II) catalysed hydrogenation of α,β-unsaturated carboxylic acids takes place with excellent enantioselectivity (Fig. 7.13 and Table 7.2) to give either **7.28** or **7.29** depending on which enantiomer of the catalyst is used.

This type of asymmetric hydrogenation was used to prepare the anti-inflammatory agent (S)-naproxen **7.30** in excellent yield and an enantiomeric excess of 97 per cent (Fig. 7.14).[11]

Allylic alcohols are also reduced with high enantioselectivity in the presence of BINAP–ruthenium(II) catalysts. The catalysts are recoverable and given some optimization the substrate:catalyst ratio can be as high as 50 000:1, which make this a most efficient process.

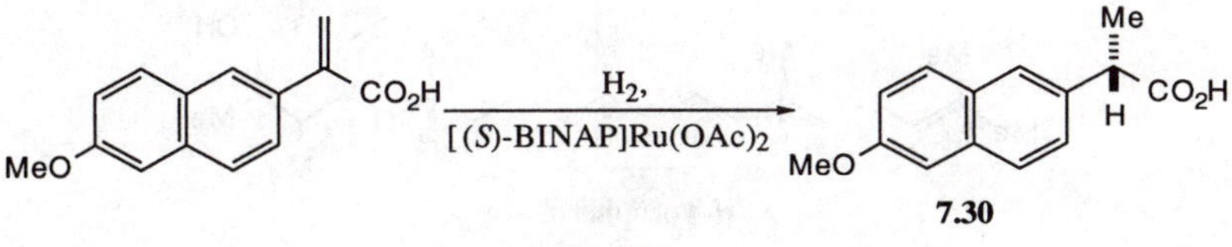

Fig. 7.14

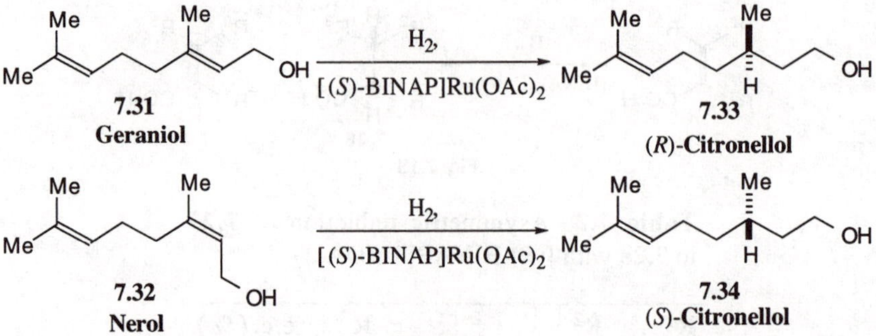

Fig. 7.15

This reaction is chemoselective as well as enantioselective, in that isolated carbon–carbon double bonds are not reduced. This reaction is exemplified by the hydrogenation of the geometric isomers geraniol **7.31** and nerol **7.32** (Fig. 7.15).[12] The isolated double bond is not reduced, and the absolute configuration of the citronellol depends on the geometry of the alkene that is reduced. The use of (*R*)-BINAP in place of (*S*)-BINAP would also reverse the absolute configuration of the product. It follows that the same enantiomer of the product, say (*R*)-citronellol **7.33**, could be obtained from either isomer double bond geometry, by reduction as shown, or by reduction of nerol **7.32** with the (*R*)-BINAP–ruthenium catalyst.

This powerful asymmetric hydrogenation methodology has been used in a total synthesis of the side chain of α-D-tocopherol (vitamin E) **7.35** (Fig. 7.16).[12] Moreover, the chiral centre present in the reduction substrate **7.36** was obtained by another asymmetric synthesis method using a BINAP–ruthenium(II) complex as an isomerization catalyst (Chapter 8).

The BINAP–ruthenium(II) catalyst system can also be used for kinetic resolution of racemic allylic alcohols, as illustrated by the reduction of **7.37** (Fig. 7.17).[13]

So far in this chapter, only the reduction of alkenes has been considered. An equally large body of work exists concerning the enantioselective reduction of carbonyl compounds, mainly prochiral ketones, to the corresponding alcohol.

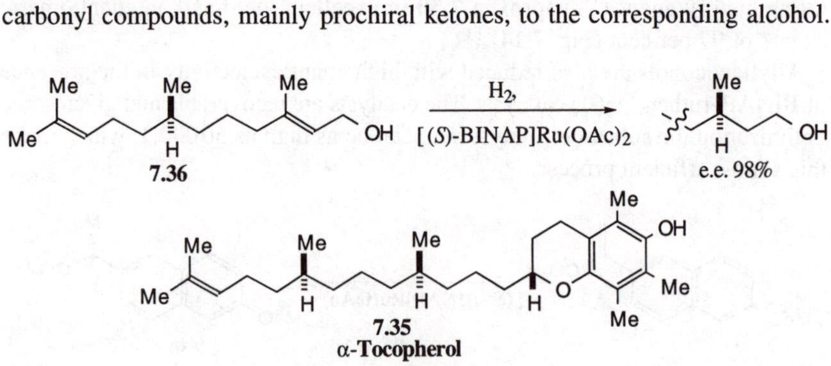

7.35
α-Tocopherol
Fig. 7.16

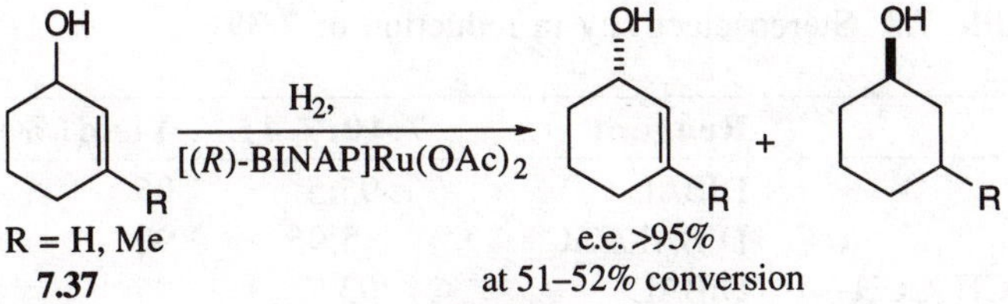

R = H, Me
7.37

e.e. >95%
at 51–52% conversion

Fig. 7.17

Selected examples of this important class of asymmetric reductions will now be discussed.

In principle, a chiral auxiliary attached to the carbonyl group should allow for asymmetric reduction of a carbonyl group. One of the most successful approaches of this type uses sulphoxide **7.38** as the stereochemical control element. Both enantiomers of this are available, although due to the high level of stereochemical control which is possible, only one is actually needed. Sulphoxide **7.38** reacts readily with esters to produce β-ketosulphoxides **7.39** (Fig. 7.18), and the reduction of these systems takes place with high stereoselectivity (Table 7.3). Moreover, reduction with diisobutylaluminium hydride (DIBAL) gives **7.40**, whereas use of lithium aluminium hydride or DIBAL in the presence of zinc chloride takes place with the opposite stereochemistry to give **7.41**.[14] This reversal of stereoselectivity is thought to be due to the β-ketosulphoxides **7.39** adopting a conformation similar to **7.42** in the absence of chelating metals. This conformation minimizes dipole–dipole repulsion between the two oxygen functionalities, and hydride attack from the less hindered face of the carbonyl results in **7.40**. When chelating metals (lithium or zinc) are present, the carbonyl group is chelated (conformation **7.43**) and reduced from the less hindered (opposite) face (Fig. 7.18).

The products from these reductions are versatile synthetic intermediates. The sulphoxide group can be reduced off with either Raney nickel, or as in the case

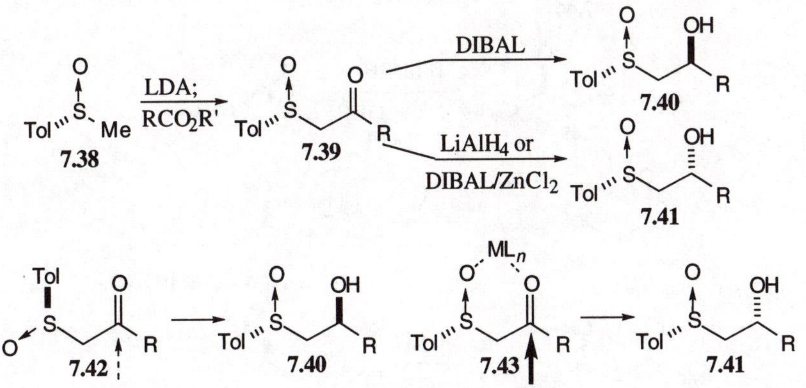

Fig. 7.18

Table 7.3 Stereoselectivity in reduction of **7.39**

R	Reagent	7.40:7.41	Yield (%)
Ph	DIBAL	>95:5	95
Ph	DIBAL/ZnCl$_2$	>5:95	90
PhCH$_2$CCH$_2$	DIBAL	93:7	95
PhCH$_2$CCH$_2$	DIBAL/ZnCl$_2$	>5:95	95
n-C$_8$H$_{17}$	DIBAL	95:5	95
n-C$_8$H$_{17}$	DIBAL/ZnCl$_2$	>5:95	92

of allylic alcohols such as **7.44**, with lithium in ethylamine.[15] The sulphoxide group can also be converted into a leaving group which allows for the conversion of β-hydroxysulphoxides into epoxides as shown for **7.45** (Fig. 7.19).[14]

Reduction of β,δ-diketosulphoxides takes place at the non-enolized carbonyl group (**7.46**) and reduction of the remaining carbonyl group can be controlled as shown in Fig. 7.19. Reductive removal of the sulphoxide group then provides a route to *anti*-1,3-diols.[16]

Impressive as this sulphoxide approach is, there is the drawback that the chiral controller is destroyed on removal. An approach which avoids this problem is the reduction of ketones with enantioselective reducing agents. As usual with chiral reagents, the enantioselectivity must be high for the process to be of real value, as the products from reduction of either face of the carbonyl group are enantiomers. Nevertheless, a large body of work exists concerning such asymmetric reductions of ketones,[17] some of which is illustrated in the next part of this chapter using a few selected examples.

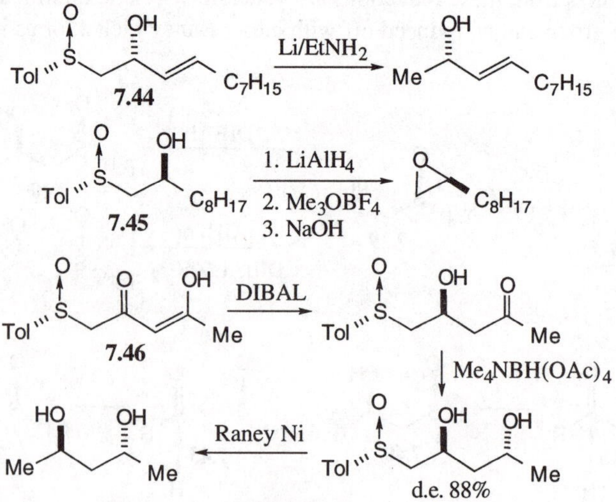

Fig. 7.19

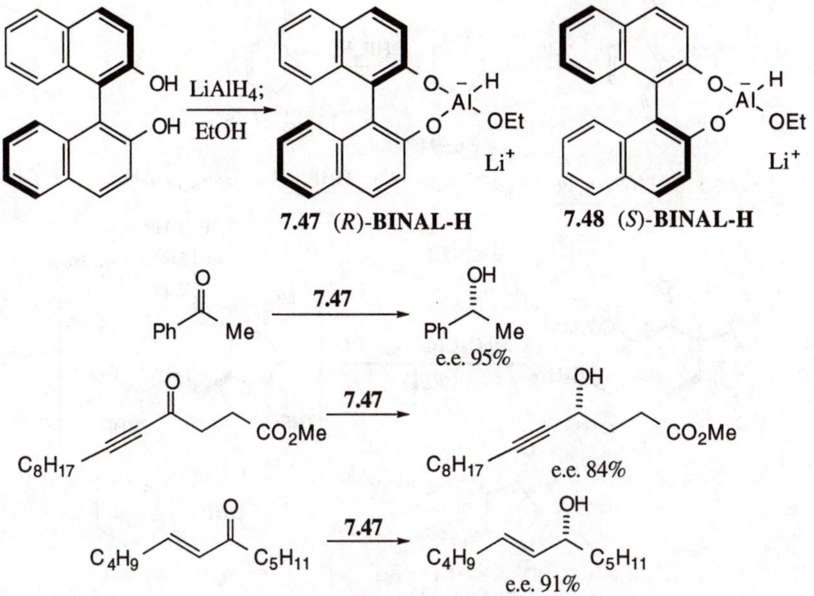

7.47 (R)-BINAL-H   7.48 (S)-BINAL-H

Fig. 7.20

Lithium aluminium hydride and sodium borohydride can both be reacted with a range of enantiomerically pure ligands to produce reagent mixtures which will carry out the enantioselective reduction of prochiral ketones. In general, reagents derived from lithium aluminium hydride give higher enantiomeric excesses, and the following discussion will be confined to such reactions.

One problem with reagents derived from lithium aluminium hydride and chiral alcohols is that some alcohols produce complexes which disproportionate rather readily to give more than one complex in solution. This problem is not encountered with enantiomeric complexes **7.47** and **7.48** derived from the corresponding binaphthols, lithium aluminium hydride, and ethanol, known as (*R*)- and (*S*)-BINAL-H respectively. These reagents reduce α,β-unsaturated ketones predictably and with high enantiomeric excess (Fig. 7.20).[18]

The BINAL reduction of prochiral ketones has been used as the cornerstone of a short, enantioselective total synthesis of the methyl ester of PGE$_1$ (Fig. 7.21), via the three-component coupling of **7.49**, **7.50**, and **7.51**.[19]

A model which is consistent with most of the reactions of the BINAL-H complex has been proposed, and is illustrated for **7.47** in Fig. 7.22. The most basic oxygen, that from the ethoxy group, acts as a bridging ligand which allows the formation of a six-membered chair transition state, in which the unsaturated group of the ketone occupies the equatorial position.

Many other ligands have been used to produce enantioselective reducing agents from lithium aluminium hydride, notably derivatives of 1,2-aminoalcohols. In general, only certain types of α,β-unsaturated ketones give

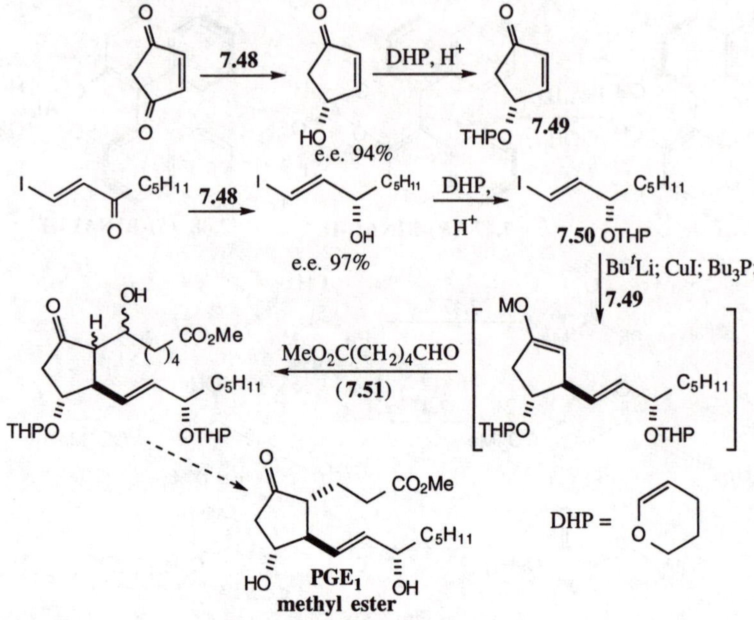

Fig. 7.21

high enantiomeric excesses with these reagents. Selected examples of the aminoalcohol controlled reduction of methyl phenyl ketone are presented in Fig. 7.23 and Table 7.4.[20]

Alkynyl ketones also tend to be good substrates for enantioselective reduction with aminoalcohol–lithium aluminium hydride complexes. For example, the

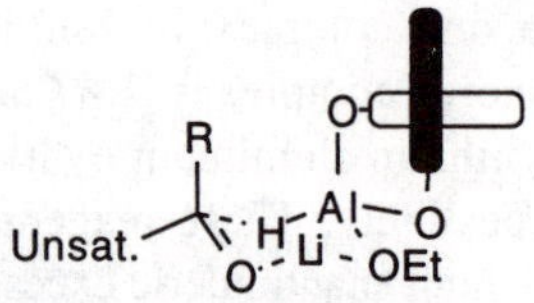

Fig. 7.22

Table 7.4   Enantioselective reduction of methyl phenyl ketone

Ligand	e.e. (%)	Product
7.52	95	(S)
7.53	92	(S)
7.54	82	(R)
7.55	68	(R)
7.56	60	(S)

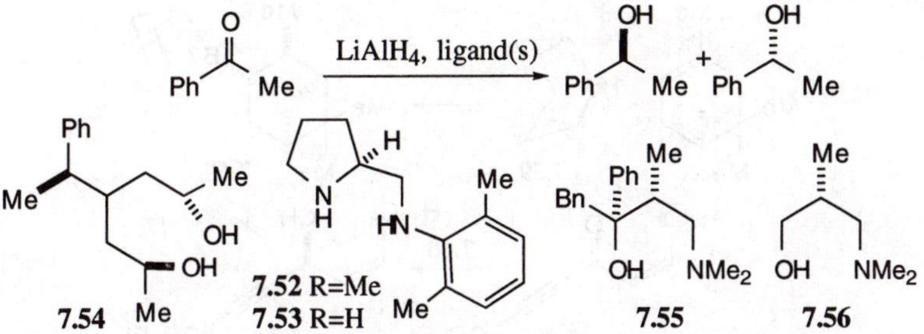

Fig. 7.23

ketone **7.57** is reduced with good enantioselectivity by the complex derived from reaction of lithium aluminium hydride, *N*-methylephedrine, and 3,5-dimethylphenol (Fig. 7.24).[21]

Allylic alcohols can be produced in high enantiomeric excess by reduction of the corresponding ketone with the complex derived from *N*-methylephedrine **7.58** alone, or with **7.58** and *N*-ethylaniline (Fig. 7.25).[22]

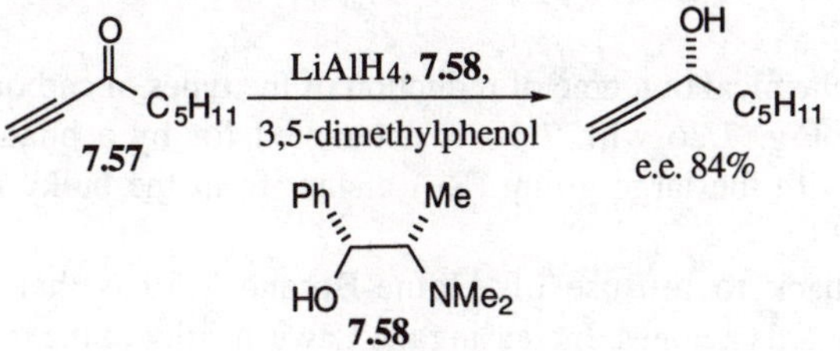

Fig. 7.24

As mentioned earlier, most complexes derived from sodium borohydride and chiral alcohols or aminoalcohols are relatively poor enantioselective reducing agents for ketones. However, there are several neutral boranes which are highly effective enantioselective reducing agents.[23]

Reaction of 9-borabicyclo[3.3.1]nonane **7.59** with (+)-pinene produces a reagent **7.60**, known as 'Alpine-Borane'. This borane reduces acetylenic

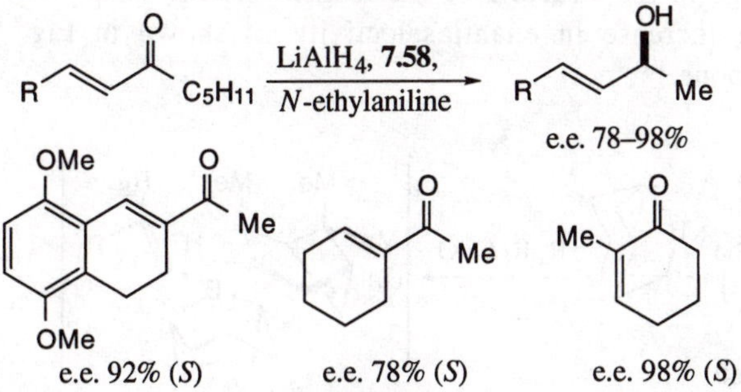

Fig. 7.25

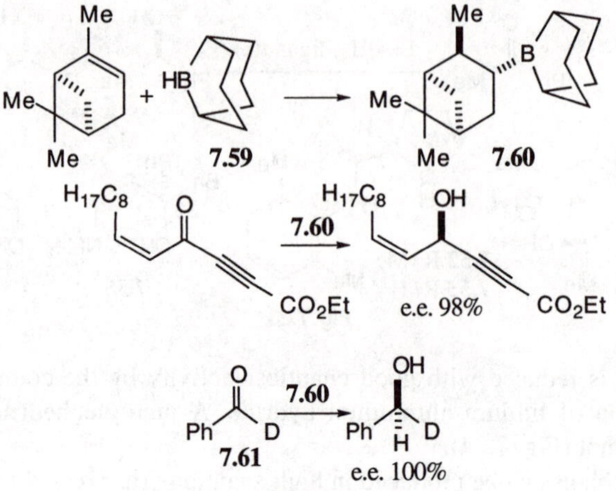

Fig. 7.26

ketones with high enantioselectivity[24] and reduction of deuterated aldehydes such as **7.61** provides the corresponding chiral labelled primary alcohols (Fig. 7.26).[25]

The stereochemical outcome of reduction of the types of carbonyl compounds illustrated in Fig. 7.26 with **7.60** is accounted for by a boat-like transition state model with the large group (R_L) distant from the bulky reducing agent (Fig. 7.27).[26]

One drawback to the use of Alpine-Borane **7.60** is that it is relatively unreactive towards ketones. Increasing the Lewis acidity of the reagent would be expected to increase this reactivity, and the most effective reagent of this type is chlorodiisopinocampheylborane (Ipc$_2$BCl) **7.62**. This is easily prepared from Ipc$_2$BH (see Chapter 6) as shown for (–)-Ipc$_2$BCl,[27] and reacts relatively rapidly with aromatic[28] and *tert*-alkyl ketones[29] to give the corresponding alcohols in high enantiomeric excess (Fig. 7.28).

A transition state model similar to that shown in Fig. 7.27 accounts for the stereoselectivity of reductions with (–)-Ipc$_2$BCl. This reducing agent is not particularly effective for the reduction of less hindered acyclic ketones, but a small change in the structure of the reagent to **7.63** (Eap$_2$BCl) results in a remarkable increase in enantioselectivity as shown in Fig. 7.29 for 3-methylbutanone.[30]

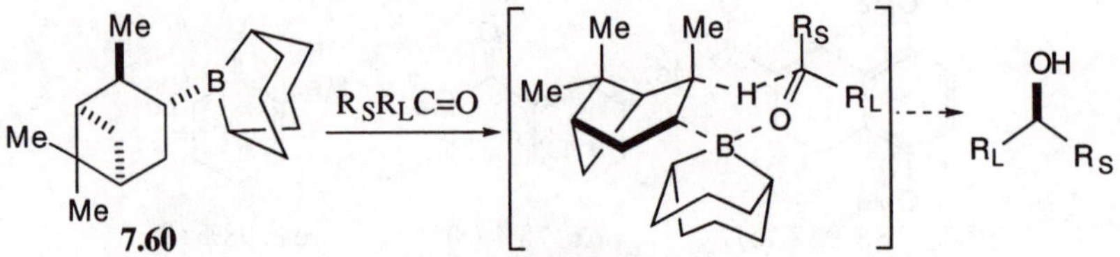

Fig. 7.27

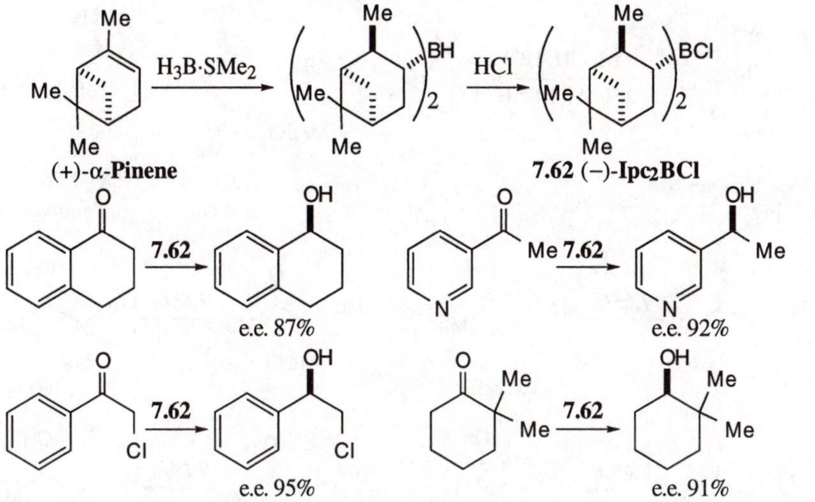

Fig. 7.28

The most demanding type of asymmetric reduction of ketones is that of acyclic, unbranched substrates. A borane reagent system based on 2,5-dimethylborolane **7.65** is capable of achieving such reductions, and of highly enantioselective reductions of other ketones (Fig. 7.30, see also Chapter 6 for other applications of this borane).[31] Although tedious to prepare, the stereoselectivity of this reagent under the appropriate reaction conditions is exceptional.

The appropriate reaction conditions referred to above involve a mixture of the borane **7.65** (1.0 equivalent) and the mesylate **7.66** (0.2 equivalent), generated by treatment of the borohydride **7.67** (1.2 equivalents) with methanesulphonic acid (1.4 equivalents). In the absence of the mesylate **7.66** reduction of butanone gives the alcohol with only 4 per cent enantiomeric excess, which rises to 80 per cent using the mixture of borane and mesylate.

It is proposed that the role of the mesylate **7.66** is that of a catalytic Lewis

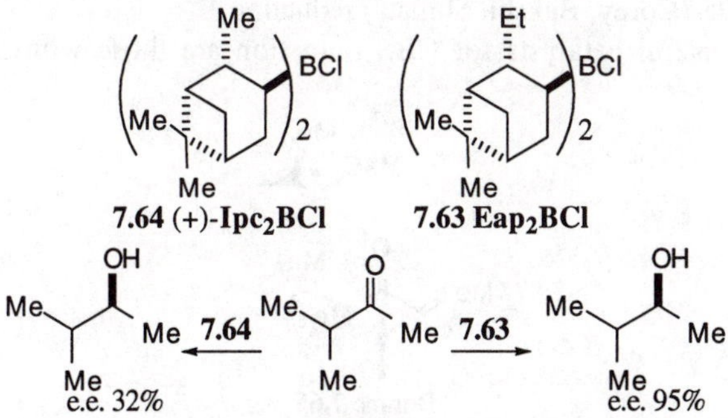

Fig. 7.29

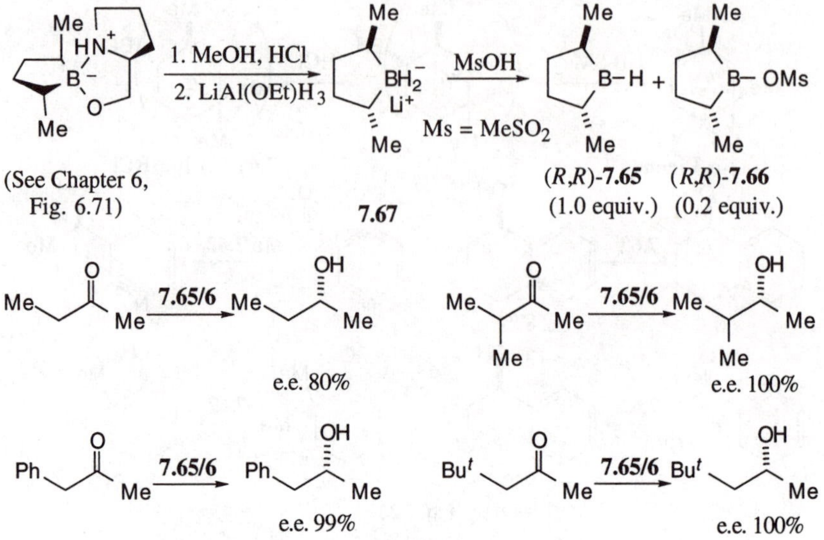

(See Chapter 6,
Fig. 6.71)

7.67

Ms = MeSO₂ → $Ms = MeSO_2$

(R,R)-**7.65** (R,R)-**7.66**
(1.0 equiv.)   (0.2 equiv.)

e.e. 80%

e.e. 100%

e.e. 99%

e.e. 100%

Fig. 7.30

acid, activating the ketone by coordination *syn* to the smaller group (Me rather than Et for butanone).[32] The borane then approaches this complex as shown in Fig. 7.31. In effect, the mesylate activates the carbonyl towards reduction, at the same time blocking approach of borane **7.65** from the lower face (Me$_A$) as represented in Fig. 7.31.

Following this line of reasoning, in principle a suitable chiral Lewis acid catalyst might be expected to cause enantioselective reduction even if the stoichiometric reducing agent is achiral. Such a reducing system would need to satisfy several criteria before it would be expected to provide good levels of enantioselectivity. In particular, the rate of reduction of uncomplexed ketone must be slow compared to that of the ketone–Lewis acid complex. An extremely powerful example of this type of system is found in the reductions of ketones with achiral boranes catalysed by oxazaborolidines **7.68** to **7.73** derived from the enantiomeric aminoalcohols **7.74** and **7.75** (Fig. 7.32), known as CBS (Corey, Bakshi, Shibata) reduction.[33]

The most useful catalysts for CBS reduction are those with alkyl groups

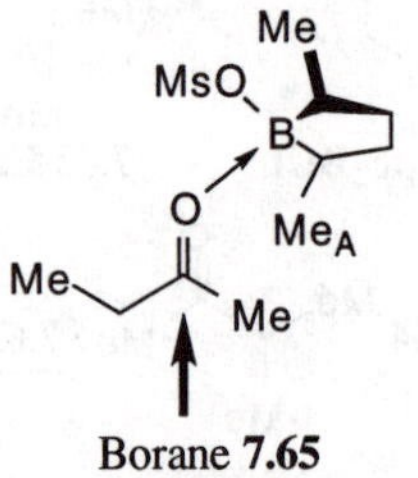

Borane **7.65**
Fig. 7.31

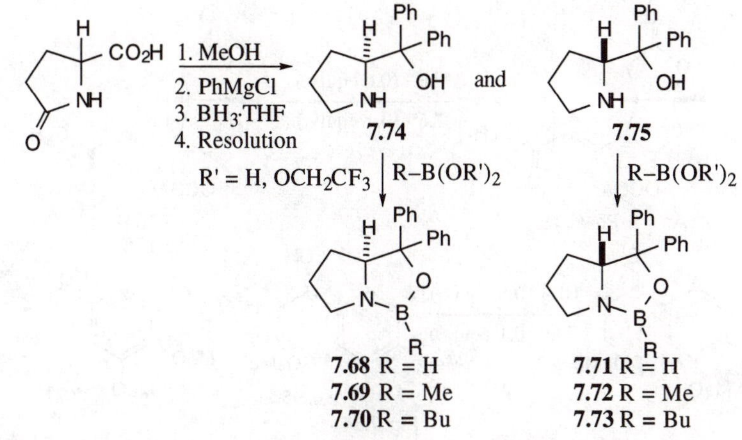

Fig. 7.32

attached to boron as these are substantially less sensitive to moisture and air than **7.68** and **7.71**, making their use much easier. Some examples of the reduction of prochiral ketones catalysed by **7.69** (Fig. 7.32) are illustrated in Fig. 7.33 and Table 7.5.

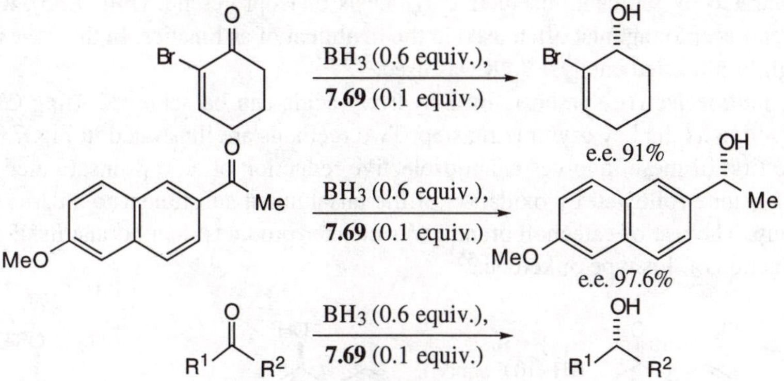

Fig. 7.33

Table 7.5 Ketone reductions catalysed by **7.69**

R^1	R^2	e.e. (%)
Ph	Me	96.5
Ph	Et	96.7
Ph	CH_2Cl	95.3
Bu^t	Me	97.3
Cyclohexyl	Me	84
Ph	$(CH_2)_3CO_2Me$	96.7

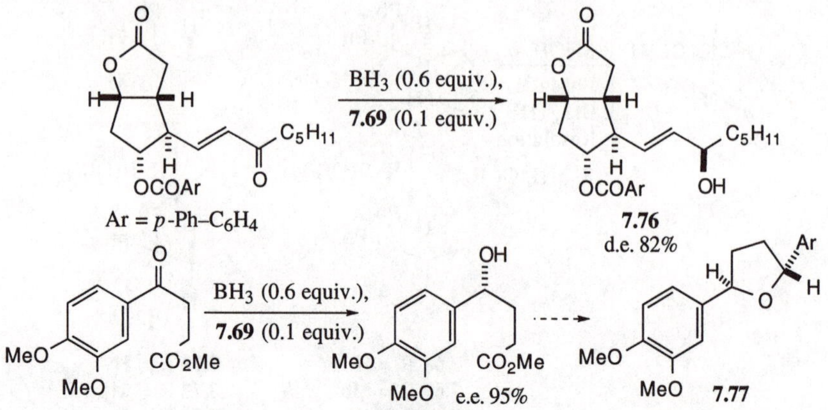

Fig. 7.34

CBS reduction has been used to accomplish chemoselective enantioselective reduction of ketones in the presence of ester and lactone functionality, as in the preparation of the prostaglandin intermediate **7.76**, and furans of general structure **7.77**, inhibitors of platelet activating factor (PAF) (Fig. 7.34).

The efficient reduction of chloromethyl ketones provides easy access to the biologically active enantiomers of several clinically useful drugs.[34] This is illustrated by the enantioselective synthesis of isoproterenol (Fig. 7.35), a β-adrenoreceptor agonist often used in the treatment of asthmatics. In this case the slightly modified catalyst **7.78** was used.[34a]

Enantioselective synthesis of α-hydroxy acids can be achieved using CBS reduction as the key asymmetric step. Two methods are illustrated in Fig. 7.36. The first of these involves enantioselective reduction of an α,β-unsaturated or aryl ketone followed by oxidation of the unsaturated substituent to a carboxyl group. The use of catecholborane gives cleaner products than borane itself for reduction of this type of ketone.[35]

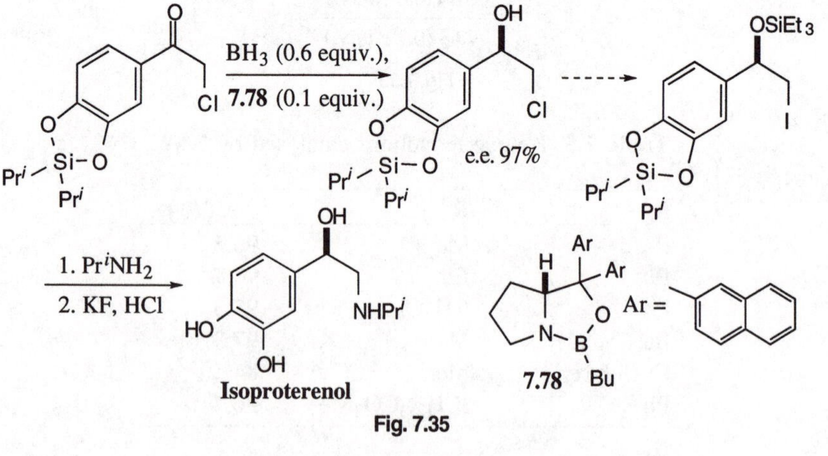

Isoproterenol

Fig. 7.35

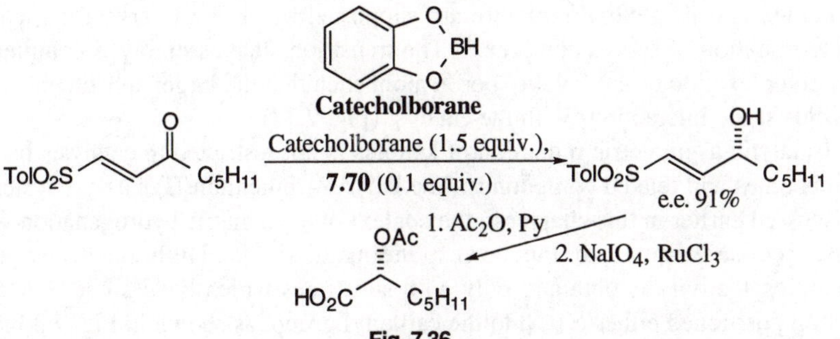

Fig. 7.36

The second route to α-hydroxy acids,[36] which can also be used to prepare α-amino acids,[37] (Fig. 7.37) relies on the enantioselective reduction of trichloromethyl ketones. The reduction system is the same as that used in the previous method, but the following sequence is quite different and utilizes the particular reactivity of trichloromethyl carbinols under basic conditions. Suitable conditions result in the formation of the intermediate dichloroepoxide, which reacts with nucleophiles as shown by the examples in Fig. 7.37.

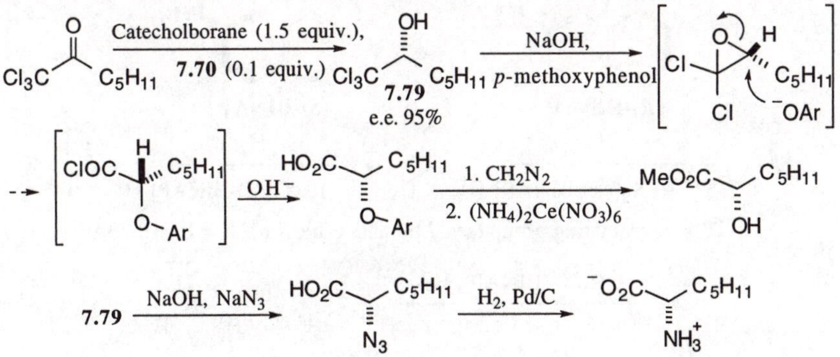

Fig. 7.37

A mechanistic scheme which is consistent with the stereochemical results of reductions catalysed by oxazaborolidines such as those used in CBS reductions is outlined in Fig. 7.38.[33,38] The existence of an initial borane–catalyst

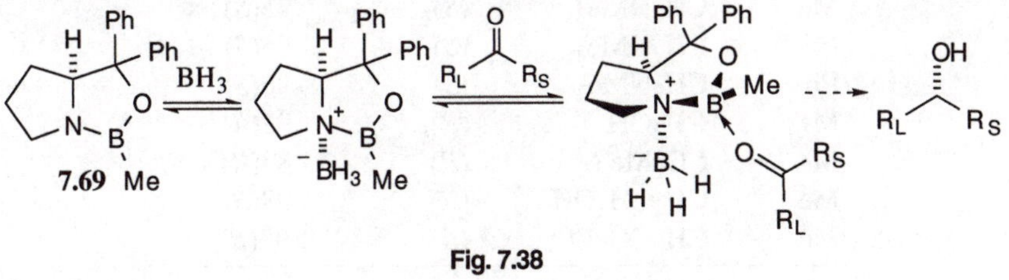

Fig. 7.38

complex and its general structure are supported by an X-ray crystal structure determination of such a complex.[39] The transition state assembly is completed by complexation of the 'cyclic' boron atom such that the larger substituent (R_L) avoids steric interactions with the catalyst (Fig. 7.38).

Catalytic asymmetric reduction of ketones is not restricted to catalysts based on boranes and related compounds. The BINAP–ruthenium(II) catalyst systems discussed earlier in this chapter in the context of asymmetric hydrogenation will also reduce ketones to the corresponding alcohols. High reactivity and enantioselectivity is obtained only with substrates which have a coordinating group positioned either α or β to the carbonyl group, as shown in Fig. 7.39 and Table 7.6.[40]

A coordinating group must also be present for the asymmetric reduction of alkyl aryl ketones using these complexes. In the absence of such a group the reduction is very slow. However, as an *o*-bromo group is sufficient and can be removed easily, this is not necessarily a barrier to the use of this extremely

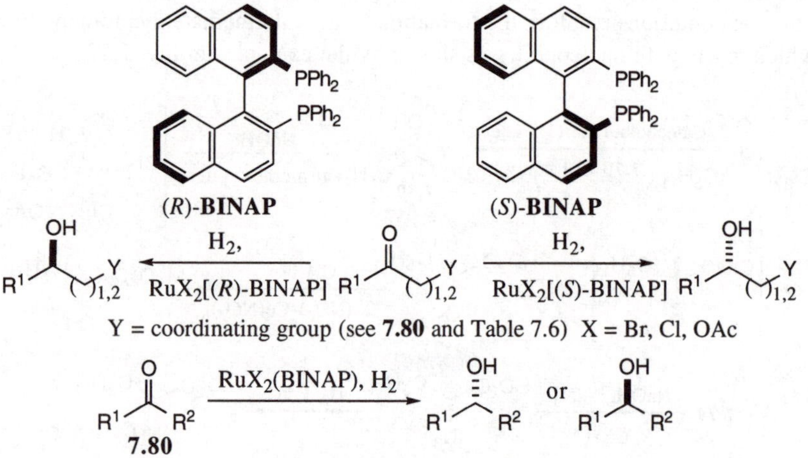

Fig. 7.39

Table 7.6 Reduction of **7.80** catalysed by BINAP-Ru(II) complexes

R^1	R^2	BINAP	e.e. (%)
Me	CH_2NMe_2	(*S*)-	96(*S*)
Pr^i	CH_2NMe_2	(*S*)-	95(*S*)
Ph	CH_2NMe_2	(*S*)-	96(*S*)
Me	CH_2OH	(*R*)-	92(*R*)
Me	CO_2Me	(*R*)-	83(*R*)
Me	CH_2CH_2OH	(*R*)-	98(*R*)
Me	CH_2CO_2Et	(*R*)-	>99(*R*)

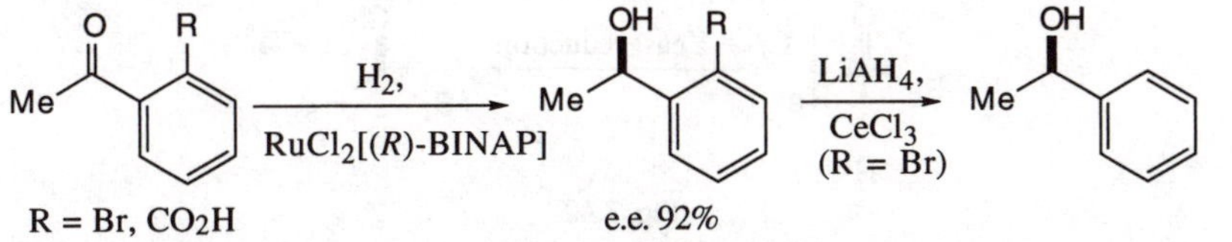

Fig. 7.40

powerful methodology for the asymmetric synthesis of alkyl aryl carbinols (Fig. 7.40).

The coordinating group itself can be a carbonyl group which can undergo enantioselective reduction. In these cases, reductions of α- or β-diketones, the reductions are extremely enantioselective and diastereoselective, with little if any of the corresponding *meso*-diol being formed (Fig. 7.41). The reductions of **7.81** and **7.82** are even more outstanding. In the former case reduction is highly regioselective, and in the latter three chiral centres are produced with very high stereoselectivity in the sense shown (Fig. 7.41).

The catalysts discussed in this chapter bind both substrate and reducing agent, and reaction only takes place when both are coordinated. The asymmetry of the catalyst then results in control of the absolute stereochemistry of the reduction product. This is analogous to the way in which enzymes function, and given that redox reactions are widespread in Nature, it is not surprising to find that certain types of enzymes are useful for the enantioselective reduction of ketones.

The enzymes which are useful for the reduction of ketones are oxidoreductases, and it is possible to use either pure enzymes or whole cells which possess the required enzymatic activity.[41] One example of each of these approaches will now be discussed, following some general considerations.

One of the problems with using enzymes is that by their very nature, they tend to be substrate selective. Fortunately, the asymmetric reduction of many types of ketone can be achieved with yeast enzymes, usually by fermentation methods, and with the purified enzyme horse liver alcohol dehydrogenase

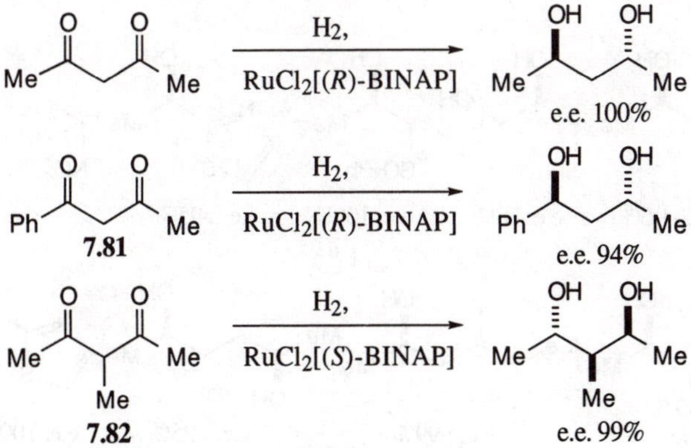

Fig. 7.41

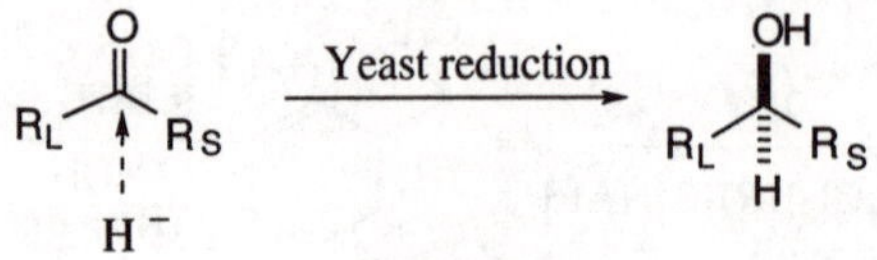

Fig. 7.42

(HLADH). Another problem associated with the use of oxidoreductases is the need for coenzymes, cosubstrates which undergo chemical transformation in the reaction. In principle, these are required in stoichiometric amounts, which is not a problem in fermentations as they are constantly being produced by the cells. If an isolated enzyme is used then the appropriate coenzyme needs to be used, but cost and availability usually rule out the use of stoichiometric quantities, and some protocol has to be adopted which allows for coenzyme regeneration during the reaction. A detailed discussion of this is not appropriate in this chapter, and coenzyme recycling, where required, will be assumed to be possible.

A great deal of work has been carried out on the enantioselective reduction of ketones with fermenting yeast and selected examples of the enantioselective reduction of ketones are discussed below.[41,42] If the two ketone substituents are classified as 'large' (R_L) and 'small' (R_S), yeast reduction often gives the product of reduction from the lower face of the carbonyl group as represented in Fig. 7.42. Nevertheless, there are exceptions to this rule, and the stereochemistry can be changed simply by including other additives in the fermentation mixture or by varying the structure of the ketone (*vide infra*).

Representative examples of alcohols which have been prepared by this type of yeast reduction are provided in Fig. 7.43, and all conform to the rule (often referred to as Prelog's rule). A wide range of substituents can be accommodated including aryl groups,[43] sulphides,[44] sulphones,[45] thiazoles,[46] and nitro groups. α-Diketones,[47] β-diketones,[48] γ-diketones,[49] and α-keto esters[50] can also be reduced in high enantiomeric excess.

The reduction of β-keto esters with yeast has received much attention.[51] Although Prelog's rule often applies, the absolute configuration and

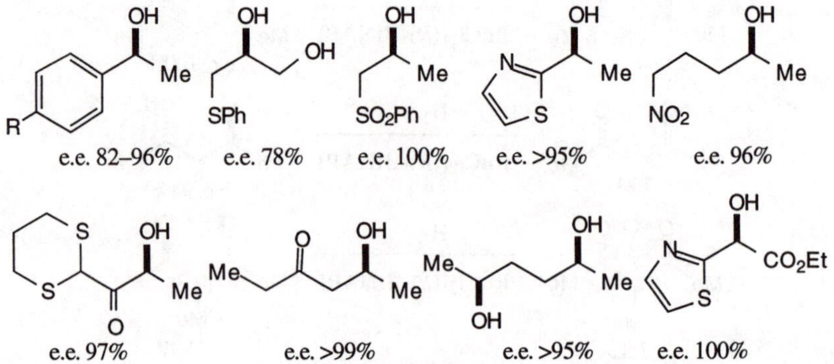

e.e. 82–96%   e.e. 78%   e.e. 100%   e.e. >95%   e.e. 96%

e.e. 97%   e.e. >99%   e.e. >95%   e.e. 100%

Fig. 7.43

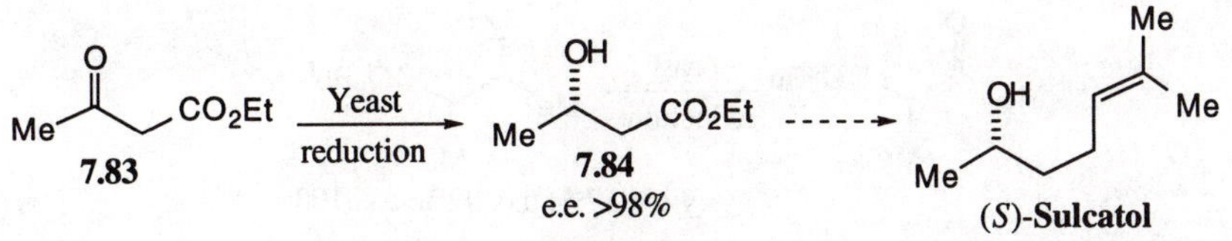

Fig. 7.44

enantiomeric excess of the product can depend on the size and nature of the ketone substituents, and on the ester used. Instances are also known in which variables such as the glucose concentration, the pH, and the physiological state of the cells influence the reduction stereoselectivity. The origin of these effects, and the effect of additives on the stereoselectivity, can be traced to the presence in the yeast cells of several oxidoreductases, which possess different enantioselectivity and substrate selectivity.

The reduction of ethyl acetoacetate **7.83** has been thoroughly studied and the product **7.84** has been used in numerous syntheses such as that of (S)-sulcatol (Fig. 7.44).[51]

It is possible to carry out the enantioselective reduction of many other β-keto esters with high enantioselectivity, and a selection of these is presented in Fig. 7.45 and Table 7.7.[52]

β-Keto esters with a substituent between the carbonyl groups are chiral, and the two enantiomers are usually in relatively rapid equilibrium via the enol. In this situation it is possible to achieve high yields and enantioselectivity in yeast reduction if one of the two enantiomers is a relatively poor substrate.[53] Fortunately this situation often pertains and it is possible to reduce both cyclic

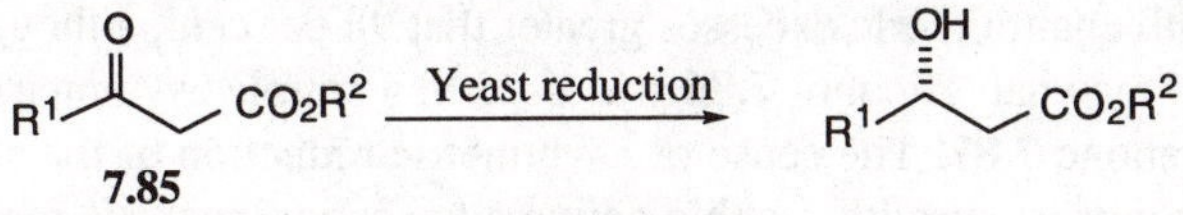

Fig. 7.45

Table 7.7 Yeast reduction of **7.85**

R^1	R^2	Yield (%)	e.e. (%)
Me	Bu^t	61	85(S)
Ph	Et	70	100(S)
Et	C_8H_{17}	75	95(R)
$(CH_2)_2CH=CH_2$	H	35	99(R)
$(CH_2)_2CH=CH_2$	Me	30	92(R)
EtS	Me	55	70(R)
CH_2Br	Bn	40-50	100(S)
CH_2N_3	Bn	70-80	95(R)
CH_2CH_2OBn	CH_2Bu^t	35	96(S)

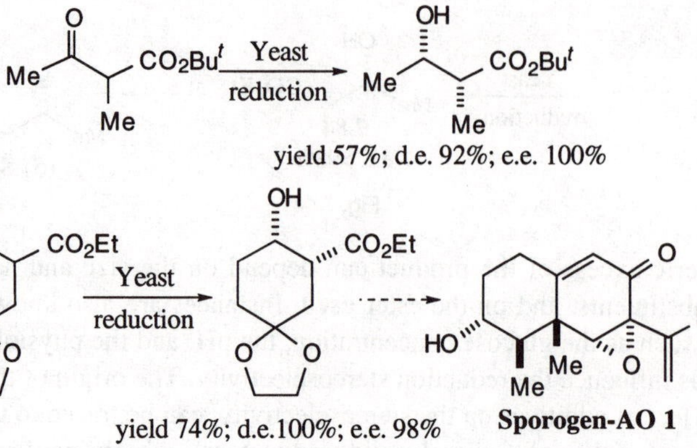

yield 57%; d.e. 92%; e.e. 100%

yield 74%; d.e.100%; e.e. 98%   **Sporogen-AO 1**

Fig. 7.46

and acyclic systems very effectively,[54] as illustrated by the examples in Fig. 7.46.

A heteroatom such as sulphur[55] or nitrogen[56] can also be incorporated into the cyclic part of the β-keto ester without diminishing the efficiency of the yeast reduction, which extends the potential applications of this type of enantioselective reaction (Fig. 7.47).

The most commonly used pure oxidoreductase is horse liver alcohol dehydrogenase (HLADH). Although this enzyme needs coenzyme recycling, protocols have been developed which make such reductions possible on a scale which is synthetically useful. The examples of enantioselective reductions of prochiral diketones shown in Fig. 7.48 illustrate the selectivity and predictability of this type of process.[57] All the examples shown provide products with enantiomeric excesses greater that 98 per cent, although the yield does vary somewhat. Alcohol **7.86** was used in a simple asymmetric synthesis of the twistanone **7.87**. The sense of asymmetric reduction of the substrates in Fig. 7.48 is consistent with a cubic active site section analysis method which has been proposed for reductions catalysed by HLADH.[58]

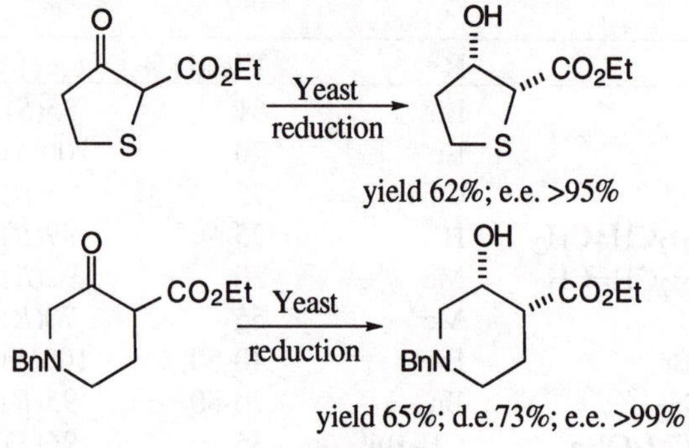

yield 62%; e.e. >95%

yield 65%; d.e.73%; e.e. >99%

Fig. 7.47

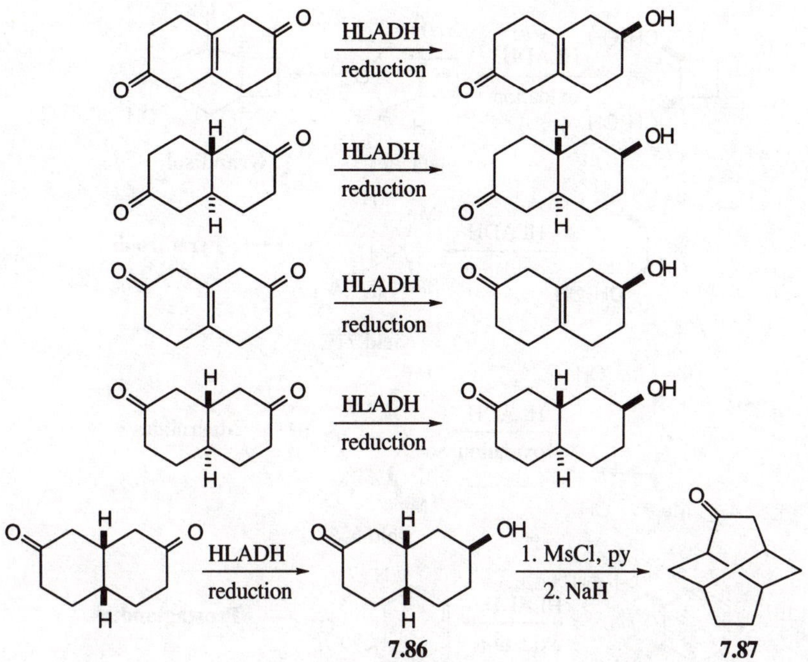

Fig. 7.48

As the name implies, oxidoreductases can be used to catalyse oxidations as well as reductions. For the enantioselective oxidation of alcohols the most studied system involving a pure enzyme is HLADH operating under conditions suitable for oxidation rather than reduction. *Meso*-diols can be oxidized to aldehydes or lactones in high enantiomeric excess using this approach (Fig. 7.49).[59]

This approach, the enantioselective oxidation of *meso*-diols with HLADH, has been used to prepare a variety of lactones in essentially 100 per cent enantiomeric excess (Fig. 7.50). Such lactones represent useful intermediates for the asymmetric synthesis of various natural products including grandisol,[58] pyrethroids,[60] macrolides,[61] and prostaglandins.[62]

While the preceding enzyme-catalysed oxidation of prochiral diols is both

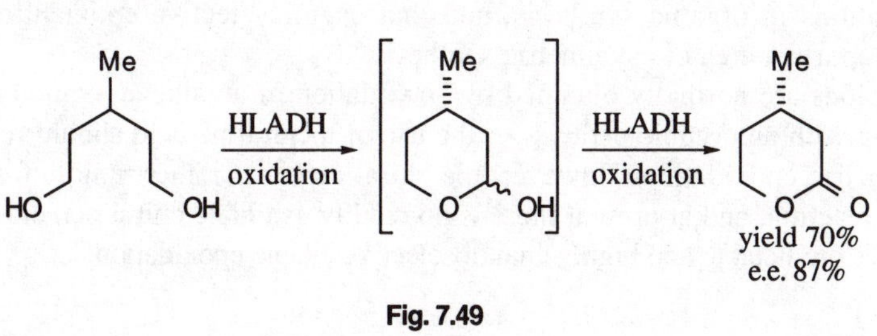

Fig. 7.49

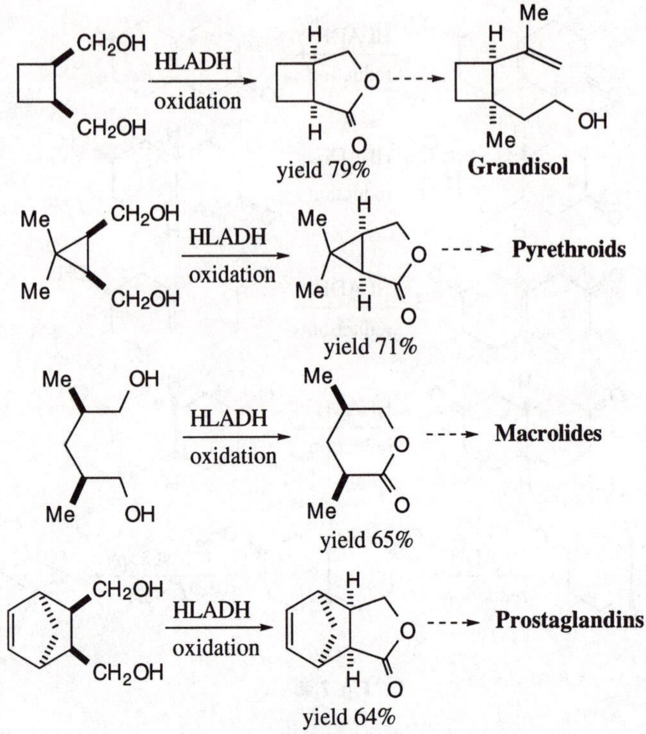

Fig. 7.50

interesting and useful in the asymmetric synthesis of several classes of natural products, a much more general and widely applicable asymmetric oxidation is that of alkenes to epoxides. Epoxides have numerous uses in synthesis and the alkenes required as starting materials are easy to obtain, either from commercial sources or by standard synthetic procedures.

The value of epoxides in synthesis is due largely to their opening by nucleophilic reagents, which almost always takes place with high stereoselectivity, and often with high regioselectivity. The range of effective nucleophiles is wide, and unlike nucleophilic displacement at secondary or tertiary carbons, it is rare for epoxide opening is to be complicated by elimination. It is to be expected therefore that epoxides find widespread applications in organic synthesis, and that enantioselective epoxidation is a most important area of asymmetric synthesis.

Epoxides are normally obtained by epoxidation of an alkene, commonly by reaction with an organic peracid, so the use of a chiral peracid should result in asymmetric epoxidation. However, the chiral centre is rather remote from the site of reaction, and at present there is no readily available chiral peracid which provides predictable and highly enantioselective alkene epoxidation.

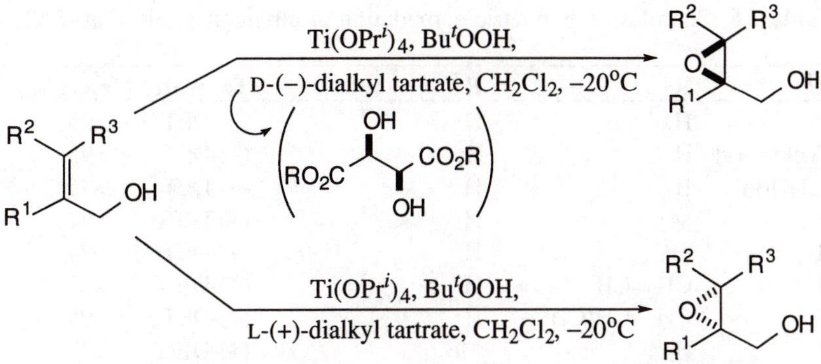

Tartrates usually diethyl (DET) or diisopropyl (DIPT) esters.
'Catalytic conditions' use 5 to 10 per cent tartrate and
$Ti(OPr^i)_4$, with a tartrate:$Ti(OPr^i)_4$ ratio 1.1:1 to 1.2:1

Fig. 7.51

The most versatile and widely used method for enantioselective alkene epoxidation is that due to Sharpless, commonly referred to as 'Sharpless epoxidation'.[63] The reagents involved in this oxidation, *tert*-butyl hydroperoxide (oxidant), titanium tetraisopropoxide (Lewis acid), and a diester of (+)- or (−)-tartaric acid (the 'source of chirality'), are readily available and easily handled.

Sharpless epoxidation is general for allylic alcohols ('isolated' alkenes are not oxidized), highly predictable, and often highly enantioselective. The selectivity for allylic alcohols renders the reaction chemoselective in that only a double bond which is adjacent to a hydroxyl-bearing carbon is epoxidized. Other double bonds present in the substrate are not oxidized. The final advantage of this oxidation lies in the fact that in most cases it is possible to run the reaction under catalytic conditions.[64] In summary, the Sharpless methodology is a general, highly chemoselective, enantioselective oxidation which uses an asymmetric catalyst system derived from readily available components, and is easy to perform (Fig. 7.51). It is hardly surprising that it has had a major impact on asymmetric organic synthesis, and a considerable part of the rest of this chapter will be devoted to this remarkable reaction.

The oxidation of a prochiral allylic alcohol, or an allylic alcohol in which the chirality is relatively remote from the double bond undergoing epoxidation, is the most general version of this reaction, and will be referred to as 'Sharpless asymmetric epoxidation'. A very wide range of allylic alcohols of all

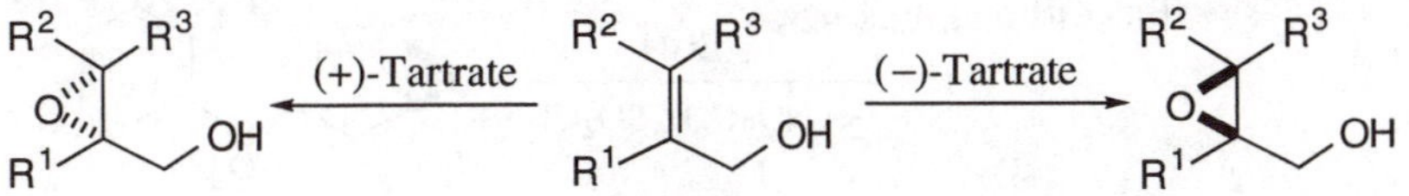

Fig. 7.52

Table 7.8 Sharpless asymmetric epoxidation of allylic alcohols (Fig. 7.52)

R^1	R^2	R^3	Tartrate	e.e. (%)
Me	H	H	(−)-DET	>95
Cyclohexyl	H	H	(+)-DET	>95
CH_2OBn	H	H	(−)-DET	>95
H	Me	H	(+)-DIPT	92
H	Pr^i	H	(+)-DET	94
H	$CH_2=CH$	H	(+)-DIPT	>91
H	$CH_2=CH(CH_2)_3$	H	(+)-DET	95
H	H	Me	(+)-DIPT	92
H	H	$CH_2CH=CHC_5H_{11}$	(+)-DMT	94
H	H	CH_2OBn	(−)-DET	92
Me	Me	H	(+)-DET	94
$-(CH_2)_4-$		H	(+)-DET	93
Me	CH_2OBn	H	(−)-DIPT	90
Me	H	Bu^n	(+)-DET	89
H	$Me_2CH=CHCH_2$	Me	(+)-DET	95
H	$TBSOCH_2CH_2$	Me	(−)-DET	95
Me	Ph	Bn	(+)-DIPT	94
$-(CH_2)_{10}-$		Me	(+)-DIPT	94

substitution patterns has been used in this reaction, and a small selection of the various structural types is shown in Fig. 7.52 and Table 7.8.[65]

The use of only small quantities of tartrate and titanium(IV) isopropoxide under 'catalytic conditions' (Fig. 7.51) has several practical benefits. The most important of these are that the much simplified isolation procedure allows the preparation of small, water-soluble epoxides, and that it is possible to derivatize or open the epoxy alcohol *in situ* (Fig. 7.53).[66]

The absolute configuration of the epoxide produced by Sharpless asymmetric epoxidation can be predicted using the simple model illustrated in Fig. 7.54. If the substrate is drawn as shown, the oxygen is delivered from above the plane if

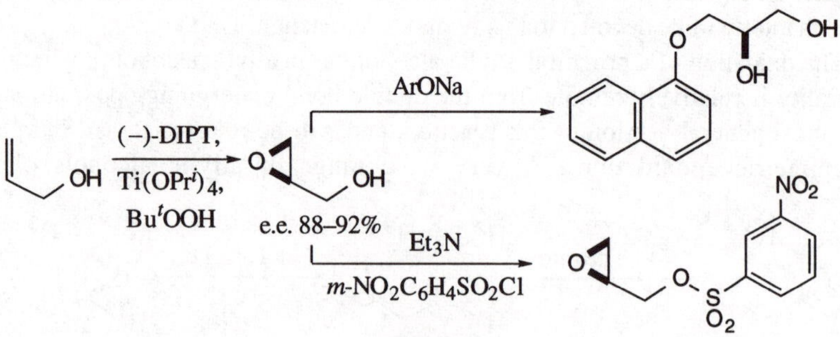

Fig. 7.53

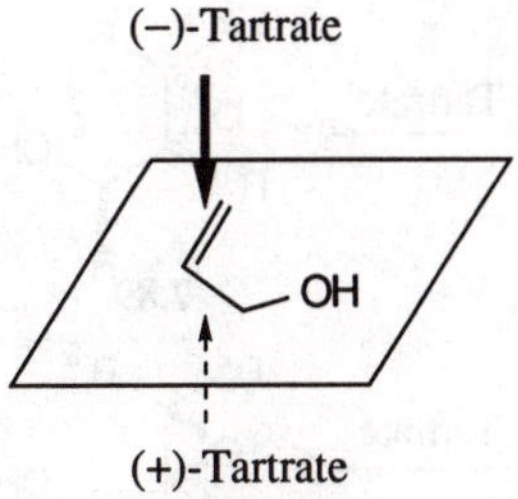

(−)-Tartrate

(+)-Tartrate

Fig. 7.54

a (−)-tartrate ester is used, and from below with the enantiomeric tartrate. This extremely reliable predictive model is one of the great strengths of Sharpless asymmetric epoxidation, with over three hundred examples all conforming to this model.[67]

Sharpless epoxidation of allylic alcohols which carry a substituent at C-1 (Fig. 7.55) is also a most useful procedure, although the situation is now complicated by the fact that almost all such alcohols are chiral. Nevertheless the simple model shown above (Fig. 7.54) can be applied and the stereoselectivity predicted.

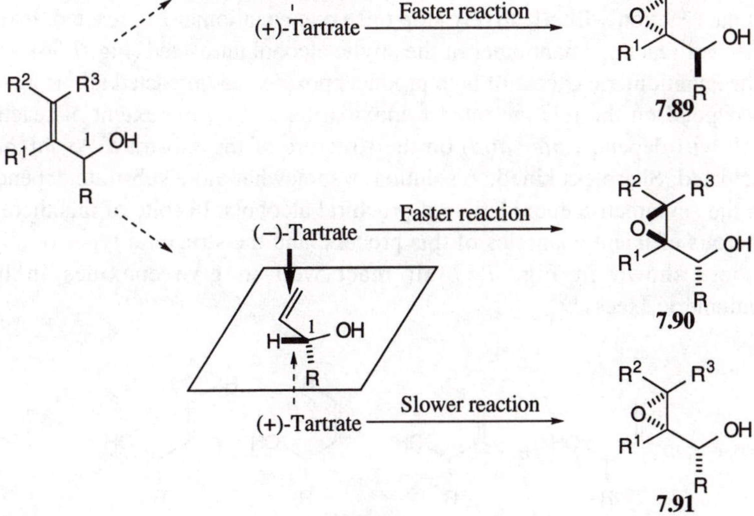

Fig. 7.55

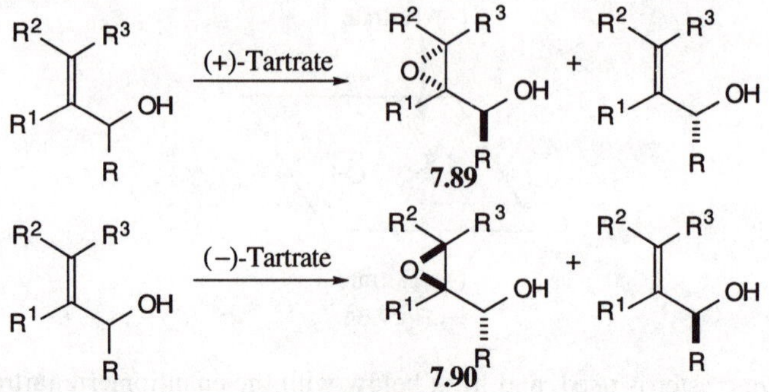

Fig. 7.56

The catalytically active epoxidation species is discriminating enough to allow it to be used in the kinetic resolution of racemic alcohols of this type, and once again the simple model can be used to predict which enantiomer of the substrate will react more rapidly with the enantiomeric epoxidation catalysts (Fig. 7.55).[68] One prediction arising from the model shown (Fig. 7.55), which is borne out in practice, is that of the four possible epoxide products **7.88 – 7.91**, the two *syn*-isomers (**7.88** and **7.91**) should be formed in smaller quantities than the corresponding *anti*-isomers (**7.89** and **7.90**). The reaction is diastereoselective in addition to being enantioselective.

For a given tartrate enantiomer, one of the enantiomers of the racemic allylic alcohol will react more rapidly. In the case of (−)-tartrates, epoxide **7.90** will be formed rapidly from the enantiomer shown, whereas the other enantiomer of the substrate reacts slowly. If there is only enough oxidant (*tert*-butyl hydroperoxide) added to consume all of the faster reacting enantiomer, and the rate difference is sufficiently high (a ratio of rates of ~25:1 is usually enough), then the reaction will effectively stop once one enantiomer has reacted, leaving the slower reacting enantiomer of the allylic alcohol unreacted (Fig. 7.56).

The enantiomeric excess of both product epoxide and unreacted allylic alcohol will depend on the relative rate of epoxidation and on the extent of reaction, which will depend (*inter alia*) on the structure of the substrate. As might be anticipated, Sharpless kinetic resolution is somewhat more substrate dependent than the asymmetric epoxidation of prochiral alcohols. In spite of this there are numerous efficient examples of this process, and the structural types of allylic alcohols shown in Fig. 7.57 all react well to give epoxides in high enantiomeric excess.[69]

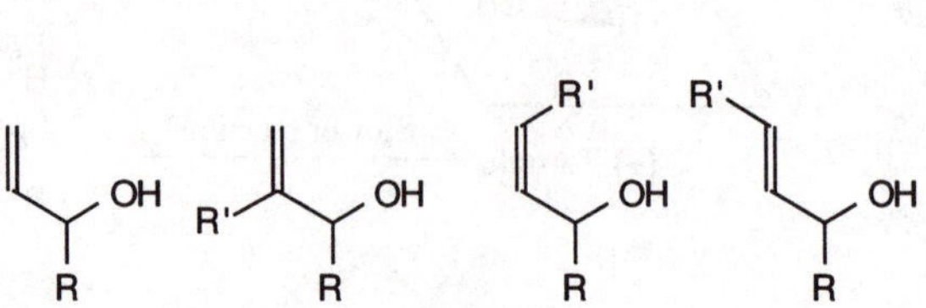

Fig. 7.57

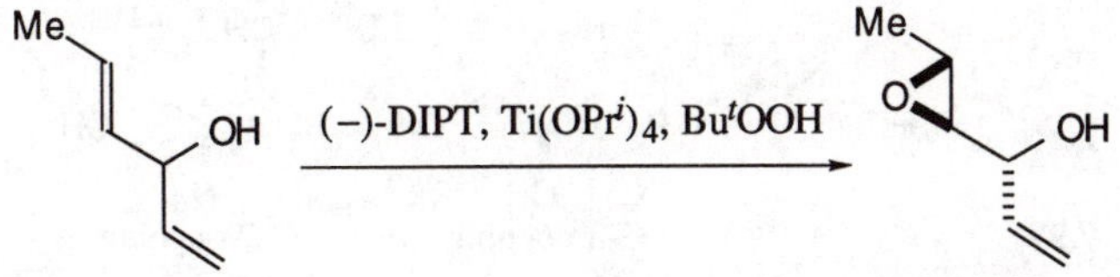

(±)-**7.92**

7.93
yield 35%; e.e. >95%

Fig. 7.58

A remarkable example of Sharpless kinetic resolution is that of alcohol **7.92** (Fig. 7.58). Both double bonds have the potential to undergo Sharpless epoxidation, but the more nucleophilic of the two reacts more rapidly so that epoxide **7.93** is formed in high enantiomeric excess and good yield (the maximum yield is 50 per cent as in all resolutions).[70]

There are numerous examples of the use of Sharpless asymmetric epoxidation and kinetic resolution in the asymmetric synthesis of various types of targets. An example of the power of this approach can be found in the preparation of the C-29 to C-37 sequence **7.94** of the polyether antibiotic X-206 (see Chapter 4, Fig. 4.15). The chiral centres at C-34 and C-35 are established in a kinetic resolution step (**7.95** to **7.96**). Protection, selective epoxide opening (see below), and further reactions provide the substrate **7.97** for asymmetric epoxidation.[71] The epoxide undergoes cyclization under the reaction conditions to give **7.94**.

As illustrated by the reaction sequence shown in Fig. 7.59, the great value of epoxides in organic synthesis is linked to their selective opening with nucleophiles. In order to illustrate the factors which are important in this, and the control that is possible, reactions of a typical type of acyclic *trans*-2,3-epoxy alcohol **7.98** (Fig. 7.60) will be considered. Epoxide opening generally follows an S_N2 pathway, and as such, electron-withdrawing groups adjacent to

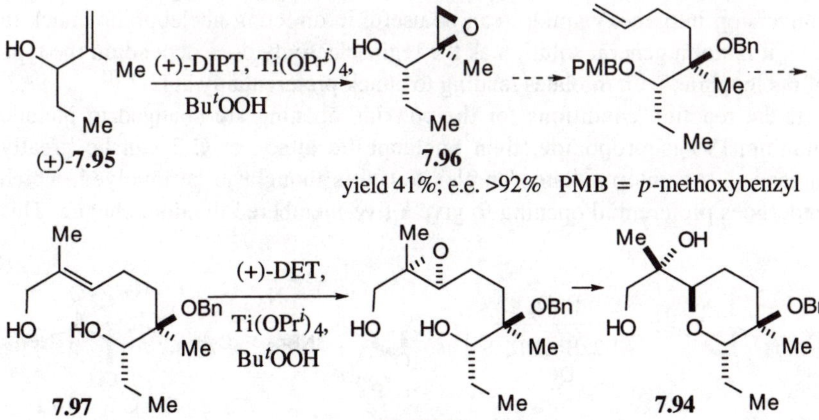

(+)-**7.95**

7.96
yield 41%; e.e. >92% PMB = *p*-methoxybenzyl

7.97 **7.94**

Fig. 7.59

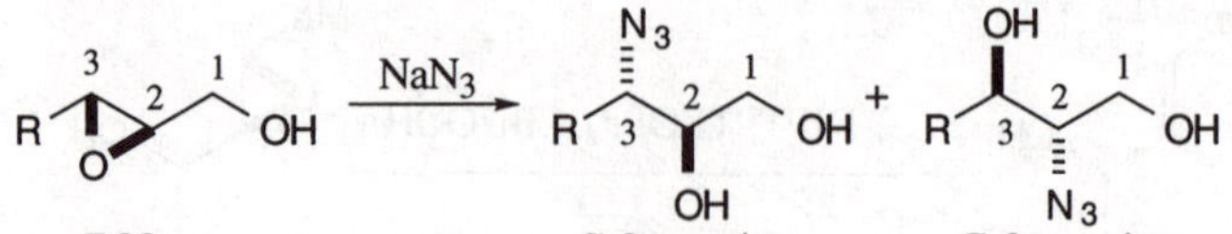

Fig. 7.60

Table 7.9 Regioselectivity in opening of **7.98**

R	C-3:C-2
C_7H_{15}	3.5:1
Cyclohexyl	1.7:1
$BnOCH_2$	1:1
Bu^t	C-2 only

the carbon undergoing substitution cause a decrease in rate. The electron withdrawing effect of the C-1 hydroxyl group slows down attack at C-2 (Fig. 7.60). Therefore, opening of the epoxide at C-3 is favoured electronically, but the regioselectivity is also sensitive to steric effects. As the bulk of the substituent at C-3 increases, the proportion of attack at this position decreases (Table 7.9). 'Simple' opening of these epoxides is a balance of steric and electronic effects.[72]

The tendency for epoxide opening at C-3 in systems such as **7.98** can be increased by manipulation of the C-1 substituent or by changing the conditions of the reaction. For example, if the C-1 primary hydroxyl is oxidized to a carboxylic acid and converted to an amide such as **7.99**, the proportion of opening at C-3 is increased (Fig. 7.61).[72]

This approach was used in a short synthesis of bestatin (Fig. 7.62), a natural enzyme inhibitor with antitumour activity. Although this approach involving conversion into epoxy amides can be useful in directing nucleophilic attack to C-3, it is not a general solution as the regioselectivity does depend on the type of nucleophile, with thiolates tending to attack preferentially at C-2.[70]

If the reaction conditions for the epoxide opening are changed to include titanium(IV) isopropoxide, then nucleophilic attack at C-3 can be greatly enhanced. An intermediate complex **7.100** is thought to be involved, which undergoes preferential opening to give a five-membered titanium chelate. This

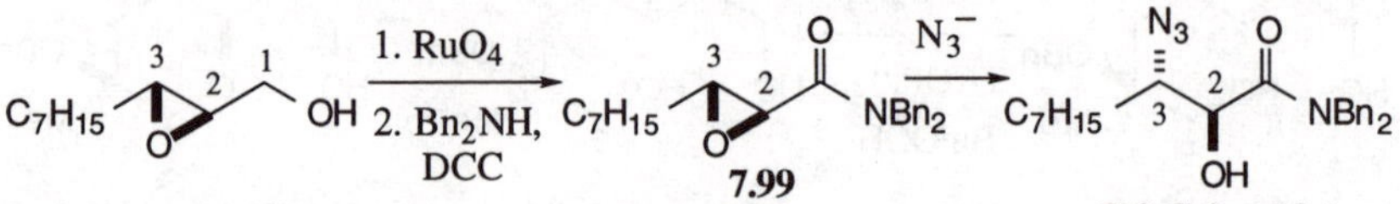

Fig. 7.61

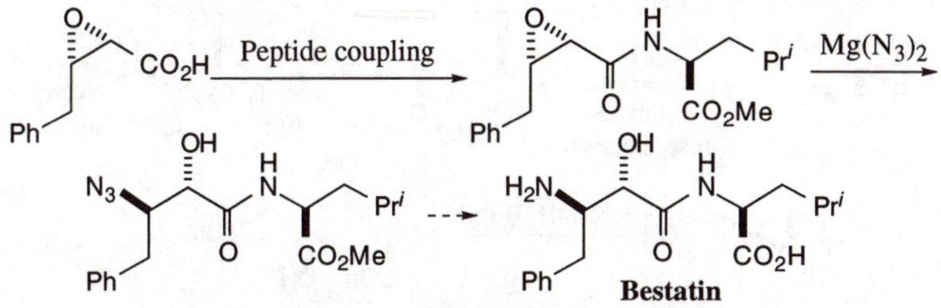

Fig. 7.62

modification is successful with both 2,3-epoxy alcohols and acids (Fig. 7.63), but is restricted in that the nucleophile must be relatively unreactive towards titanium(IV) isopropoxide.[73]

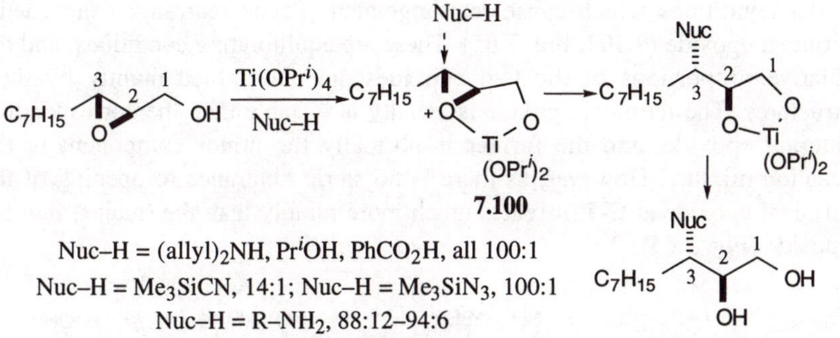

Nuc–H = (allyl)$_2$NH, Pr^{i}OH, PhCO$_2$H, all 100:1

Nuc–H = Me$_3$SiCN, 14:1; Nuc–H = Me$_3$SiN$_3$, 100:1

Nuc–H = R–NH$_2$, 88:12–94:6

Fig. 7.63

In spite of the electronic bias against opening of 2,3-epoxy alcohols, it is possible to introduce some nucleophiles at this position by tethering the nucleophile to the C-1 hydroxyl group. This can be carried out either 'temporarily' as in the case of Red-Al (Na(MeOCH$_2$CH$_2$O)$_2$AlH$_2$),[74] or 'permanently' as shown for the cyclizations of the carbamate **7.101** (Fig. 7.64).[75]

It is possible to carry out apparent displacement of the C-1 hydroxyl group under the correct conditions.[72] This is achieved by carrying out the reaction

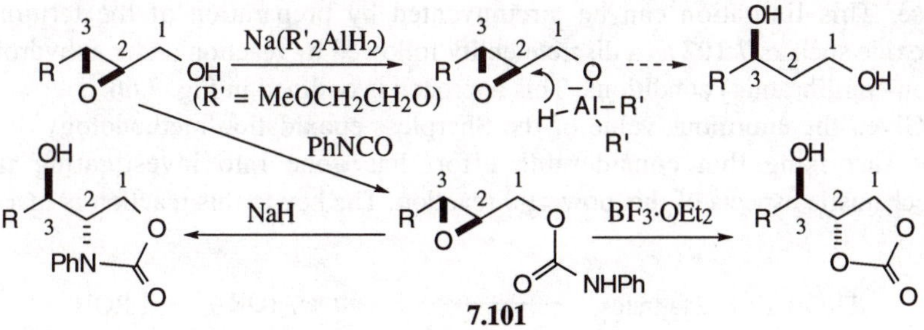

7.101

Fig. 7.64

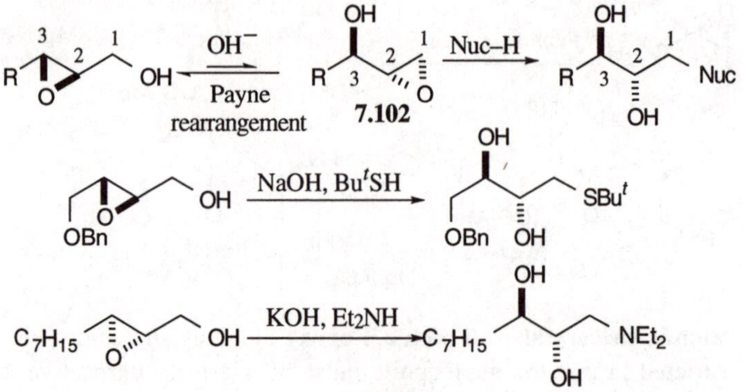

Fig. 7.65

under conditions which cause rearrangement (Payne rearrangement) to the terminal epoxide (**7.102**, Fig. 7.65). These are equilibrating conditions, and the relative proportions of the two epoxides are determined mainly by their structures. The terminal epoxide is usually less stable than the corresponding internal epoxide, and the former is normally the minor component in the reaction mixture. However, as there is no steric hindrance to opening of the terminal epoxide at C-1, it reacts much more rapidly than the (major) internal epoxide (Fig. 7.65).[76]

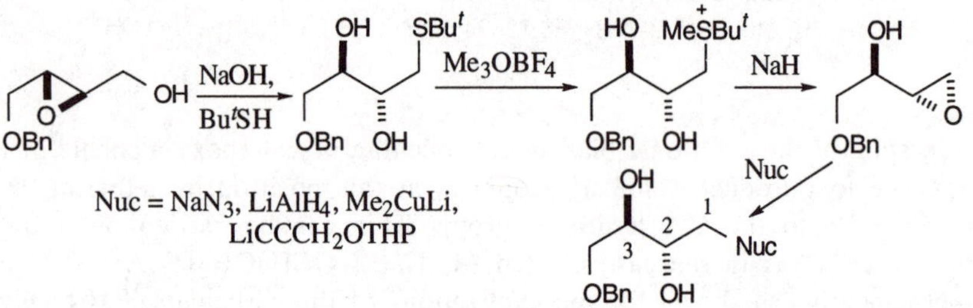

Nuc = NaN$_3$, LiAlH$_4$, Me$_2$CuLi,
         LiCCCH$_2$OTHP

Fig. 7.66

As can be seen from the examples in Fig. 7.65, the range of nucleophiles which can be used in this process is limited to those compatible with aqueous base. This limitation can be circumvented by preparation of the terminal epoxide such as **7.102** as a discrete entity followed by reaction under anhydrous (non-equilibrating) conditions. This approach is outlined in Fig. 7.66.[77]

Given the enormous value of the Sharpless epoxidation methodology, it is not surprising that considerable effort has gone into investigating the mechanistic aspects of this powerful reaction. The key to this reaction is that in

$$Ti(OR)_4 + 2\,tartrate \rightleftharpoons Ti(tartrate)_2(OR)_2 + 2\,ROH$$

Fig. 7.67

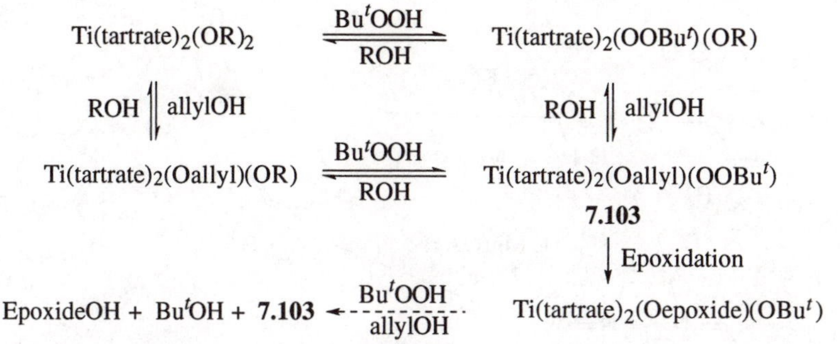

$$Ti(tartrate)_2(OR)_2 \underset{ROH}{\overset{Bu^tOOH}{\rightleftharpoons}} Ti(tartrate)_2(OOBu^t)(OR)$$

$$ROH \updownarrow allylOH \qquad\qquad ROH \updownarrow allylOH$$

$$Ti(tartrate)_2(Oallyl)(OR) \underset{ROH}{\overset{Bu^tOOH}{\rightleftharpoons}} Ti(tartrate)_2(Oallyl)(OOBu^t)$$

7.103

$$\downarrow Epoxidation$$

$$EpoxideOH + Bu^tOH + \mathbf{7.103} \xleftarrow[allylOH]{Bu^tOOH} Ti(tartrate)_2(Oepoxide)(OBu^t)$$

Fig. 7.68

solution, titanium(IV) alkoxides undergo rapid ligand exchange with other alcohols. On reaction with a tartrate ester, the equilibrium lies on the side of the chelate (Fig. 7.67).[78]

Ligand exchange continues, with *tert*-butyl hydroperoxide and the allylic alcohol replacing the remaining alkoxides (RO^-). It is only when the catalyst is 'fully loaded' **7.103** that the epoxidation reaction occurs (Fig. 7.68). After transfer of the oxygen, the titanium complex containing product epoxide and *tert*-butoxide undergoes further ligand exchanges until it is 'fully loaded' again and ready for the next epoxidation.

The 'fully loaded' catalyst **7.103** is thought to be a dimeric species with a structure similar to that shown in Fig. 7.69, a structure which is consistent with kinetic and spectroscopic studies.

In the Sharpless epoxidation, the substrate must possess a group capable of coordination to the catalyst, almost always a hydroxyl group, and without such a group asymmetric epoxidation does not take place. Asymmetric epoxidation of 'isolated' alkenes lacking this type of coordination site is possible but is considerably less advanced than Sharpless epoxidation. Nevertheless, one group of asymmetric catalysts are showing great potential; these are manganese(III) complexes prepared as outlined in Fig. 7.70.[79]

A range of such complexes has been prepared and several of these show useful

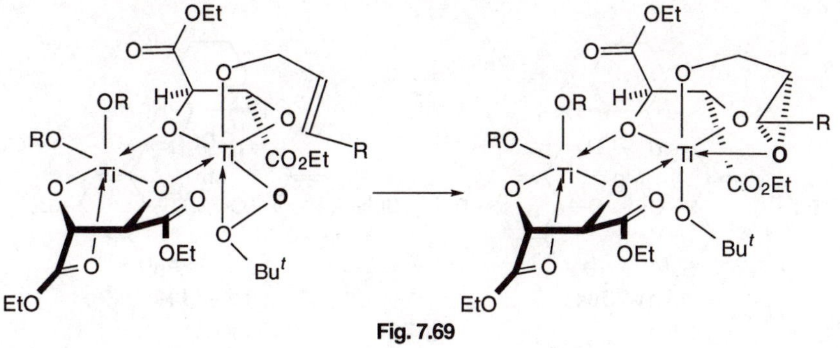

Fig. 7.69

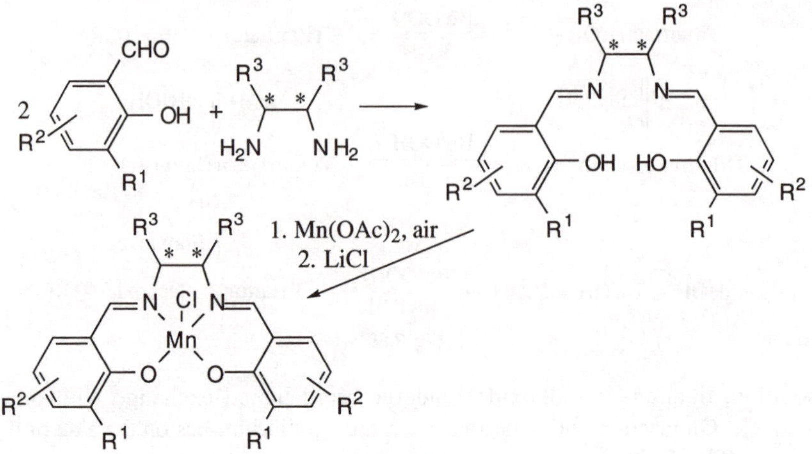

Fig. 7.70

levels of enantioselectivity. To illustrate this approach to the enantioselective epoxidation of isolated alkenes, reactions catalysed by the enantiomeric complexes (R,R)-**7.104** and (S,S)-**7.104** will be considered (Fig. 7.71).[80]

Some examples of asymmetric epoxidation using these catalysts are presented in Fig. 7.72. High enantiomeric excess is obtained with *cis*-alkenes, with *trans*-isomers being poor substrates. The reactions are easy to carry out, and use diluted household bleach (NaOCl) as the stoichiometric oxidant.

Substrate selectivity studies and the effect of structural changes in the ligands have led to a model which accounts for the observed stereochemical aspects of this reaction. It is thought that a manganese(IV) oxo species is the oxidant, and that only *cis*-alkenes can approach properly. The bulky *tert*-butyl groups are most important for high enantioselectivity, and are considered to prevent approach of the substrate from directions other than that indicated in Fig. 7.73. The alkene is thought to approach as shown with the large group being better accommodated in the space remote from the axial hydrogen H* (Fig. 7.73).[80]

Whatever the detailed mechanism of the process,[81] it is an extremely easy and a potentially very valuable method for asymmetric synthesis epoxidation, and

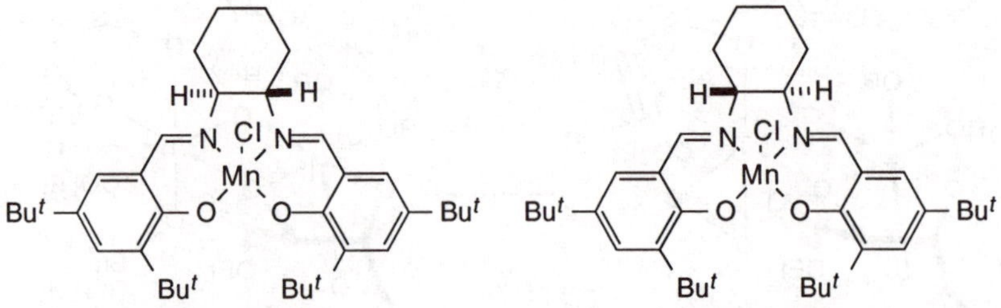

(S,S)-**7.104** (R,R)-**7.104**

Fig. 7.71

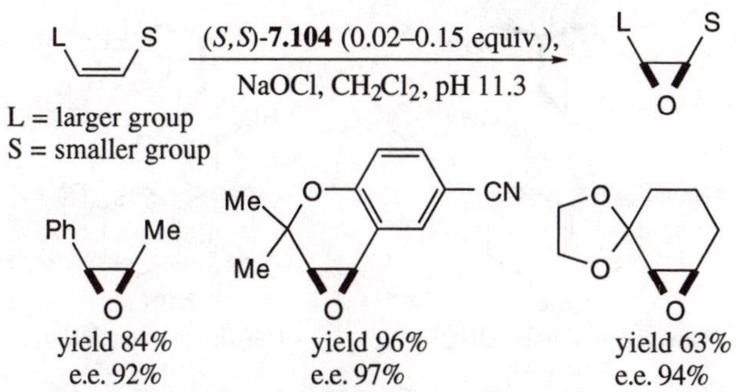

Fig. 7.72

will undoubtedly find widespread use in organic synthesis. A simple example of this is shown in Fig. 7.74, in the synthesis of **7.105**, the side chain of taxol, a powerful antileukaemic and tumour-inhibiting natural product.

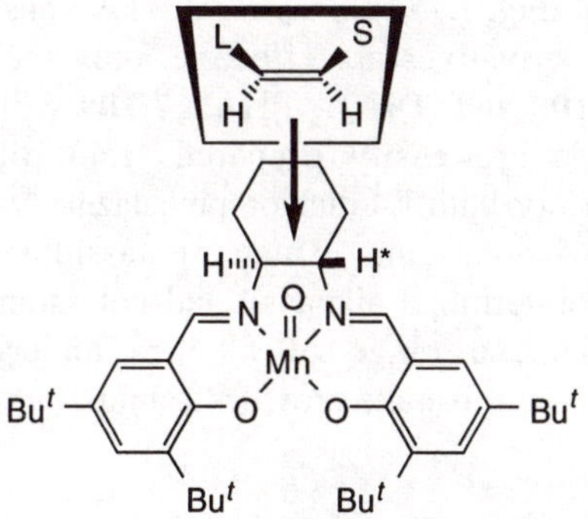

Fig. 7.73

In this synthesis, the stoichiometric reagents are very simple (e.g. bleach, hydrogen, and ammonia) and the enantiomeric excess is high. Although some *trans*-epoxide is produced, probably via an alternative epoxidation pathway, the overall yield is good (25 per cent).[82]

The final type of asymmetric oxidation which will be covered in this chapter is asymmetric dihydroxylation. A range of catalysts is available which provide

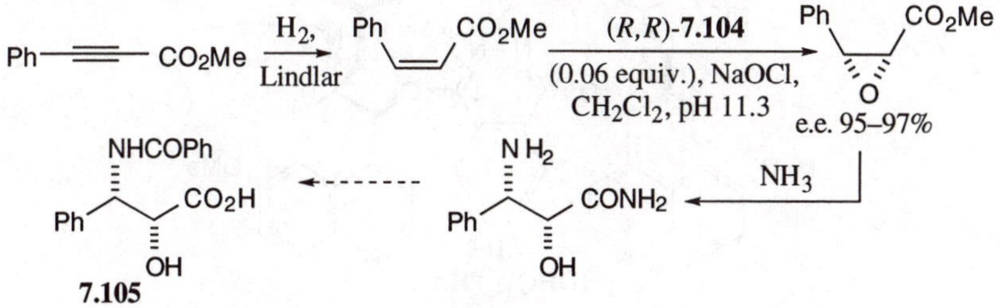

Fig. 7.74

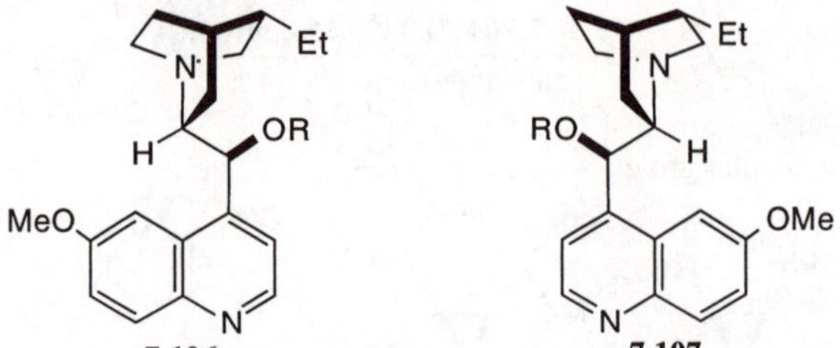

7.106
Dihydroquinidine (DHQD)
derivatives

7.107
Dihydroquinine (DHQ)
derivatives

Fig. 7.75

cis-diols from various alkenes in high enantiomeric excess.[63a] The ligand class which will be discussed is based on derivatives of dihydroquinidine **7.106** and dihydroquinine **7.107** (Fig. 7.75). These two catalysts systems give opposite asymmetric induction, and although diastereoisomers they are often referred to as 'pseudo-enantiomeric'. Fig. 7.75 emphasizes this relationship.

Although various derivatives are effective, only the phthalazine ligands (DHQD)$_2$-PHAL **7.108** and (DHQ)$_2$-PHAL **7.109** will be considered (Fig. 7.76). These ligands are easily prepared from dihydroquinidine and dihydroquinine by reaction with 1,4-dichlorophthalazine.[83]

In the presence of potassium osmate, potassium carbonate, methane sulphonamide (for non-terminal alkenes), and potassium ferricyanide as the stoichiometric oxidant, a range of alkenes undergo enantioselective dihydroxylation. Potassium osmate provides osmium tetroxide *in situ*, and the

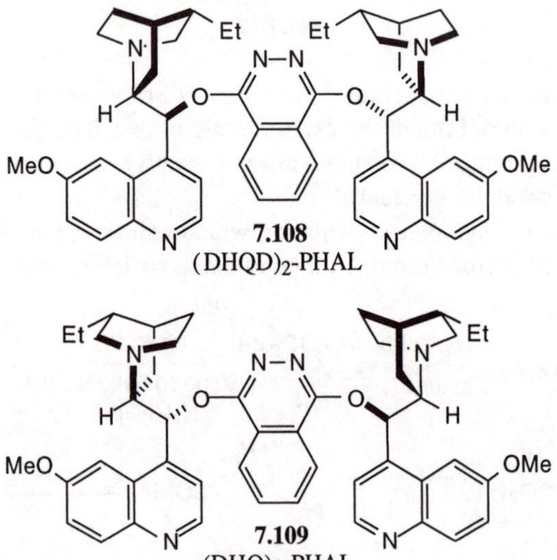

7.108
(DHQD)$_2$-PHAL

7.109
(DHQ)$_2$-PHAL

Fig. 7.76

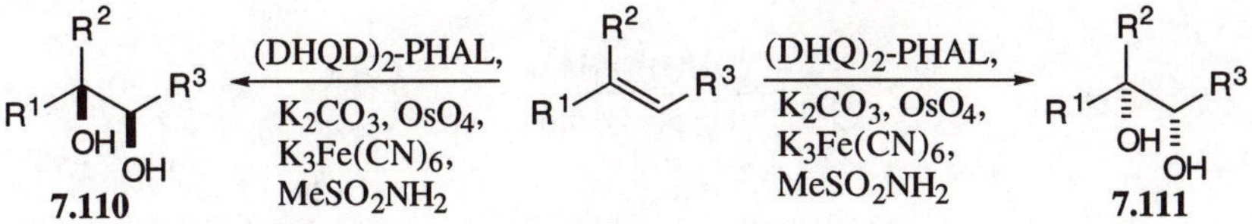

Fig. 7.77

Table 7.10 Enantioselectivity of asymmetric dihydroxylation of alkenes (Fig. 7.77)

R^1	R^2	R^3	7.110[a]	7.111[a]
Bu^n	Me	Me	98	95
Ph	$-(CH_2)_4-$		99	98
Bu^n	H	Bu^n	97	93
C_5H_{11}	H	CO_2Et	99	96
Ph	H	Ph	>99	>99
C_5H_{11}	Me	H	78	76
Ph	Me	H	94	93
C_8H_{17}	H	H	84	80
Ph	H	H	97	97
CO_2Bn	H	H	77	70

[a]Per cent e.e. of diol **7.110** or **7.111**.

complex between this and the ligand is responsible for enantioselective osmylation of the alkene. Hydrolysis of the resulting osmate ester is accelerated by methane sulphonamide, providing the diol and a reduced osmium species which is re-oxidized by the stoichiometric oxidant. Representative examples are shown in Fig. 7.77 and Table 7.10.[84]

In spite of much effort, a detailed mechanistic pathway has not yet been proved, but based on the results of the asymmetric dihydroxylation of many

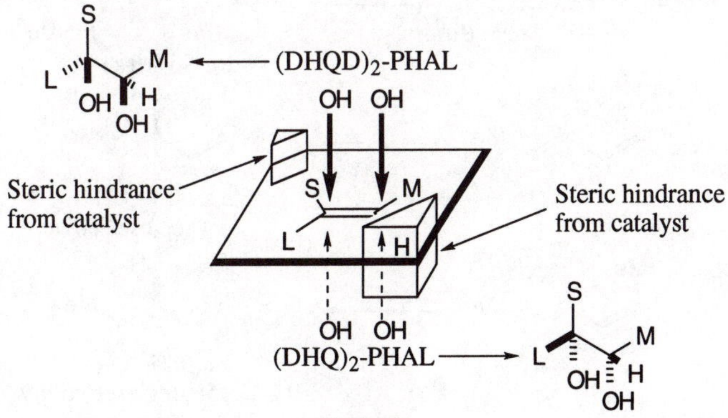

Fig. 7.78

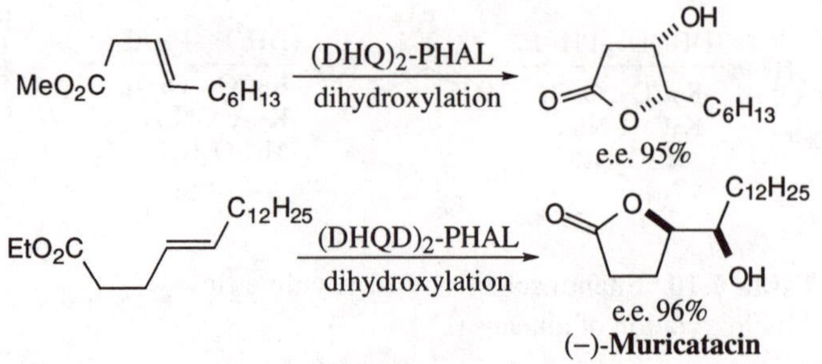

Fig. 7.79

alkenes, a simple predictive model has been developed.[83]

This is shown schematically in Fig. 7.78, and relies on the classification of the alkene substituents as small (S), medium (M), and large (L). The model can then be applied as shown. The observation that *trans*-alkenes are much better substrates than *cis*-alkenes is also consistent with this simple model.

A large range of alkenes has been used as substrates in this reaction, and no attempt will be made to review these reactions. Instead, examples which involve regioselectivity and chemoselectivity will be considered, along with selected synthetic applications of asymmetric dihydroxylation controlled by ligands **7.108** and **7.109**.

Asymmetric dihydroxylation of β,γ- and γ,δ-unsaturated esters provides a simple method for the preparation of lactones; in both cases only the γ-lactone is formed (Fig. 7.79). This approach was used in an asymmetric total synthesis of the natural product (−)-muricatacin (Fig. 7.79).[85]

Selective asymmetric dihydroxylation of conjugated enynes, dienes, and trienes is possible. In general, an alkene is oxidized in preference to an alkyne, and in a polyene it is the most electron-rich double bond that is the major or sole site of reaction (Fig. 7.80).[86] The sense of the enantioselective

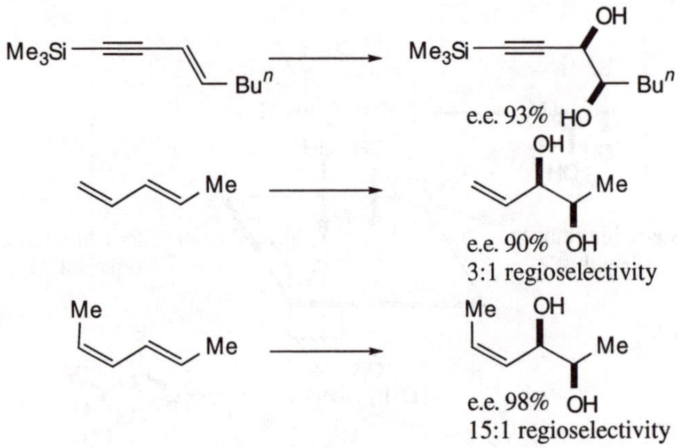

Fig. 7.80

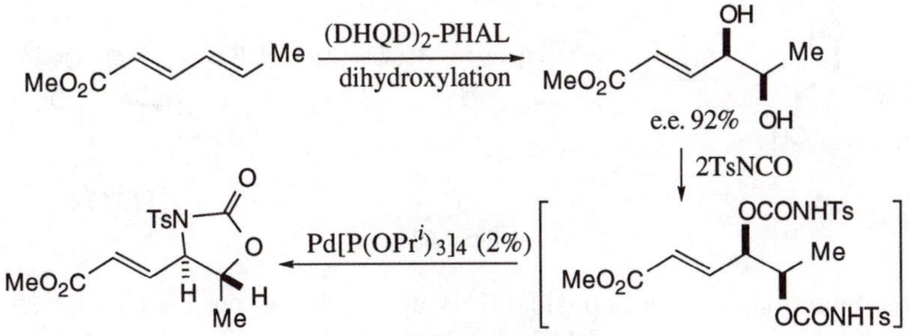

Fig. 7.81

dihydroxylation of these substrates is consistent with the simple model illustrated in Fig. 7.78, and as expected from this model, a *trans*-alkene is oxidized in preference to a *cis*- alkene.

Products of the chemoselective dihydroxylations illustrated in Fig. 7.80 are potentially very useful in organic synthesis. A simple example of this potential is the conversion of diols derived from the asymmetric dihydroxylation of dienes into oxazolidinones (Fig. 7.81).[87]

This method for asymmetric dihydroxylation, as it is very easy to carry out, highly enantioselective, and predictable, is bound to have a major impact in the area of enantioselective organic synthesis. 1,2-Diols themselves are versatile functional groups and often easily manipulated, but their synthetic potential is increased enormously by conversion to epoxides and cyclic sulphates (Fig. 7.83), which react similarly to epoxides with nucleophiles.

1,2-Diols derived from this type of dihydroxylation can be converted into epoxides using the procedure illustrated in Fig. 7.82. A mixture of acetoxy halides is often formed, but this is of no consequence as both cyclize to the

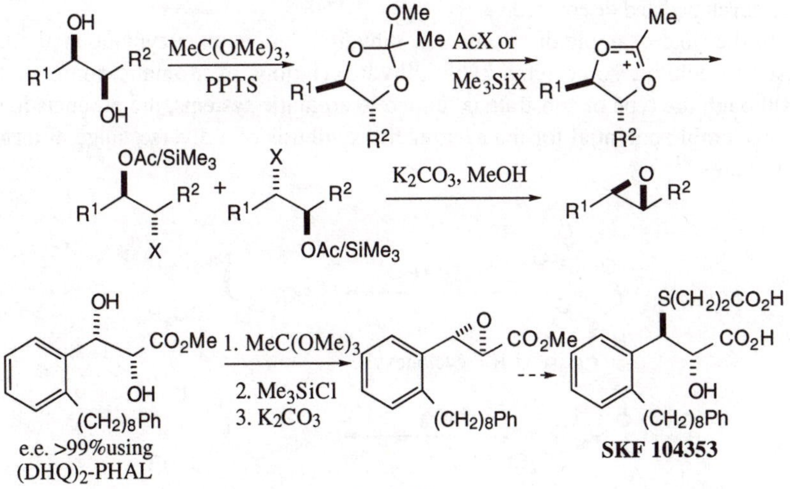

Fig. 7.82

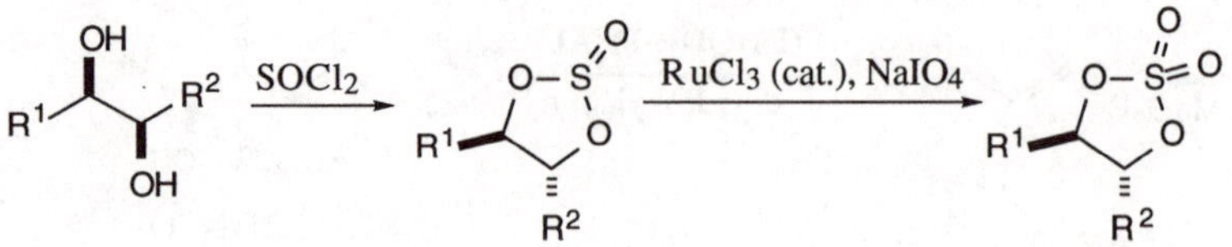

Fig. 7.83

same enantiomer of the epoxide. This approach has been used in a short enantioselective synthesis of the leucotriene antagonist SKF 104353.[88]

The conversion of 1,2-diols into epoxides effectively allows nucleophilic displacement to be carried out at one of the hydroxyl-bearing carbons. This is also possible if the diol is converted into the cyclic sulphate **7.112**, which is readily achieved using the two-step procedure shown in Fig. 7.83.[89]

These cyclic sulphates are similar to epoxides in that they undergo nucleophilic displacements readily (usually S_N2) but there are some significant differences. As with epoxides, usually the less hindered position reacts more rapidly, but unlike epoxides there is a strong tendency for selective displacement α to a carbonyl group (Fig. 7.84).[90]

The other important difference in reactivity between epoxides and cyclic sulphates is that it is possible to carry out displacement at both positions in a cyclic sulphate. This principle is illustrated by the generalized 'one-pot' reaction sequence leading to aziridines shown in Fig. 7.85.

The various methods discussed in this chapter for the asymmetric oxidation of alkenes have focused on selected versatile and powerful methods of some generality. Given the ease of carrying out such reactions using enantioselective catalysis and the versatility of epoxides and 1,2-diols in organic synthesis, the importance of these methods can only increase as more and more applications are developed and discovered.

In the final example of asymmetric dihydroxylation, an enzymatic method is used to achieve some remarkable dihydroxylations of aromatic compounds. Although the type of substrate is limited to aromatic systems, the products have considerable potential for the asymmetric synthesis of a diverse range of target structures.[91]

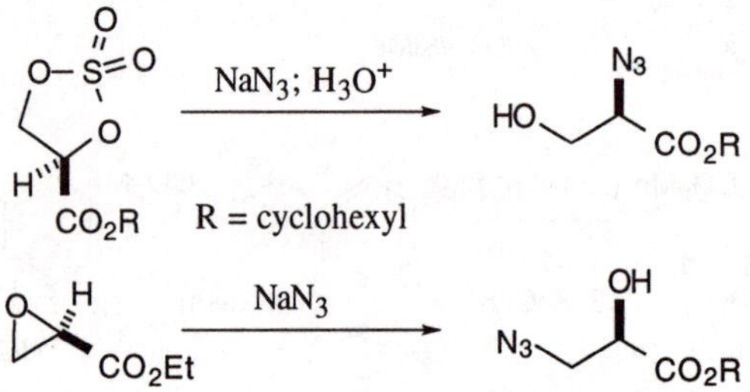

Fig. 7.84

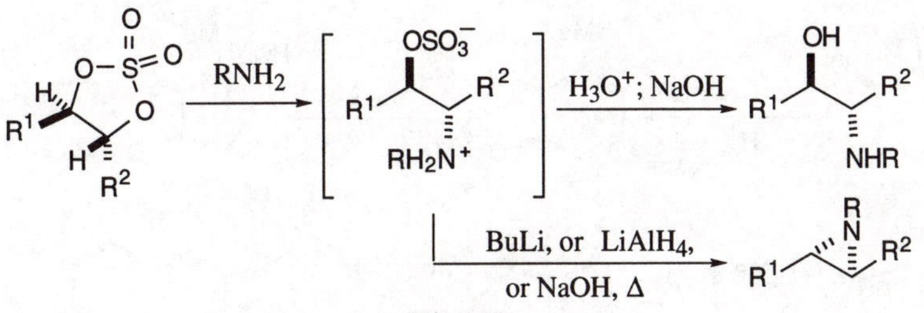

Fig. 7.85

One pathway by which microorganisms process aromatic compounds is through enzyme-catalysed dihydroxylation followed by re-aromatization. By using mutant organisms in which the re-aromatization stage is blocked, the intermediate cyclohexadiene diol accumulates and can be isolated in high yield (Fig. 7.86).

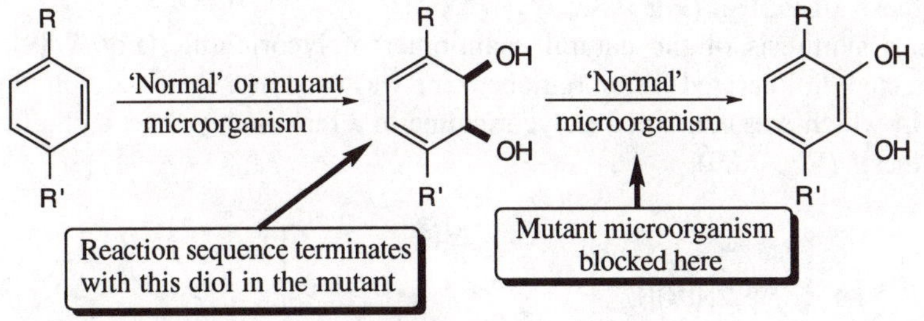

Fig. 7.86

Mutant strains of *Pseudomonas putida* are often used and oxidations using these systems are usually highly site-specific but will accept a wide range of types of substitution on the aromatic ring. Many different cyclohexadiene diols have been prepared in essentially 100 per cent enantiomeric excess using these microorganisms, and a few representative examples of the product diols are shown in Fig. 7.87.[92]

The usefulness and versatility of this type of intermediate is illustrated by the following two examples of their use in the synthesis of two quite different classes of compound.

The diol obtained by microbial dihydroxylation of toluene (Fig. 7.87),

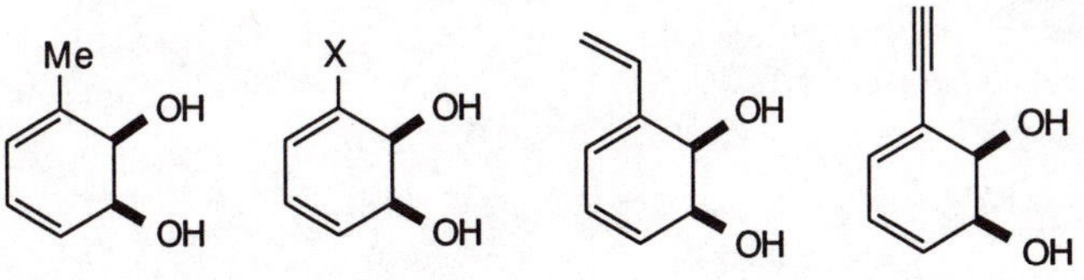

X = Cl, Br, I

Fig. 7.87

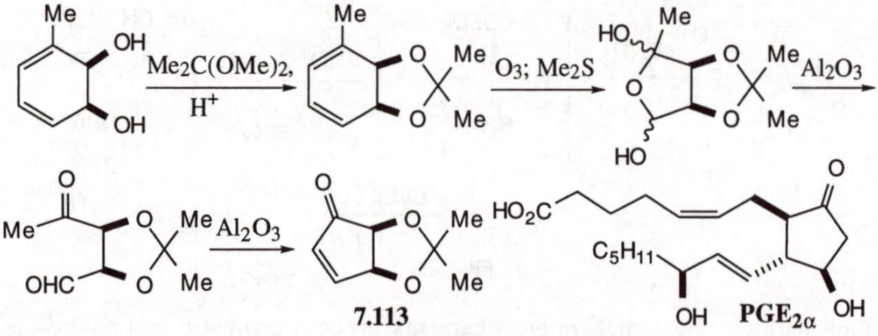

Fig. 7.88

protected as its acetonide, has been converted to the cyclopentenone **7.113** by ozonolysis of both double bonds and subsequent intramolecular aldol condensation. This aldol condensation product is a known intermediate for the synthesis of $PGE_{2\alpha}$ (Fig. 7.88).[93]

In a synthesis of the natural enantiomer of lycoricidine (Fig. 7.89), the protected diol derived from bromobenzene was used in a cycloaddition to give **7.114** which was converted to lycoricidine in a remarkably short and efficient synthesis (Fig. 7.89).

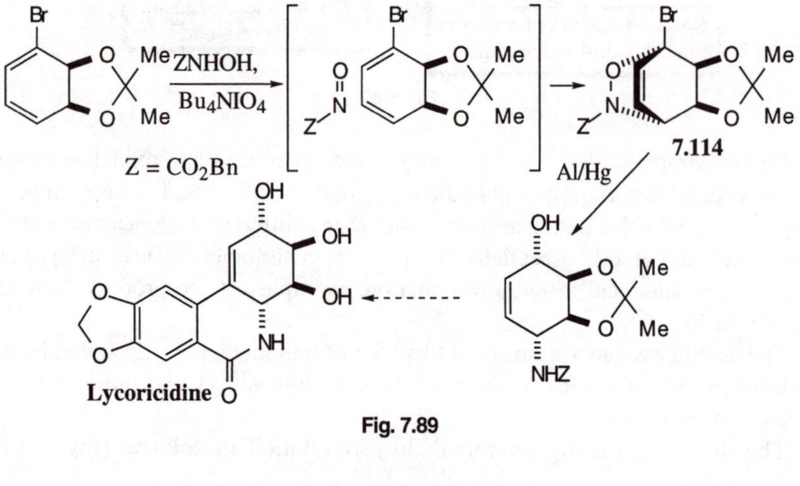

Fig. 7.89

References

1. Ojima, I., Clos, N., and Bastos, C. (1989), *Tetrahedron*, **45**, 6901, and references cited therein.
2. For more extensive lists, see Ref. 1.
3. Kagan, H. B. (1985), in *Asymmetric Synthesis*, (ed. J.D. Morrison), Vol. 5, pp. 1–40, Academic Press, New York.
4. a) Halpern, J. (1983), *Pure and Appl. Chem.*, **55**, 99; b) Halpern, J. (1985), in *Asymmetric Synthesis*, (ed. J.D. Morrison), Vol. 5, pp. 41–70, Academic Press, New York; c) Landis, C. R. and Halpern, J. (1987), *J. Amer. Chem. Soc.*, **109**, 1746.
5. Taken from Koenig, K. E. (1985), in *Asymmetric Synthesis*, (ed. J.D. Morrison), Vol. 5, pp. 71-101, Academic Press, New York.
6. Knowles, W. S. (1983), *Acc. Chem. Res.*, **16**, 106.
7. Ojima, I. (1984), *Pure and Appl. Chem.*, **56**, 99.
8. Ref. 4b, pp. 44–45.
9. Hayashi, T., Kawamura, N., and Ito, Y. (1987), *J. Amer. Chem. Soc.*, **109**, 7876.
10. Noyori, R., Ohta, M., Hsiao, Y., Kitamura, M., Ohta, T., and Takaya, H. (1986), *J. Amer. Chem. Soc.*, **108**, 7117.
11. Ohta, T., Takaya, H., Kitamura, M., Nagai, K., and Noyori, R. (1987), *J. Org. Chem.*, **52**, 3174.
12. Takaya, H., Ohta, T., Sayo, N., Kumobayashi, H., Akutagawa, S., Inoue, S.-I., Kasahara, I., and Noyori, R. (1987), *J. Amer. Chem. Soc.*, **109**, 1596.
13. Kitamura, M., Kasahara, I., Manabe, K., Noyori, R., and Takaya, H. (1988), *J. Org. Chem.*, **53**, 708.
14. Solladié, G., Demailly, G., and Greck, C. (1985), *Tetrahedron Lett.*, **26**, 435.
15. Solladié, G., Demailly, G., and Greck, C. (1985), *J. Org. Chem.*, **50**, 1552.
16. Solladié, G. and Ghiatou, N. (1992), *Tetrahedron Lett.*, **33**, 1605.
17. ApSimon, J. W. and Lee Collier, T. (1986), *Tetrahedron*, **42**, 5157; Singh, V. K. (1993), *Synthesis*, 605.
18. Noyori, R., Tomino, I., Tanimoto, Y., and Nishizawa, M. (1984), *J. Amer. Chem. Soc.*, **106**, 6709; Noyori, R., Tomino, I., Yamada, M., and Nishizawa, M. (1984), *J. Amer. Chem. Soc.*, **106**, 6717.
19. Suzuki, M., Morita, Y., Koyano, H., Koga, M., and Noyori, R. (1990), *Tetrahedron*, **46**, 4809.
20. Taken from Table 1, Ref. 17.
21. Vigeron, J. P. and Bloy, V. (1980), *Tetrahedron Lett.*, **21**, 1735.
22. Terashima, S., Tanno, N., and Koga, K. (1980), *J. Chem. Soc., Chem. Commun.*, 1026.
23. For a comparative review, see Brown, H. C. and Ramachandran, P. V. (1991), *Pure and Appl. Chem.*, **63**, 307.
24. Midland, M. M., McDowell, D. C., Hatch, R. L., and Tramontano, A. (1980), *J. Amer. Chem. Soc.*, **102**, 867; Midland, M. M., Tramontano, A., Kazubski, A., Graham, R. S., Tsai, D. J. S., and Cardin, D. B. (1984), *Tetrahedron*, **40**, 1371; Midland, M. M. and Graham, R. S. (1984), *Organic Syntheses*, **63**, 57.
25. Midland, M. M., Greers, S., Tramontano, A., and Zderic, S.A. (1979), *J. Amer. Chem. Soc.*, **101**, 2352.

26. Midland, M. M. and Zderic, S.A. (1982), *J. Amer. Chem. Soc.*, **104**, 525.
27. Reviews of this and related reagents can be found in Refs. 17 and 23.
28. Chandrasekharan, J., Ramachandran, P. V., and Brown, H. C. (1985), *J. Org. Chem.*, **50**, 5446.
29. Brown, H. C., Chandrasekharan, J., and Ramachandran, P. V. (1986), *J. Org. Chem.*, **51**, 3394.
30. Brown, H. C., Ramachandran, P. V., Teodorovic, A. V., and Swaminathan, S. (1991), *Tetrahedron Lett.*, **32**, 6691.
31. Masamune, S., Kennedy, R. M., Petersen, J. S., Houk, K. N., and Wu, Y.-D. (1986), *J. Amer. Chem. Soc.*, **108**, 7404.
32. Masamune, S., Kim, B.-M., Petersen, J. S., Sato, T., and Veenstra, S. J. (1985), *J. Amer. Chem. Soc.*, **107**, 4549.
33. Corey, E. J., Bakshi, R. K., and Shibita, S. (1987), *J. Amer. Chem. Soc.*, **109**, 5551; Corey, E. J., Bakshi, R. K., Shibita, S., Chen, C-P., and Singh, V. K. (1987), *J. Amer. Chem. Soc.*, **109**, 7925; For an improved preparation of the catalyst, see Corey, E. J. and Link, J. O. (1992), *Tetrahedron Lett.*, **33**, 4141.
34. a) Corey, E. J. and Link, J. O. (1990), *Tetrahedron Lett.*, **31**, 601; b) Corey, E. J. and Link, J. O. (1991), *J. Org. Chem.*, **56**, 442.
35. Corey, E. J. and Bakshi, R. K. (1990), *Tetrahedron Lett.*, **31**, 611.
36. Corey, E. J. and Link, J. O. (1992), *Tetrahedron Lett.*, **33**, 3431.
37. Corey, E. J. and Link, J. O. (1992), *J. Amer. Chem. Soc.*, **114**, 1906.
38. Corey, E. J., Link, J. O., and Bakshi, R. K. (1992), *Tetrahedron Lett.*, **33**, 7107; Jones, D. K. and Liotta, D. C. (1993), *J. Org. Chem.*, **58**, 799.
39. Corey, E. J., Azimioara, M., and Sarshar, S. (1992), *Tetrahedron Lett.*, **33**, 3429.
40. Kitamura, M., Ohkuma, T., Inoue, S., Sayo, N., Kumobayashi, H., Akutagawa, S., Ohta, T., Takaya, H., and Noyori, R. (1988), *J. Amer. Chem. Soc.*, **110**, 629.
41. Jones, J. B.(1985), in *Asymmetric Synthesis*, (ed. J.D. Morrison), Vol. 5, pp. 309–344, Academic Press, New York.
42. Csuk, R. and Glänzer, B. I. (1991), *Chem. Rev.*, **91**, 49.
43. Eichberger, G., Faber, K., and Griengl, H. (1985), *Monatsh. Chem.*, **116**, 1233.
44. Fujisawa, T., Itoh, T., Nakai, M., and Sato, T. (1985), *Tetrahedron Lett.*, **26**, 771; Indahl, S. R. and Scheline, R. R. (1986), *Appl. Microbiol.*, **16**, 667.
45. Iruchijma, S. and Kojima, N. (1978), *Agric. Biol. Chem.*, **42**, 451; Crumbie, R. L., Deol, B. S., Nemorin, J. E., and Ridley, D. D. (1978), *Aust. J. Chem.*, **31**, 1965; Kozikowski, A. P., Mugrage, B. B., Li, C. S., and Felder, L. (1986), *Tetrahedron Lett.*, **27**, 4817.
46. Guanti, G., Banfi, L., and Narisano, E. (1986), *Tetrahedron Lett.*, **27**, 3547.
47. Fujisawa, T., Kojima, E., Itoh, T., and Sato, T. (1985), *Tetrahedron Lett.*, **26**, 6089.
48. Bolte, J., Gourcy, J. G., and Veschambre, H. (1986), *Tetrahedron Lett.*, **27**, 565; Fauve, A. and Veschambre, H. (1988), *J. Org. Chem.*, **53**, 5215.
49. Lieser, J. K. (1983), *Synth. Commun.*, **13**, 765; Short, R. P., Kennedy, R. M., and Masamune, S. (1989), *J. Org. Chem.*, **54**, 1755.
50. Deol, B. S., Ridley, D. D., and Simpson, G. W. (1976), *Aus. J. Chem.*, **29**, 2459.
51. Ref. 42, pp.65–71; Servi, S. (1990), *Synthesis*, 1.

52. Taken from Schemes 72 and 74, Ref. 42.
53. Shieh, W.-R. and Sih, C. J. (1993), *Tetrahedron Asymmetry*, **4**, 1259.
54. Nakamura, K., Miyai, T., Nagar, A., Oka, S., and Ohno, A. (1989), *Bull. Chem. Soc. Jpn.*, **62**, 1179; Kitihara, T., Kurata, H., and Mori, K. (1988), *Tetrahedron*, **44**, 4339.
55. Hoffmann, R. W., Helbig, W., and Ladner, W. (1982), *Tetrahedron Lett.*, **23**, 3479.
56. Seebach, D., Rogo, S., Maetzke, T., Braunschweiger, H., Cercus, J., and Krieger, M. (1987), *Helv. Chim. Acta*, **70**, 1605.
57. Dodds, D. R. and Jones, J. B. (1988), *J. Amer. Chem. Soc.*, **110**, 577, and references cited therein.
58. Jakovac, I. J., Goodbrand, H. B., Lok, K. P., and Jones, J. B. (1982), *J. Amer. Chem. Soc.*, **104**, 4659.
59. Irwin, A. J. and Jones, J. B. (1977), *J. Amer. Chem. Soc.*, **99**, 556; Jones, J. B. and Lok, K. P. (1979), *Can. J. Chem.*, **57**, 1025.
60. Jakovac, I. J., Ng, G., Lok, K. P., and Jomes, J. B. (1980), *J. Chem. Soc., Chem. Commun.*, 515.
61. Collum, D. B., McDonald, J. H., and Still, W. C. (1980), *J. Amer. Chem. Soc.*, **102**, 2118.
62. Jones, J. B., Finch, M. A. W., and Jakovac, I. J. (1982), *Can. J. Chem.*, **60**, 2007.
63. a) Sharpless, K. B. (1991), in *Comprehensive Organic Synthesis*, (eds. B.M. Trost and I. Fleming), Vol. 7, pp. 389–436, Pergamon, Oxford; b) Rossiter, B. E. (1985), in *Asymmetric Synthesis*, (ed. J.D. Morrison), Vol. 5, pp. 194–246, Academic Press, New York; c) Finn, M. G. and Sharpless, K. B. (1985), in *Asymmetric Synthesis*, (ed. J.D. Morrison), Vol. 5, pp. 247–308, Academic Press, New York.
64. Gao, Y., Hanson, R. M., Klunder, J. M., Ko, S. Y., Masamune, H., and Sharpless, K. B. (1987), *J. Amer. Chem. Soc.*, **109**, 5765.
65. Compiled from Ref. 63a.
66. Klunder, J. M., Ko, S. Y., and Sharpless, K. B. (1986), *J. Org. Chem.*, **51**, 3710; Ko, S. Y. and Sharpless, K. B. (1986), *J. Org. Chem.*, **51**, 5413.
67. See Ref. 63a, pp. 390–393.
68. Martin, V. S., Woodward, S. S., Katsuki, T., Yamada, Y., Ikeda, M., and Sharpless, K. B. (1981), *J. Amer. Chem. Soc.*, **103**, 6237.
69. See Ref. 63a for a compilation.
70. Sharpless, K. B., Behrens, C. H., Katsuki, T., Lee, A. W. M., Martin, V. S., Takatani, M., Viti, S. M., Walker, F. J., and Woodward, S. S. (1983), *Pure and Appl. Chem.*, **55**, 589.
71. Evans, D. A., Bender, S. L., and Morris, J. (1988), *J. Amer. Chem. Soc.*, **110**, 2506.
72. For a thorough discussion, see Behrens, C. H. and Sharpless, K. B. (1983), *Aldrichimica Acta*, **16**, 67.
73. Caron, M. and Sharpless, K. B. (1985), *J. Org. Chem.*, **50**, 1557; Canas, M., Poch, M., Verdaguer, X., Moyano, A., Pericàs, M. A., and Reira, A. (1991), *Tetrahedron Lett.*, **32**, 6931.

74. a) Minami, N., Ko, S. S., and Kishi, Y. (1982), *J. Amer. Chem. Soc.*, **104**, 1109; b) Ma, P., Martin, V. S., Masamune, S., Sharpless, K. B., and Viti, S. M. (1982), *J. Org. Chem.*, **47**, 1378; c) Finan, J. M. and Kishi, Y. (1982), *Tetrahedron Lett.*, **23**, 2719; d) Viti, S. M. (1982), *Tetrahedron Lett.*, **23**, 4541; e) Nicolaou, K. C. and Venishi, J. (1982), *J. Chem. Soc.*, *Chem. Commun.*, 1292; f) Mubarak, A. M. and Brown, D. M. (1982), *J. Chem. Soc.*, *Perkin Trans. 1*, 809; g) Takano, S., Kasahara, C., and Ogasawara, K. (1983), *Chem. Lett.*, 175.
75. a) Corey, E. J., Hopkins, P. B., Munroe, J. E., Marfat, A., and Hashimoto, S. (1980), *J. Amer. Chem. Soc.*, **102**, 7986; b) Roush, W. R. and Brown, R. J. (1982), *J. Org. Chem.*, **47**, 1371.
76. Katsuki, T., Lee, A. W. M., Ma, P., Martin, V. S., Masamune, S., Sharpless, K. B., Tuddenham, D., and Walker, F. J., (1982), *J. Org. Chem.*, **47**, 1373; Wrobel, J. E. and Ganem, B. (1983), *J. Org. Chem.*, **48**, 3761.
77. Taken from Ref. 72, Scheme 11.
78. Finn, M. G. and Sharpless, K. B. (1991), *J. Amer. Chem. Soc.*, **113**, 106; Finn, M. G. and Sharpless, K. B. (1991), *J. Amer. Chem. Soc.*, **113**, 113.
79. Zhang, W., Loebach, J. L., Wilson, S. R., and Jacobsen, E. N. (1990), *J. Amer. Chem. Soc.*, **112**, 2801; Zhang, W. and Jacobsen, E. N. (1991), *J. Org. Chem.*, **56**, 2296; Irie, R., Ito, Y., and Katsuki, T. (1991), *Tetrahedron Lett.*, **32**, 6891.
80. Jacobsen, E. N., Zhang, W., Muci, A. R., Ecker, J. R., and Deng, L. (1991), *J. Amer. Chem. Soc.*, **113**, 7036.
81. Fu, H., Lock, G. C., Zhang, W., Jacobsen, E. N., and Wong, C.-H. (1991), *J. Org. Chem.*, **56**, 6497.
82. Deng, L. and Jacobsen, E. N. (1992), *J. Org. Chem.*, **57**, 4320.
83. Amberg, W., Bennani, Y. L., Chada, R. K., Crispino, G. A., Davis, W. D., Hartung, J., Jeong, K.-S., Ogino, Y., Shibita, T., and Sharpless, K. B. (1993), *J. Org. Chem.*, **58**, 844, and references cited therein.
84. Sharpless, K. B., Amberg, W., Bennani, Y. L., Crispino, G. A., Hartung, J., Jeong, K.-S., Kwong, H.-L., Morikawa, K., Wang, Z.-M., Xu, D., and Zhang, X.-L. (1992), *J. Org. Chem.*, **57**, 2768.
85. Wang, Z.-M., Zhang, X.-L., and Sharpless, K. B. (1992), *Tetrahedron Lett.*, **33**, 6407.
86. Xu, D., Crispino, G. A., and Sharpless, K. B. (1992), *J. Amer. Chem. Soc.*, **114**, 7570; Jeons, K.-S., Sjö, P., and Sharpless, K. B. (1992), *Tetrahedron Lett.*, **33**, 3833.
87. Xu, D. and Sharpless, K. B. (1993), *Tetrahedron Lett.*, **34**, 951.
88. Kolb, H. C. and Sharpless, K. B. (1992), *Tetrahedron*, **48**, 10515.
89. For a review, see Lohray, B. B. (1992), *Synthesis*, 1035.
90. Lohray, B. B. and Sharpless, K. B. (1989), *Tetrahedron Lett.*, **30**, 2623.
91. For a review, see Widdowson, D. A., Ribbons, D. W., and Thomas, S. D. (1990), *Janssen Chim. Acta*, **8**, 3.
92. Hudlicky, T., Luna, H., Olivio, H. F., Andersen, C., Nugent, T., and Price, J. D. (1991), *J. Chem. Soc.*, *Perkin Trans. 1*, 2907.
93. Hudlicky, T., Luna, H., Barbieri, G., and Kwart, L. D. (1988), *J. Amer. Chem. Soc.*, **110**, 4735.

8 Rearrangements

The two main areas which will be considered in this chapter are sigmatropic rearrangements[1] and alkene isomerizations.[2] There are numerous examples of the application of these two classes of reaction to the asymmetric synthesis of organic compounds, and in both cases a limited selection of examples will be presented in order to illustrate the principles and applications of these methods.

Of most importance in asymmetric synthesis are 2,3- and 3,3-sigmatropic rearrangements typified by the Wittig[3] and Claisen[4] rearrangements respectively (Fig. 8.1). Taking these two as typical examples of their classes, some of the important general features of these reactions will be considered before specific examples of their use in asymmetric synthesis are discussed.

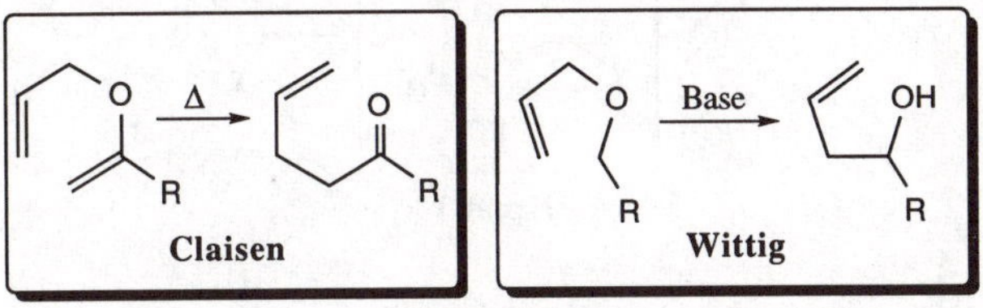

Fig. 8.1

The Claisen rearrangement and its variants usually proceed through a six-membered chair transition state (Fig. 8.2),[1] analogous to the Zimmerman–Traxler model which is so useful in accounting for the stereochemical outcome of aldol reactions (Chapter 5). Just as in aldol reactions this type of transition state leads to a number of important characteristics.

The newly formed double bond is almost always produced as the (*E*)-isomer. This follows from the preference for the larger group to be equatorial in the chair transition state, as does the high level of the transfer of chirality which is usually observed (Fig. 8.3). Precursor **8.1** produces the (*E*)-alkene **8.2** as a result of the strong preference for the group **x** to adopt an equatorial position. This same preference results in rearrangement of **8.3** through transition state **8.4** and hence to product **8.5**, rather than to **8.6** which would involve transition state **8.7** in which **x** is axial.

The tendency for the group **x** to adopt an equatorial position in the transition state for the rearrangement of systems such as **8.3** results in efficient chirality transfer, **8.5** being produced in preference to **8.6** which has the opposite

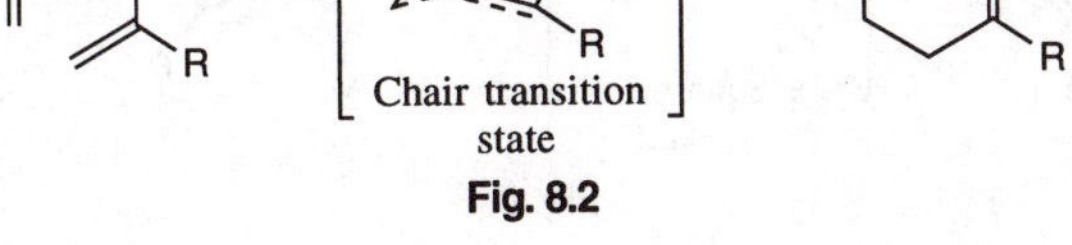

Fig. 8.2

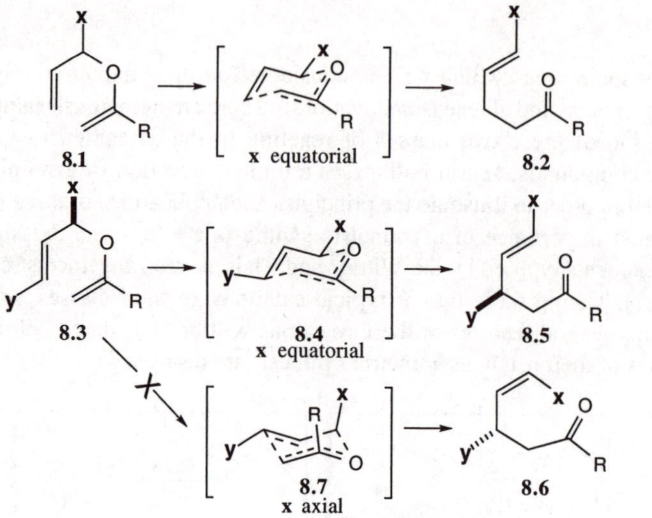
8.1
x equatorial
8.2
8.3
8.4
x equatorial
8.5
8.7
x axial
8.6

Fig. 8.3

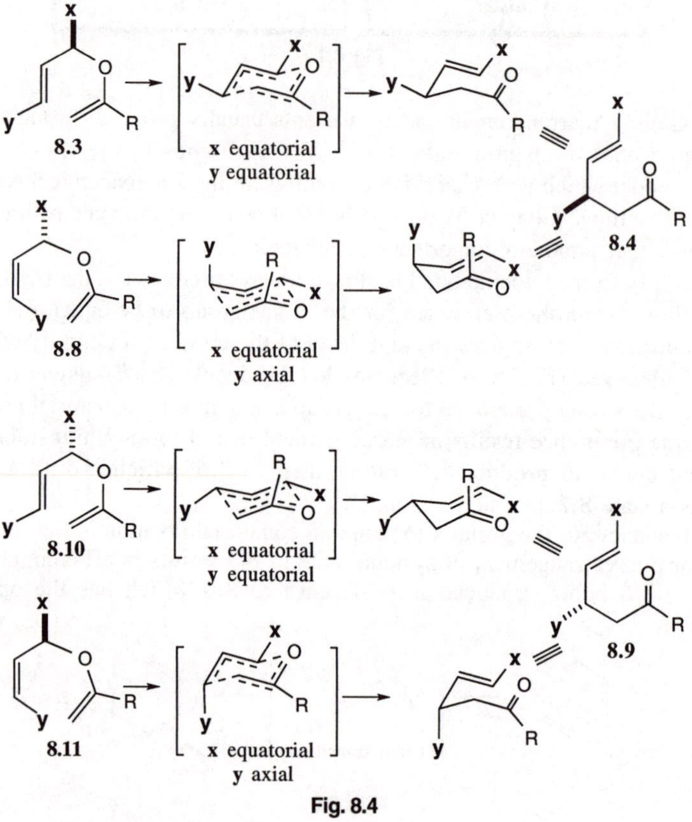
8.3
x equatorial
y equatorial
8.4
8.8
x equatorial
y axial
8.10
x equatorial
y equatorial
8.9
8.11
x equatorial
y axial

Fig. 8.4

absolute configuration at the new chiral centre (Fig. 8.3). A consequence of this is that it is usually easy to predict the absolute stereochemistry of the product in an unknown case, or to decide on the appropriate precursor to a particular product stereochemistry.

Moreover, in appropriate cases it is possible to use both enantiomers of the starting allylic alcohol to provide the same product enantiomer. This can be achieved because the geometry of the double bond in this component is also important in determining the stereochemistry of the product. For a given enantiomer of the allylic alcohol, the (*E*)- and (*Z*)-isomers will produce opposite absolute configurations of the product (Fig. 8.4). It follows that **8.3** and **8.8** will both give the same product **8.4**, and the enantiomer **8.9** will be produced from **8.10** and **8.11** (Fig. 8.4).

It follows from the preceding analysis that if the vinyl ether unit carries a substituent **z**, then the geometry of this vinyl ether unit should have an important role in determining the stereochemistry of the product. Precursors **8.12** and **8.13** which differ only in the geometry of the vinyl ether give rise to opposite enantiomers **8.14** and **8.15** (Fig. 8.5).

If both double bonds in the precursor carry substituents **y** and **z**, then for a given enantiomer of the allyl ether portion, the relative and absolute stereoichemistry of the product is determined by the configuration of both double bonds. The various possibilities and the appropriate transition state models are illustrated in Fig. 8.6.

The terms which are widely used to describe the relative stereochemistry of the diastereoisomeric products *threo*-**8.17** and *erythro*-**8.17** (Fig. 8.6) are different to those which are usually used currently to describe the relative stereochemistry of aldol products, where the *syn*- and *anti*- convention (Chapter 5) has largely replaced the descriptors *threo*- and *erythro*-. Care must be exercised in the description of the geometry of the enolate as either (*E*) or (*Z*). The situation is straightforward in most cases, as the 'internal' oxygen atom will take precedence over the substituent R. This is not the case for the widely used

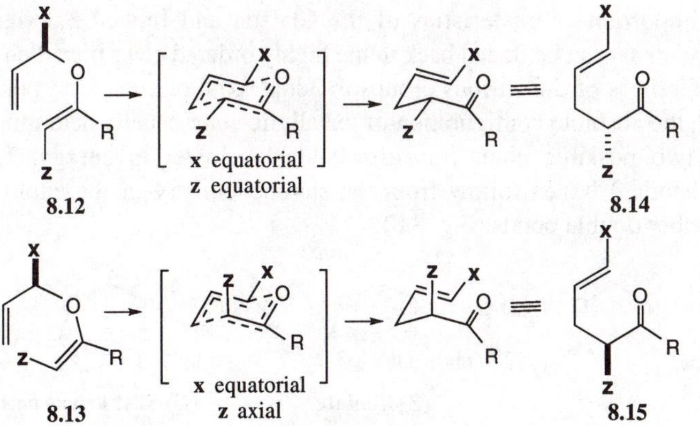

Fig. 8.5

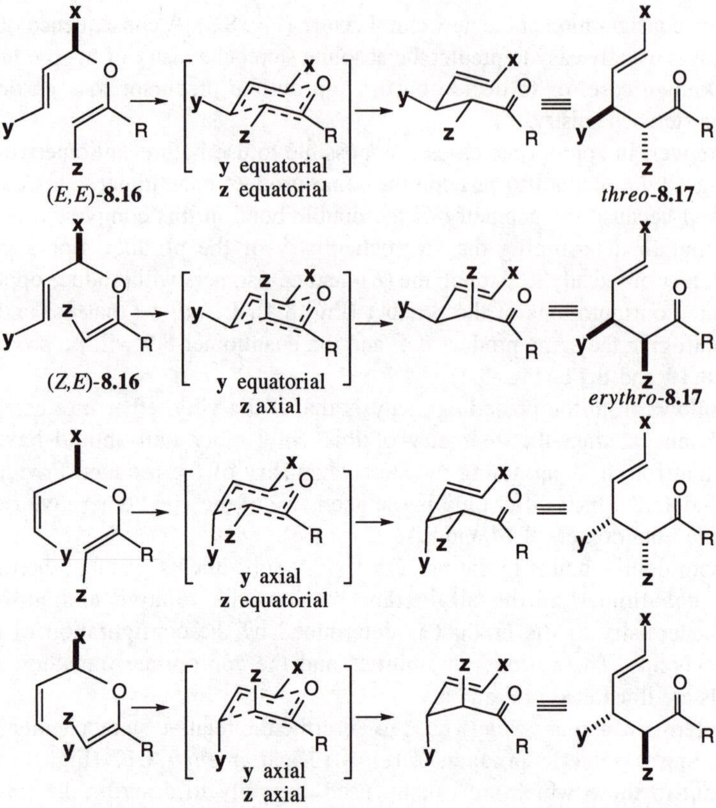

Fig. 8.6

Ireland–Claisen rearrangement in which this substituent, $OSiR_3$, takes precedence over the internal oxygen. This results in a change in stereochemical assignment (*E*) to (*Z*), even though the 'actual' stereochemistry and the relative stereochemistry of the product, remain the same (Fig. 8.7).

The important characteristics of the Claisen and related 3,3-sigmatropic rearrangements can be traced back to the highly ordered chair transition state and the preferences of the various groups to adopt axial or equatorial positions. In general, the absolute configuration of the allylic ether usually determines which of the two possible chair transition states is lower in energy. The other stereochemical issues follow from the stereochemistry of the enol ether and allylic ether double bonds (Fig. 8.8).

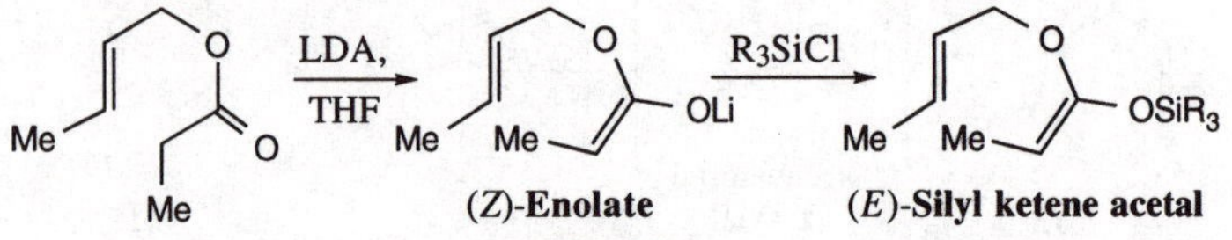

Fig. 8.7

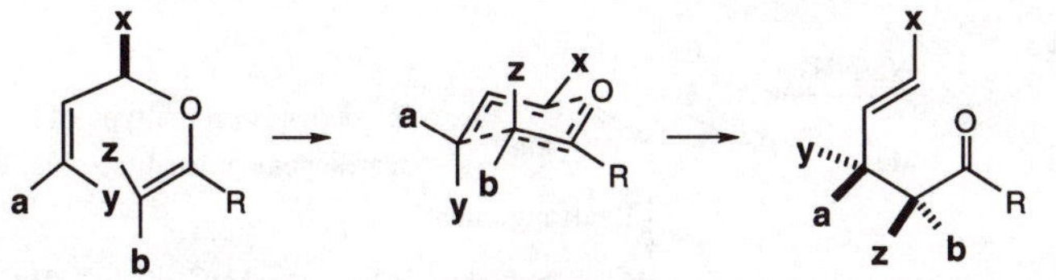

Fig. 8.8

In most rearrangements of this type the precursor is prepared from the appropriate allylic alcohol, which is often relatively easy to obtain with the correct absolute configuration and double bond geometry. The various types of rearrangement differ in the group R (Fig. 8.8) and the way in which the enol ether is introduced.

Some of most common variants are illustrated in Fig. 8.9, although this is by no means a complete list. Which of these variants is used in a particular situation depends on a number of factors, the most obvious being that different carbonyl functional groups are produced. The conditions also vary from mildly acidic (Claisen,[5] Johnson[6]), through neutral (Eschenmoser[7]), to basic (Ireland[8]). All these would usually be considered when deciding which one to use in a particular synthetic scheme. Often the most important factor is the stereochemistry of the desired product, especially if there is the necessity of preparing either an *erythro*- or a *threo*-isomer in high diastereoisomeric and enantiomeric excess.

There are numerous examples of the use of this type of rearrangement for the transfer of chirality in which the vinyl ether is unsubstituted and there is therefore no question of *erythro/threo* isomers being produced. A few illustrative examples of this type of reaction will now be considered.

The synthesis of the side chain of α-tocopherol provides an excellent demonstration of the control which is posssible using the Claisen

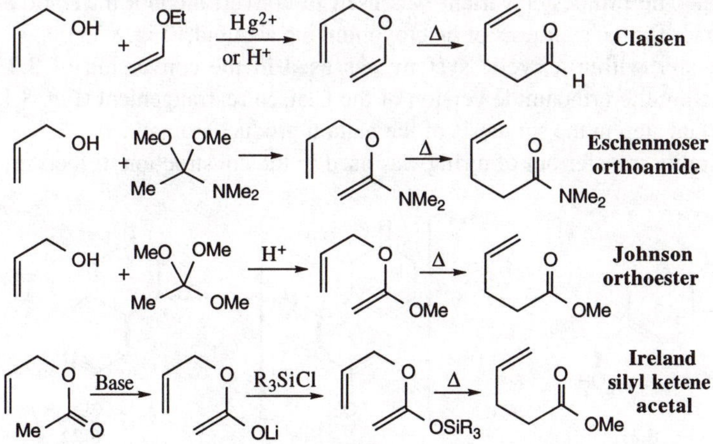

Fig. 8.9

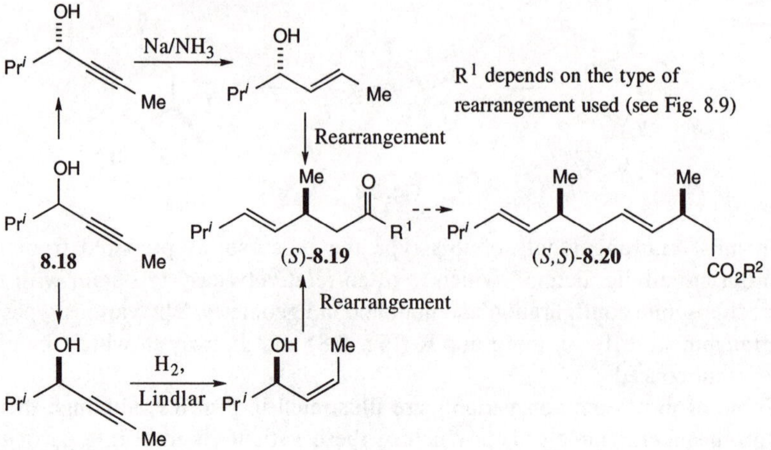

Fig. 8.10

rearrangement, and its variations, in acyclic systems (Fig. 8.10).[9] The starting material **8.18** was resolved, but *both* enantiomers could be used in the synthesis of the *same* enantiomer of the product **8.19** (cf. Fig. 8.4). The (S)-enantiomer of **8.18** was converted to the (E)-alkene and (R)-**8.18** was reduced to the (Z)-alkene. Both of these were subjected to all four of the Claisen rearrangement conditions shown in Fig. 8.9 and both gave (S)-**8.19**. All these Claisen rearrangements were shown to proceed with 97–99 per cent transmission of chirality. This overall sequence was repeated to provide (S,S)-**8.20**, and although two separation steps are required, the sequence **8.18** to **8.20** is rendered efficient as all the isomers produced are converted to the same enantiomer of the desired product.

The Claisen rearrangement has been widely used in the transfer of chirality in cyclic systems. The transfer can take place into, within, and out of the cyclic system. The orthoester variant was used to convert alcohol **8.21** into **8.22**, a precursor for the synthesis of heteroyohimbine alkaloids (Fig. 8.11).[10]

Transfer within a cyclic system was used in the conversion of **8.23** into **8.24** using the orthoamide version of the Claisen rearrangement (Fig. 8.12), an important step in the synthesis of the natural product aromatin.[11]

Chirality transfer out of a ring was used in the construction of the sterol side

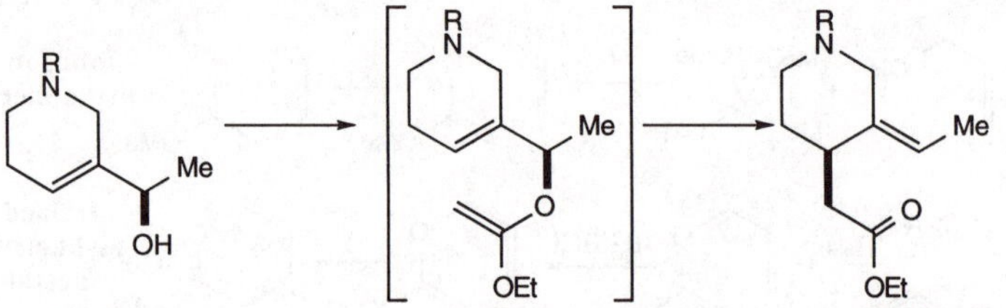

Fig. 8.11

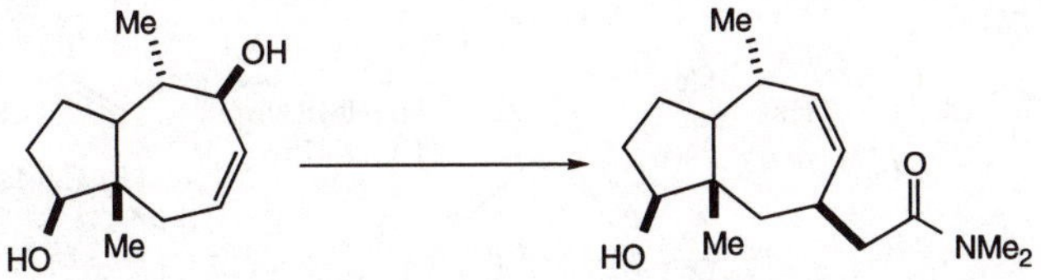

Fig. 8.12

chain outlined below (Fig. 8.13). This utilizes the Carroll rerrangement, a reaction in which the enol of an allylic β-keto ester undergoes a Claisen-type rearrangement with subsequent decarboxylation of the resulting β-keto acid.

In all the examples used so far, the question of formation of *erythro/threo* isomers has not arisen. As outlined in Fig. 8.6, these isomers arise from either of the two possible configurations of a substituted enolic double bond. It follows that to obtain high *erythro/threo* selectivity the enol ether must be formed with high stereoselectivity.

The desired level of control is relatively difficult to achieve for the enol ether, orthoester, and orthoamide rearrangements. Although some methods are available, a more general solution to this problem is found in the rearrangement of silyl ketene acetals, often referred to as the Ireland–Claisen rearrangement (Fig. 8.9).[4,8] The final part of the discussion of Claisen rearrangements will be concerned with this topic.

The stereochemical outcome of enolization of esters has been studied thoroughly,[12] and a detailed discussion of the results is beyond the scope of this volume. A simplified scheme is presented in Fig. 8.14. Enolization in THF

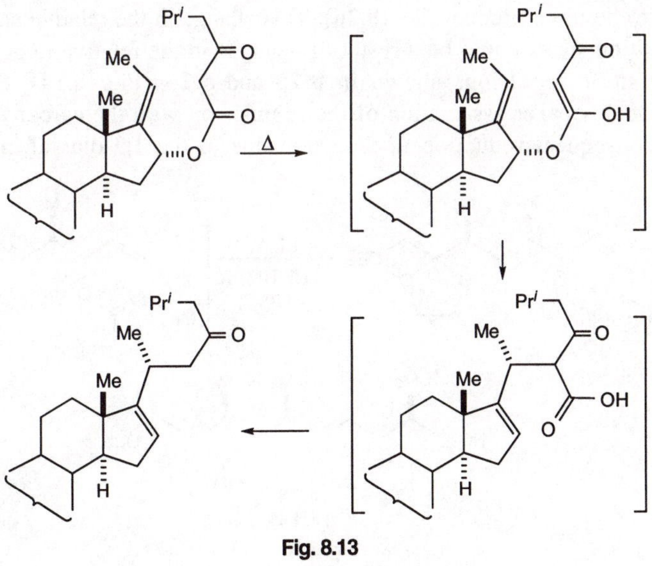

Fig. 8.13

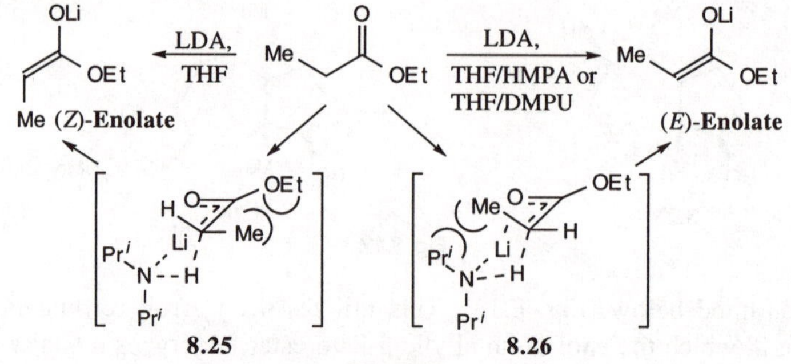

Fig. 8.14

alone generally produces an enolate mixture rich in the (Z)-lithium enolate. In a mixed solvent system involving THF and a polar solvent capable of solvating lithium ions such as hexamethyl phosphoramide (HMPA) or *N,N*-dimethyl-*N,N*-propyleneurea (DMPU) the (E)-lithium enolate is favoured. The exact (E):(Z) ratio depends on several variables including the relative amount of base used, and the precise composition of the solvent.

It is considered that these enolizations are under kinetic control with both sets of reaction conditions, and that the difference in selectivity arises because of a difference in transition states. In the absence of strong solvation for the lithium ion, the transition state is thought to involve close coordination of the lithium ion with the carbonyl oxygen and the base. This transition state is proposed to resemble **8.25** (Fig. 8.14) in which the steric repulsion between the methyl group and the ethoxy group is less important than repulsion between the methyl group and the bulky *iso*-propyl group of the base (see **8.26**, Fig. 8.14).

In the presence of strong solvation for the lithium ion, the transition state is thought to be more product-like (**8.26**). This change in the relative energies of the transition states could be a result of changes in the relative importance of the two steric repulsions shown in **8.25** and **8.26** (Fig. 8.14). Solvation causes much weaker association of the lithium ion with the carbonyl oxygen with a consequent reduction in repulsion due to the 1,3-diaxial interaction

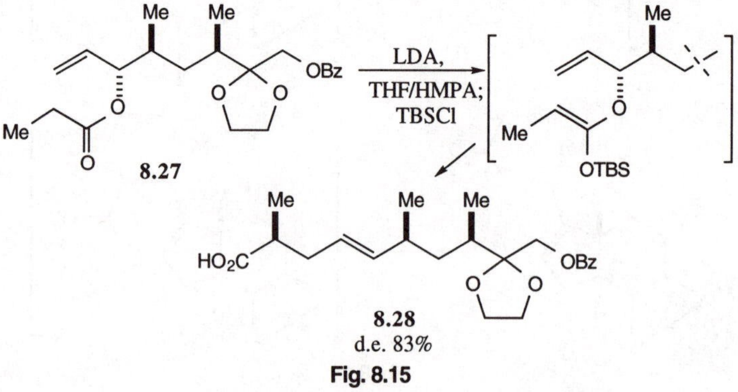

8.28
d.e. 83%
Fig. 8.15

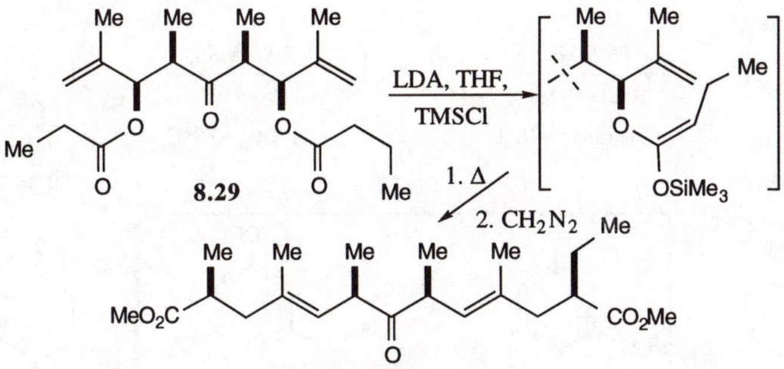

Fig. 8.16

between the methyl group and the *iso*-propyl group of the base, whereas the repulsion between the methyl group and the ethoxy group (see **8.25**, Fig. 8.14) will be essentially unchanged. Under the most favourable solvating conditions it is possible that enolization takes place by acyclic transition states, and the outcome is then likely to be influenced by the relative ground state energies of the ester conformations.

The Ireland–Claisen rearrangement has found numerous applications in organic synthesis owing to the ready availability of substrates, its efficient chirality transfer, and its predictability. Enolization of **8.27** under conditions which favour formation of the (*E*)-lithium enolate followed by silylation and rearrangement gives **8.28** with high stereochemical control (Fig. 8.15).[13]

This high level of acyclic stereocontrol is also evident in the double rearrangement of diester **8.29**, which gives **8.30** with very high diastereoselectivity (>90 per cent) (Fig. 8.16).[14]

It is also possible to carry out Ireland–Claisen rearrangements on lactones, in which the enolate is part of a ring. Often the ring size constrains the enolate geometry thereby removing one variable which affects the diastereoselectivity, and the stereochemical purity of the product can be very high. A simple example of this type of reaction is provided in Fig. 8.17, in which lactone **8.31** rearranges to the cyclohexane **8.32** very efficiently.[15]

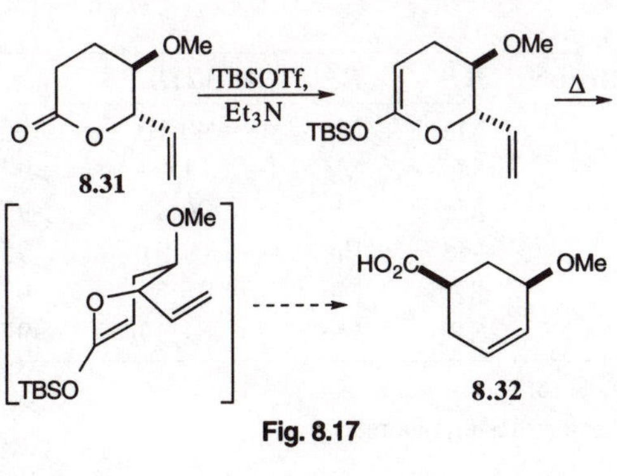

Fig. 8.17

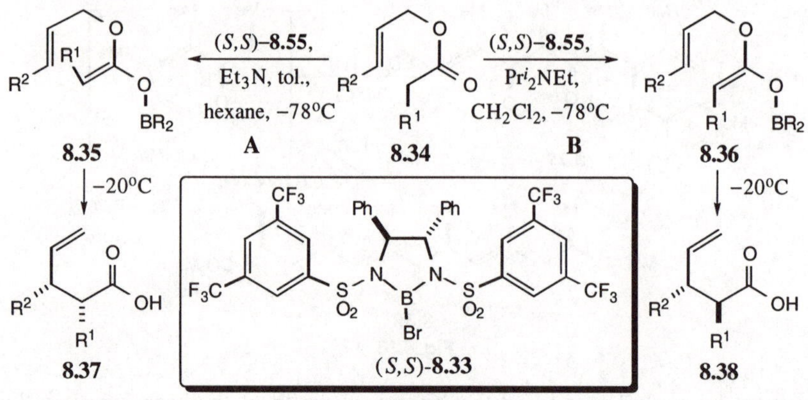

Fig. 8.18

In all of the examples of Claisen rearrangements discussed so far, the chirality has been transferred from one site in the molecule to another. Recently, a chiral reagent approach has been developed which allows the formation of products with high enantioselectivity from achiral esters.[16] The reagent, (S,S)-**8.33** (Fig. 8.18) is available in either enantiomer. It has been used previously for asymmetric aldol reactions (Chapter 5), and variants have been used in carbonyl allylation (Chapter 3) and Diels–Alder reactions (Chapter 6). Enolate geometry can be controlled simply by the reaction conditions (Fig. 8.18), and the products are usually obtained with high stereoselectivity (Table 8.1).

The transition state model illustrated in Fig. 8.19 has been proposed to account for the absolute configuration of the products of these enantioselective Ireland–Claisen rearrangements.

The [2,3]-Wittig rearrangement[3] constitutes a useful and versatile reaction for asymmetric synthesis; the general reaction is outlined in Fig. 8.20 and represents an example of a general class of [2,3]-sigmatropic rearrangements encountered in organic synthesis. The reaction usually requires a carbanion-

Table 8.1   Stereoselectivity of Claisen rearrangement of **8.34**

Conditions[a]	R^1	R^2	8.37:8.38	e.e. (%)[b]
A	Me	Me	90:10	96
B	Me	Me	1:99	>97
A	Me	Ph	96:4	>97
B	Me	Ph	9:91	95
A	SPh	Me	95:5	>97
B	SPh	Me	61:39	>97

[a]See Fig. 8.18.
[b]e.e. of major diastereoisomer.

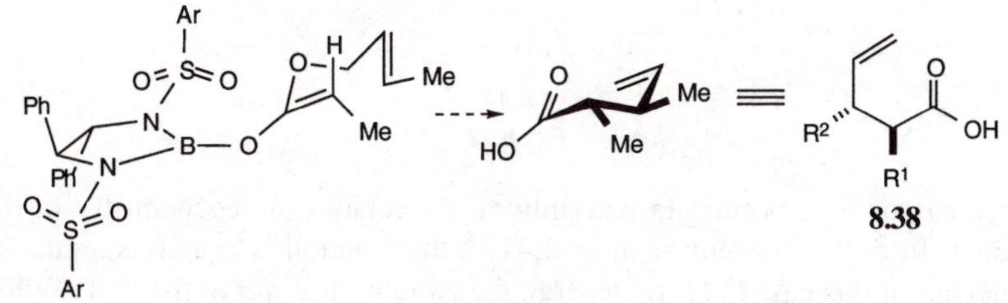

Fig. 8.19

stabilizing group X (Fig. 8.20) and like the Claisen rearrangement the questions of *erythro/threo* diastereoisomers and double bond geometry of the product must be considered, in addition to chirality transfer or asymmetric synthesis.

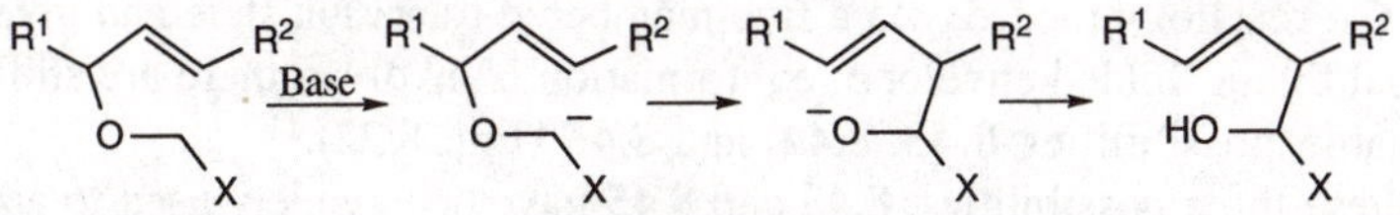

Fig. 8.20

The presence of a carbanion-stabilizing group X allows formation of the anion by deprotonation. In the absence of such a group the required carbanion can be generated by an exchange reaction. These two methods are illustrated in Fig. 8.21.

The examples shown in Fig. 8.21 also illustrate some of the general features of the [2,3]-Wittig rearrangement. Substrates such as **8.39** and **8.40**[17] often give products with high diastereoselectivity, and the double bond geometry in

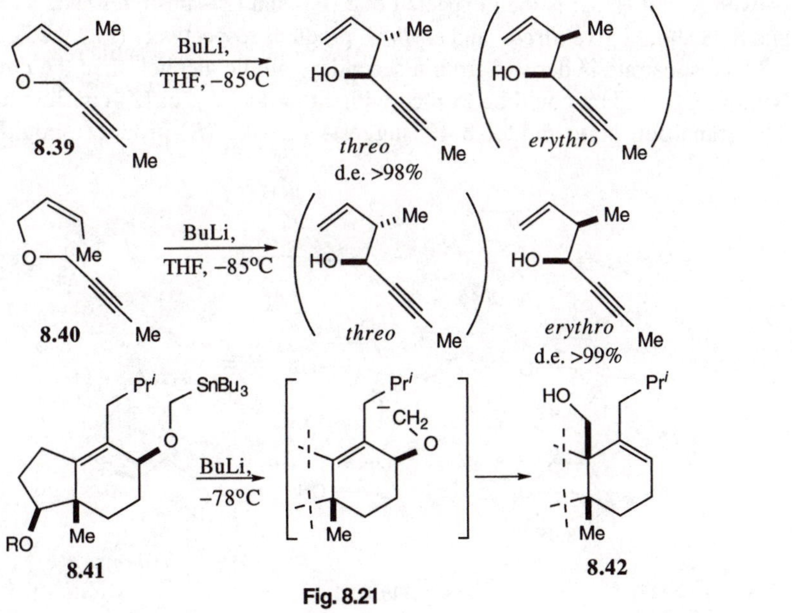

Fig. 8.21

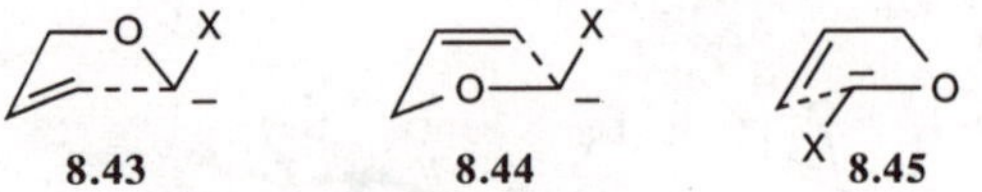

8.43   **8.44**   **8.45**

Fig. 8.22

the precursor is important in determining the relative stereochemistry of the product. In cyclic systems such as **8.41**,[18] the reaction is usually suprafacial producing in this case **8.42**. In general, the chirality transfer in the [2,3]-Wittig rearrangement is extremely high.

The stereochemical outcome of [2,3]-Wittig rearrangements is not as consistent from substrate to substrate as the Claisen rearrangement, but enough evidence is available to make a reasonable prediction in an unknown case. This substrate dependence is probably due in some part to the nature of the transition state. The reaction proceeds via a five-membered transition state and given the proposal that a 'folded envelope' conformation is involved there are still three reasonable possibilities **8.43**, **8.44**, and **8.45** (Fig. 8.22).[19]

Of these three possibilities, **8.43** and **8.45** have been widely used to account for the stereochemical outcome of [2,3]-Wittig rearrangements, with **8.45** currently being favoured. This latter transition state is thought to account for the following general observations. The sense of diastereoselection is governed largely by the substrate double bond geometry and the degree of stereoselectivity depends on the nature of X. In general, an (*E*)-substrate will give mainly *erythro*- product, and a (*Z*)-substrate mainly *threo*-. It is suggested that conformations in which X is pseudoaxial such as **8.46** (Fig. 8.23) are destabilized with respect to **8.45** as a result of pseudo-1,3-diaxial interaction betwen X and H*. It is then expected that (*E*)- and (*Z*)-substrates such as **8.47** and **8.48** should give *threo*- and *erythro*- products respectively (Fig. 8.23).

If the substrate is derived from a secondary allylic alcohol, then the double bond in the product could be formed with either the (*E*)- or (*Z*)-configuration. The transition state model **8.45** suggests that the (*E*)-product should be

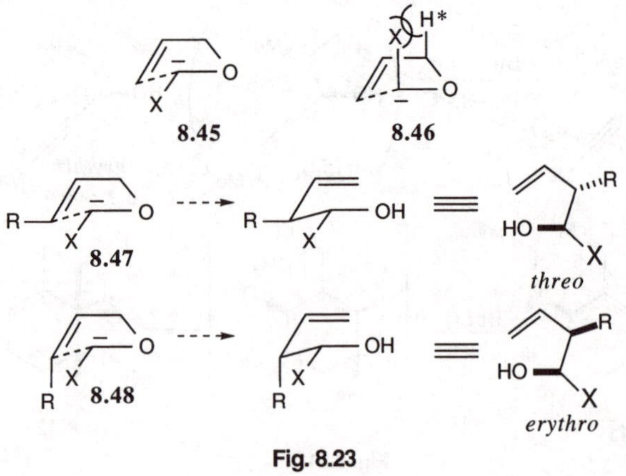

8.45   **8.46**

threo

erythro

Fig. 8.23

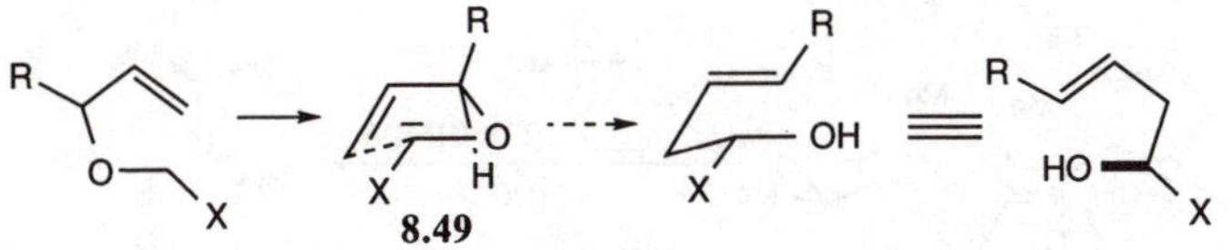

Fig. 8.24

favoured as the group R in **8.49** is likely to prefer to be '*exo*' in the transition state (Fig. 8.24). This preference for formation of an (*E*)-product is usually found to be the case.

As referred to above, the [2,3]-Wittig rearrangement can take place with extremely high levels of chirality transfer. Substrate **8.50** is converted into the (*E*)-*erythro*-product **8.51** with essentially complete transfer of chirality and very high diastereoselectivity (Fig. 8.25).[17] This observation is consistent with the transition state model discussed above.

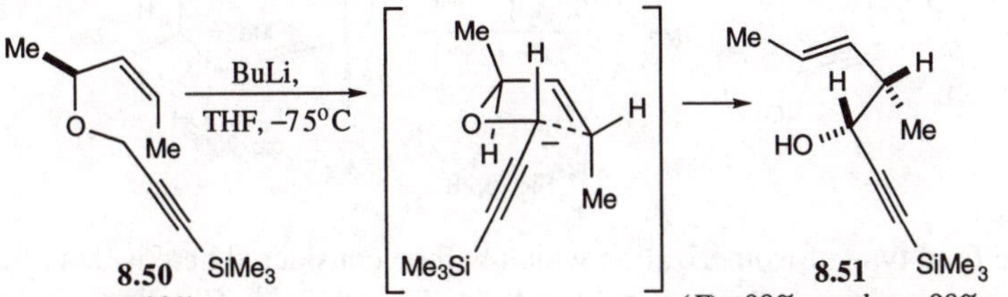

Fig. 8.25

A number of chiral auxiliary groups which can stabilize a carbanion can be used to control the absolute configuration of the [2,3]-Wittig rearrangement.[3] The results of these enantioselective [2,3]-Wittig rearrangements are somewhat variable, but reasonably high stereoselectivity has been observed in the rearrangements illustrated in Fig. 8.26.[20] In these cases the *erythro*-isomer is predominant.

Very high diastereoselectivity in favour of the *erythro*-product has also been observed in the rearrangement of **8.52** (Fig. 8.27).[21] The chiral auxiliary in

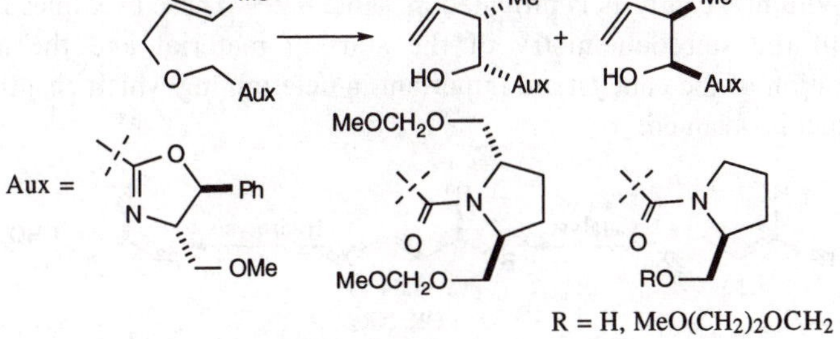

$R = H, MeO(CH_2)_2OCH_2$

Fig. 8.26

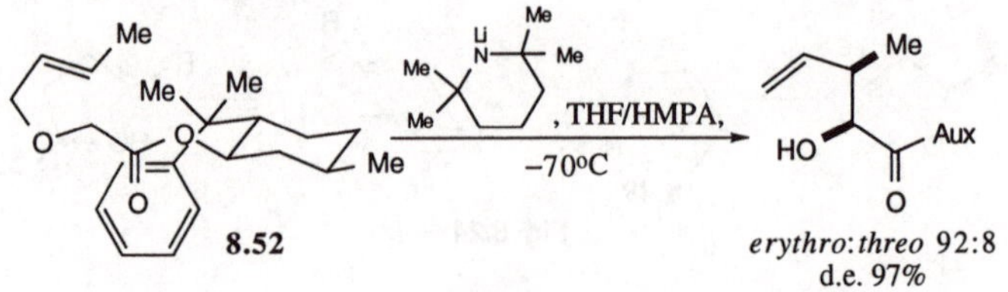

Fig. 8.27

this case is 8-phenylmenthol, which has also found use in additions to carbonyl compounds and Diels–Alder reactions (Chapters 3 and 6).

Chiral bases have also been used to effect asymmetric [2,3]-Wittig rearrangements,[22] and although high enantioselectivity has only been observed in cyclic cases (Fig. 8.28), this approach clearly holds promise for the future.

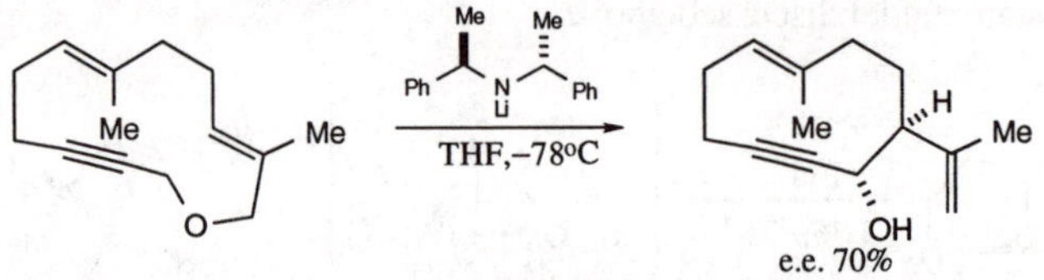

Fig. 8.28

The final type of isomerization which will be considered here is that of allylic isomerization generalized as **8.53** to **8.54** (Fig. 8.29).[2b] This isomerization is driven by the increased stability of the product **8.54** over the starting material **8.53**, and is particularly useful for synthesis as the product can be converted into an aldehyde. The absolute configuration of the product is controlled by which enantiomer of the asymmetric catalyst is used.

Although it is possible to carry out asymmetric isomerization of allylic ethers (X = OR), the enantiomeric excess is usually low.[2a] The highest levels of enantioselectivity are obtained using N,N-dialkylallylamines with the appropriate catalyst.[2b] The catalyst of choice is a rhodium(I) complex with the chelating diphosphine BINAP (Fig. 8.30). This ligand is also used in the highly enantioselective catalytic hydrogenation of alkenes (Chapter 7), with which the isomerization process shares some overall similarities.

A selection of asymmetric isomerizations of allylic amines (Fig. 8.31) using these rhodium(I) catalysts is provided in Table 8.2.[23] These examples illustrate that both the stereochemistry of the starting material and the absolute configuration of the catalyst are important in determining which enantiomer of the product is obtained.

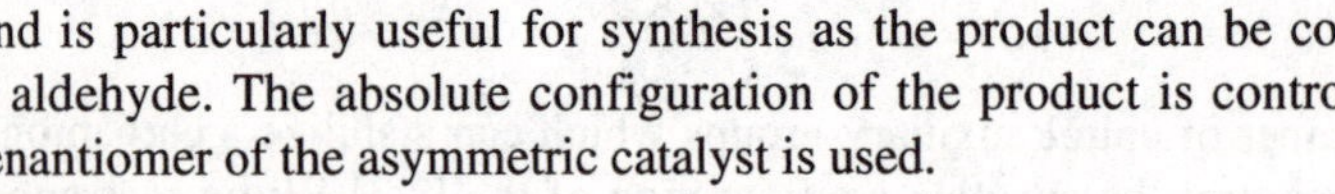

Fig. 8.29

Ligands

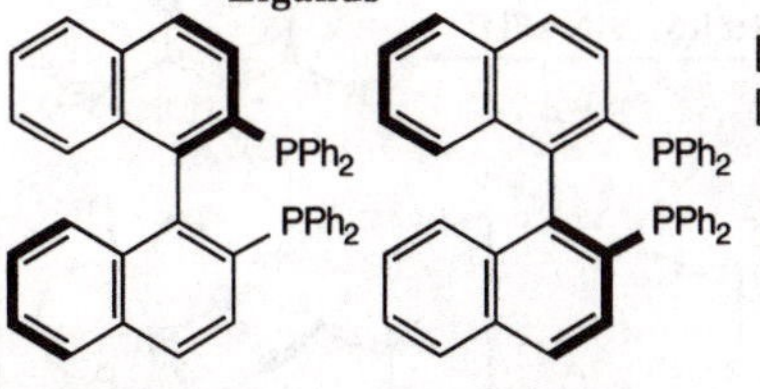

Catalysts

[Rh(BINAP)(COD)]⁺ **8.55**
[Rh(BINAP)(NBD)]⁺ **8.56**
[Rh(BINAP)₂]⁺ **8.57**
[Rh(BINAP)(solv.)]⁺ **8.58**

COD = cyclooctadiene
NBD = norbornadiene
solv. = solvent

(*R*)-**BINAP**   (*S*)-**BINAP**

Fig. 8.30

The mechanism of this process has been studied in some detail, and some of the conclusions are outlined in Fig. 8.32.[24] The reaction appears to be a clean intramolecular process, involving a suprafacial 1,3-hydrogen shift from the *transoid*-conformation. The double bond of the substrate does not appear to be coordinated to the metal in the transition state, instead this is thought to involve a square planar complex as illustrated in Fig. 8.32.[25] Transition state **8.62** is favoured for (*S*)-BINAP and **8.63** is favoured for the enantiomeric ligand.

The catalytic isomerization of **8.59** is currently used in an industrial synthesis of (−)-menthol (>1500 t/year), known as the 'Takasago Process' (Fig. 8.33). The isomerization step uses a substrate:catalyst ratio of 8000:1 at 80–100°C. The enamine is produced in very high enantiomeric excess, and the catalyst is recyclable. The 'Takasago Process' is possibly the largest production

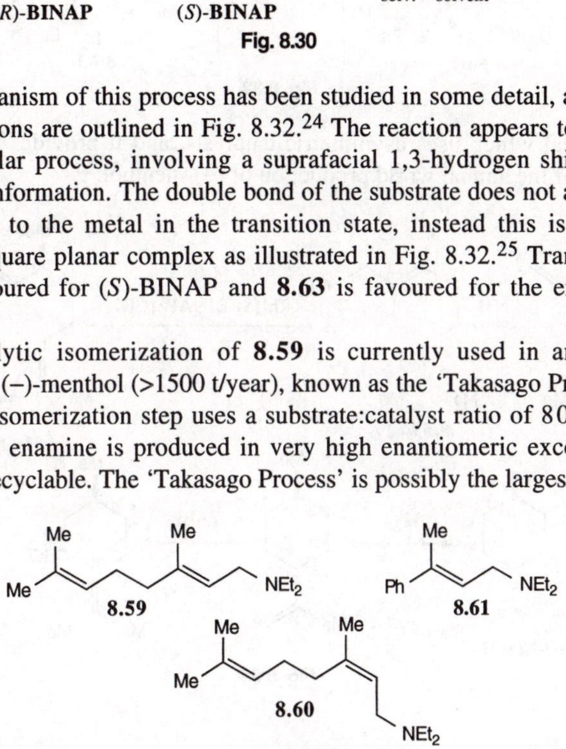

Fig. 8.31

Table 8.2 Enantioselective rearrangement of allylic amines **8.59**, **8.60**, and **8.61** (Fig. 8.31)

Substrate	Catalyst	e.e.[a]	Ratio (Temp.)[b]
8.59	(*R*)-**8.55**	96 (*S*)	100:1 (40)
8.59	(*S*)-**8.55**	99 (*R*)	8000:1 (80)
8.60	(*R*)-**8.55**	95 (*R*)	100:1 (40)
8.60	(*S*)-**8.55**	92 (*S*)	100:1 (40)
8.61	(*R*)-**8.55**	90 (*R*)	100:1 (60)

[a]e.e. (%) and absolute configuration of product.
[b]Substrate:catalyst ratio and reaction temperature (°C).

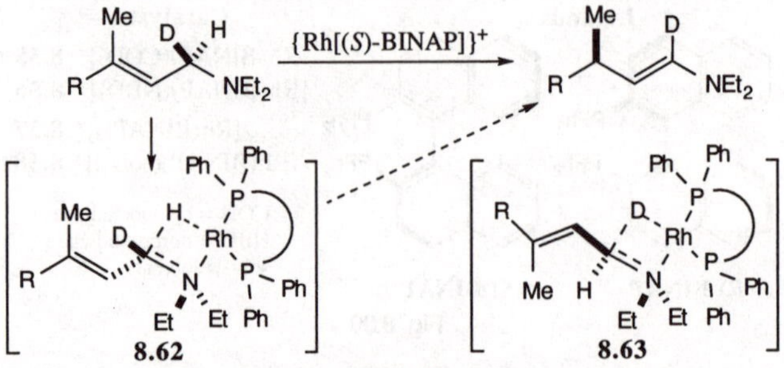

Fig. 8.32

of a chemical which uses asymmetric catalysis, and it provides a significant proportion of the annual world production of (−)-menthol.[2b]

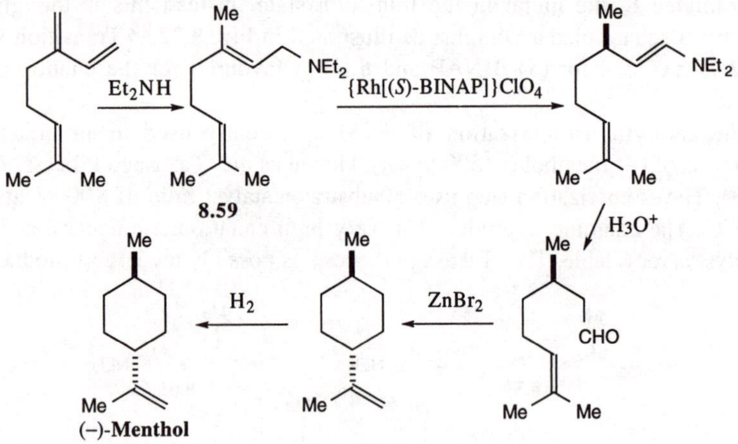

Fig. 8.33

References

1. Hill, R. K. (1984), in *Asymmetric Synthesis*, (ed. J. D. Morrison), Vol. 3, pp. 503–572, Academic Press, New York.
2. a) Otsuka, S. and Tani, K. (1985), in *Asymmetric Synthesis*, (ed. J. D. Morrison), Vol. 5, pp. 171–191, Academic Press, New York; b) Otsuka, S. and Tani, K. (1992), *Synthesis*, 665.
3. Mikami, K. and Nakai, T. (1991), *Synthesis*, 594.
4. Pereira, S. and Srebnik, M. (1993), *Aldrichimica Acta*, **26**, 17.
5. Burgstahler, A. W. and Nordin, I. C. (1961), *J. Amer. Chem. Soc.*, **83**, 198; Watanabe, W. H. and Conlon, L. E. (1957), *J. Amer. Chem. Soc.*, **87**, 2828.
6. Johnson, W. S., Wertheman, L., Bartlett, W. R., Brocksom, T. J., Li, T., Faulkner, D. J., and Petersen, M. R. (1970), *J. Amer. Chem. Soc.*, **92**, 741.
7. Felix, D., Gschwend-Steen, K., Wick, A. E., and Eschenmoser, A. (1969), *Helv. Chim. Acta*, **52**, 1031.

8. Ireland, R. E., Mueller, R. H., and Willard, A. K. (1976), *J. Amer. Chem. Soc.*, **98**, 2868.
9. Chan, K. K., Cohen, N., DeNoble, J. P., Specian, A. C., and Saucy, G. (1976), *J. Org. Chem.*, **41**, 3497.
10. Uskokovic, M. R., Lewis, R. L., Partridge, J. J., Desperaux, C. W., and Preuss, D. L. (1979), *J. Amer. Chem. Soc.*, **101**, 6742.
11. Ziegler, F. E., Fang, J.-M., and Tam, C. C. (1982), *J. Amer. Chem. Soc.*, **104**, 7174.
12. Ireland, R. E., Wipf, P., and Armstrong, J. D. (1991), *J. Org. Chem.*, **56**, 650.
13. Bartlett, P. A., Holm, K. H., and Morimoto, A. (1985), *J. Org. Chem.*, **50**, 5179.
14. Paterson, I., Hulme, A. N., and Wallace, D. J. (1991), *Tetrahedron Lett.*, **32**, 7601.
15. Schreiber, S. L. and Smith, D. B. (1989), *J. Org. Chem.*, **54**, 9.
16. Corey, E. J. and Lee, D. H. (1991), *J. Amer. Chem. Soc.*, **113**, 4026.
17. Mikami, K., Azuma, K., and Nakai, T. (1984), *Tetrahedron*, **40**, 2303.
18. Still, W. C. and Mitra, A. (1978), *J. Amer. Chem. Soc.*, **100**, 1927; Still, W. C., McDonald, J. H., Collum, D. B., and Mitra, A. (1979), *Tetrahedron Lett.*, 593; Sugimura, T. and Paquette, L. A. (1987), *J. Amer. Chem. Soc.*, **109**, 3017.
19. For a thorough discussion, see Ref. 3.
20. Uchikawa, M., Hanamoto, T., Katsuki, T., and Yamaguchi, M. (1986), *Tetrahedron Lett.*, **27**, 4577; Mikami, K., Takahashi, O., Kasuga, T., and Nakai, T. (1985), *Chem. Lett.*, 1729.
21. Takahashi, O., Mikami, K., and Nakai, T. (1987), *Chem. Lett.*, 69.
22. Marshal, J. A. and Lebreton, J. (1988), *J. Amer. Chem. Soc.*, **110**, 2925.
23. Tani, K., Yamagata, T., Akutagawa, S., Kumobayashi, H., Taketomi, T., Miyashita, A., Noyori, R., and Otsuka, S. (1984), *J. Amer. Chem. Soc.*, **106**, 5208; Tani, K. (1985), *Pure and Appl. Chem.*, **57**, 1845.
24. For a thorough discussion, see Ref. 2b.
25. Inoue, S., Takaya, H., Tani, K., Otsuka, S., Sato, T., and Noyori, R. (1990), *J. Amer. Chem. Soc.*, **112**, 4897.

9 Hydrolysis and esterification

There are two general approaches to the use of hydrolysis and esterification in asymmetric synthesis, and usually both involve the use of enzymes as asymmetric catalysts.[1] In one case the asymmetric catalyst is used in a kinetic resolution process in which one enantiomer of a racemic mixture reacts more rapidly than its antipode. In the example shown (Fig. 9.1) the lipase-catalysed hydrolysis of the (1*R*,2*R*)-enantiomer of **9.1** is rapid.[2] In effect, the (1*S*,2*S*)-enantiomer is not a substrate for the enzyme, and essentially all of the fast-reacting antipode is hydrolysed to the alcohol. As in all simple resolution procedures, the yield of an individual enantiomer cannot be greater than fifty per cent, and the two enantiomeric components, ester (1*S*,2*S*)-**9.1** and alcohol **9.2** still need to be separated.

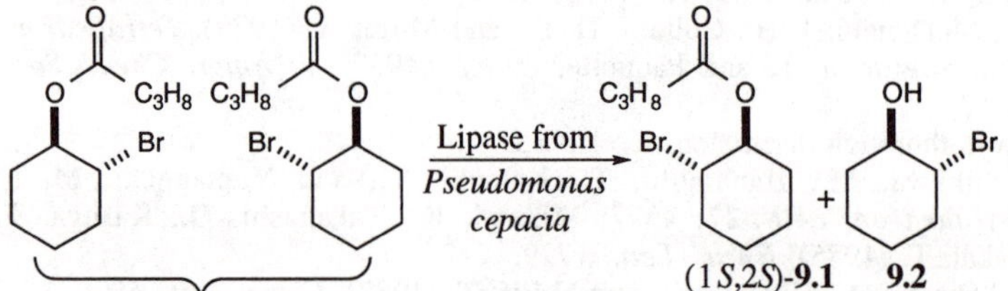

Racemate **9.1**

(1*S*,2*S*)-**9.1**
yield 46%
e.e. >99%

9.2
yield 46%
e.e. >99%

Fig. 9.1

The other approach involves using the asymmetric catalyst to 'break the symmetry' of a prochiral compound which possesses two enantiotopic groups which could react in the catalysed process. In this case the starting material is a *meso*-compound, and for high enantioselectivity to be obtained, the enzyme-catalysed reaction must occur much more rapidly at one of the two possible sites. Acetyl cholinesterase catalyses the hydrolysis of the *meso*-diacetate **9.3** to give the monoacetate **9.4** in high yield and enantiomeric excess (Fig. 9.2), with acetate group **A** being hydrolysed more rapidly than **B**.[3]

Enzymes differ somewhat from most of the reagents and catalysts which are used in organic synthesis. For example, their molecular weight is very high compared to a normal organic substrate, in Nature they usually function in an aqueous environment, and as a rule they tend to be insoluble in organic

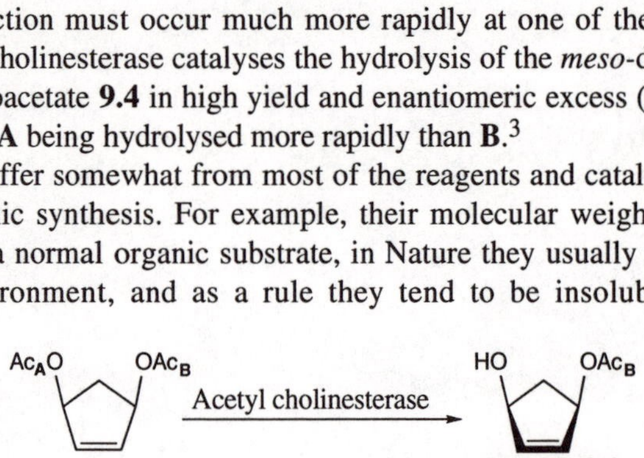

9.3

(1*R*,3*S*)-**9.4**
yield 93.6%
e.e. 96%

Fig. 9.2

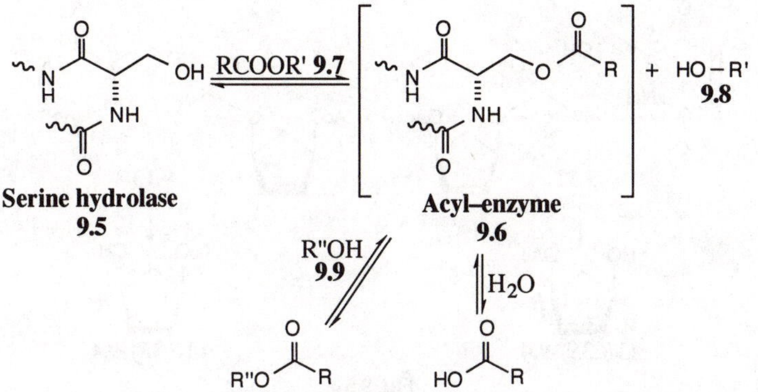

Fig. 9.3

solvents. Despite these apparent disadvantages there are numerous examples of the use of enzymes for hydrolysis and esterification. In the interest of brevity this chapter will concentrate on the principles involved, and on selected examples.

The mechanism of action of many of the hydrolase enzymes which find use in this area is outlined in Fig. 9.3 for serine hydrolases. A serine residue plays a crucial role in the process, in which the serine in the active site of the enzyme, represented by **9.5** (Fig. 9.3), attacks the ester to form an acyl–enzyme complex **9.6**. The acyl group is then transferred either to water or to an alcohol depending on whether a hydrolysis or an esterification is being carried out.

In principle, kinetic resolution might be possible if any of the components **9.7**, **9.8**, or **9.9** (Fig. 9.3) are chiral, as the enzyme is itself chiral. In the hydrolysis of **9.1** (Fig. 9.1) the alcohol portion of the ester is chiral, and although both enantiomers of **9.1** would produce the same acyl–enzyme intermediate, the transition states for the acyl transfer of each enantiomer are diastereoisomeric (Fig. 9.4). Analogous arguments can be made for the other possibilities represented in Fig. 9.3.

The enzyme-catalysed reaction of a *meso*-compound, as illustrated by the example in Fig. 9.2, is often referred to as 'asymmetrization'. In this type of reaction the enzyme distinguishes between two enantiotopic groups. Again the selectivity has its origin in diastereoselective interactions *en route* to the

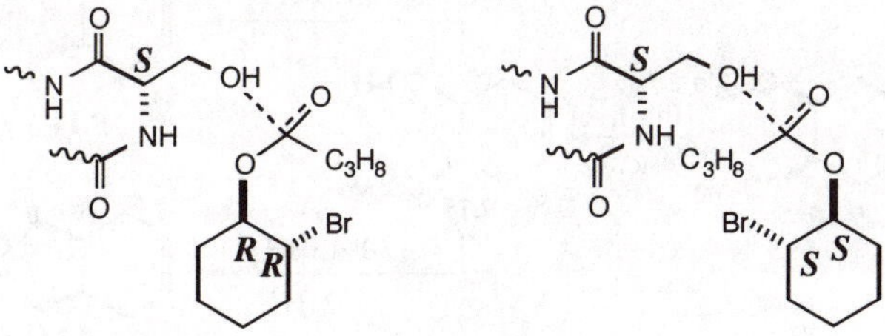

9.10   **9.11**

Fig. 9.4

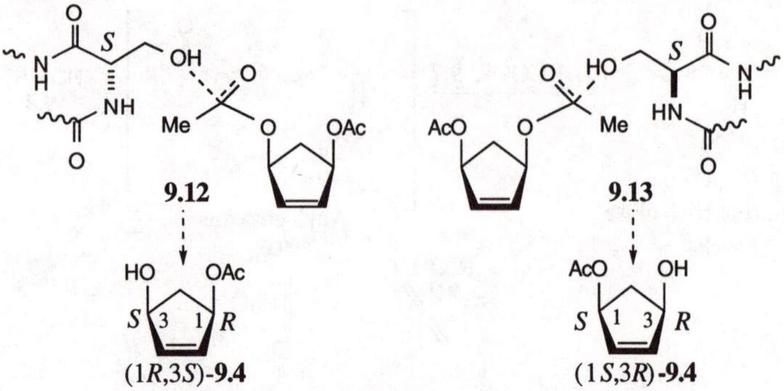

Fig. 9.5

transition state, as illustrated schematically in Fig. 9.5.

Given that the yield and enantioselectivity are high for both, in principle, asymmetrization will always be more attractive than kinetic resolution. The yield is not limited to a maximum of 50 per cent and products derived from the unwanted enantiomer do not have to be separated. In practice, many other factors influence the choice of strategy, including the ease of isolation of the desired product from the reaction mixture, and the number and type of synthetic transformations required for conversion into the target.

A potential disadvantage of asymmetrization is related to the fact that usually only the natural enantiomer of the enzyme is available, which means that only one enantiomer of the product can be produced directly. Nevertheless, due to the nature of the reactions involved, and to the symmetrical nature of the starting material, this problem is often easy to overcome using straightforward chemical manipulations. A simple example of this is illustrated by the facile synthesis of either enantiomer of lactone **9.14** from acid **9.15**,[4] obtained in high yield and enantiomeric excess by the esterase-catalysed hydrolysis of the *meso*-diester **9.16** (Fig. 9.6). This is often referred to as the '*meso*-trick'.

The enantioselectivity of an enzyme-catalysed hydrolysis or esterification is often very high, but given the limited *detailed* structural information regarding the active site itself, the absolute configuration of the product must often be based on analogy with closely related examples. This is often quite reliable, but

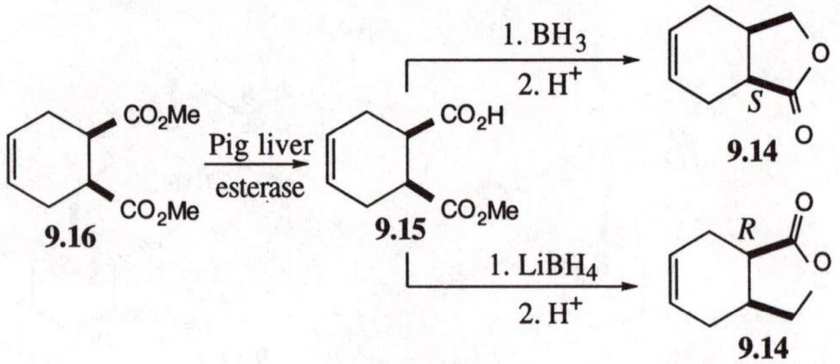

Fig. 9.6

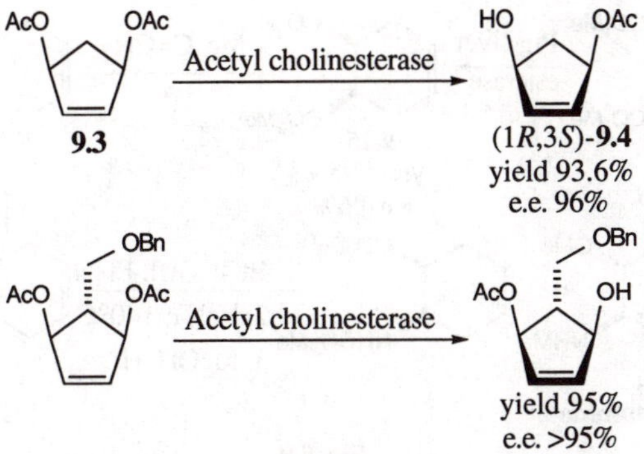

Fig. 9.7

there is always the possibility that an apparently minor structural difference could change the stereochemical outcome (Fig. 9.7),[5] presumably due to steric demands in the active site. Some simple models for this have been developed for individual enzymes and substrate types, a few of which will be considered later.

Enzyme-catalysed hydrolyses of the general types outlined so far provide an extremely valuable method for asymmetric synthesis, and appropriate enzyme systems can show predictable reactions and accept a wide variety of substrates. Moreover, such reactions are often easy to carry out on a large scale given that the enzyme is readily available. One of the most successful enzymes of this type is pig liver esterase.[6] Being readily available, cheap, and able to catalyse

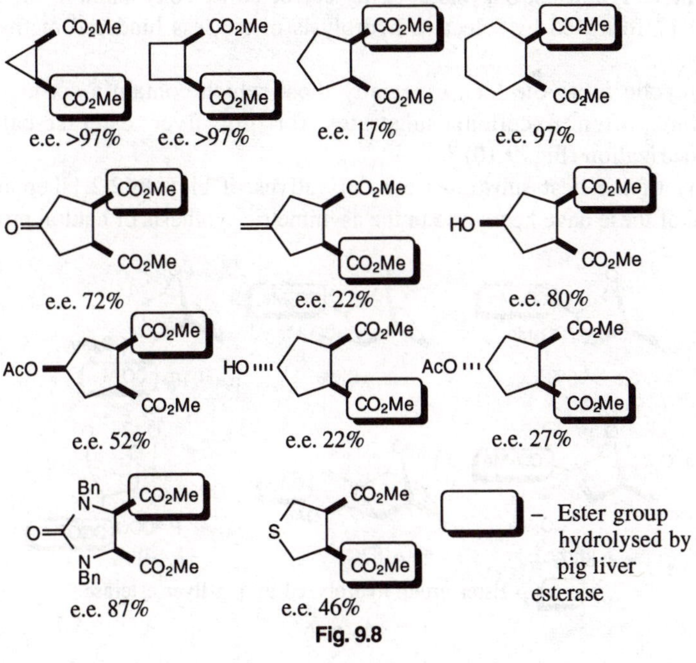

Fig. 9.8

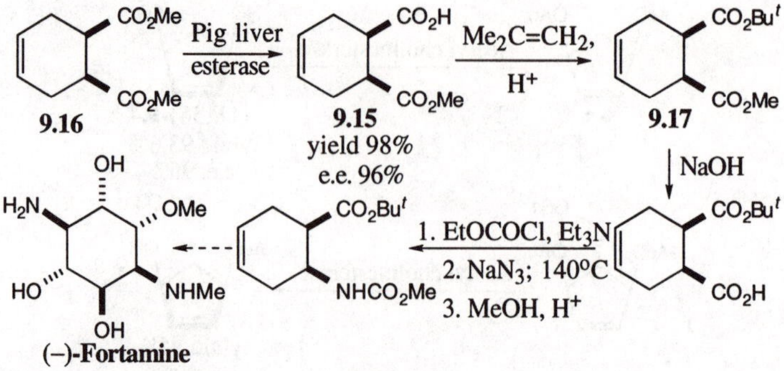

Fig. 9.9

the hydrolysis of a wide range of esters, it has found widespread use in asymmetric synthesis especially via the strategy of asymmetrization.

Pig liver esterase catalyses the enantioselective hydrolysis of a wide range of monocyclic *meso*-diesters, but care must be exercised in the prediction of the outcome of an unknown case unless very close analogies are possible with known examples, as the selected results presented in Fig. 9.8 illustrate.[7]

The particularly effective asymmetrization of diester **9.16** (Fig. 9.6) has been used in a total synthesis of (−)-fortamine[8] (Fig. 9.9), the aminocyclitol present in the natural product fortimycin and related compounds. This example illustrates the power of this approach to asymmetric synthesis, as the ester group which is hydrolysed by the enzyme is not the one which is needed for the most direct synthesis, but this is easily corrected by conversion to the *t*-butyl ester **9.17** followed by selective hydrolysis of the less hindered methyl ester (Fig. 9.9).

Polycyclic *meso*-diesters, especially those which contain a six-membered ring, are often excellent substrates for pig liver esterase-catalysed asymmetrization (Fig. 9.10).[9]

Many of the best substrates are derivatives of bicyclo[2,2,1]heptane and several of these have been used in the asymmetric synthesis of natural products.

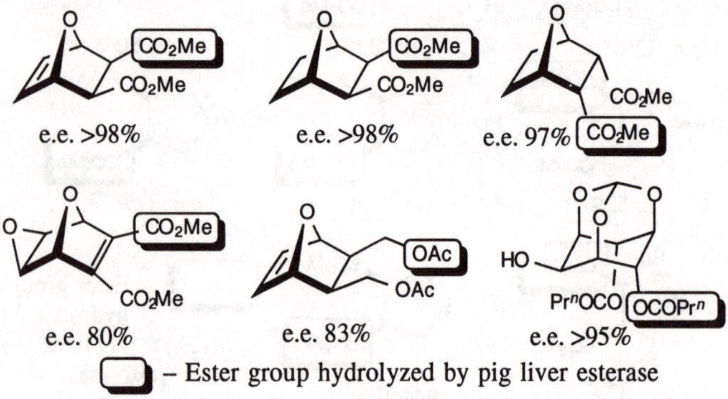

e.e. >98%   e.e. >98%   e.e. 97%

e.e. 80%   e.e. 83%   e.e. >95%

☐ – Ester group hydrolyzed by pig liver esterase

Fig. 9.10

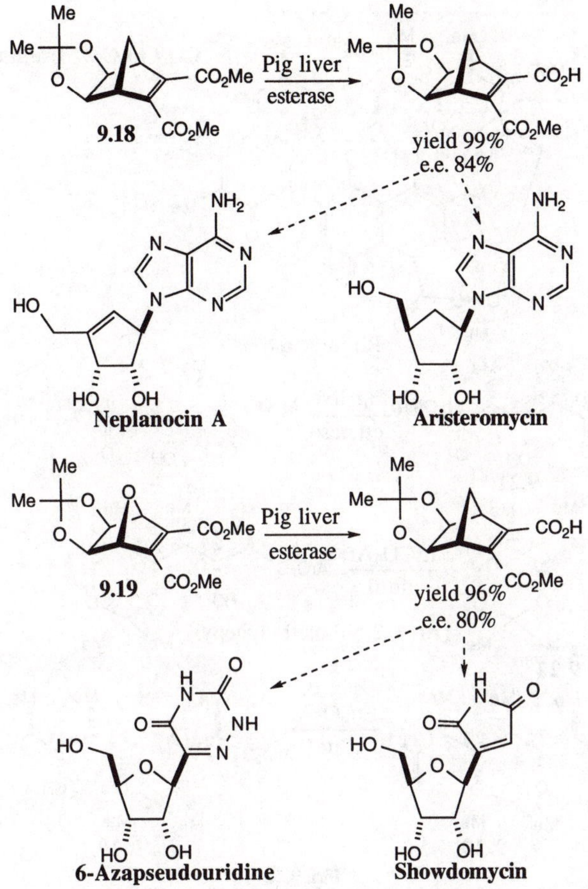

Fig. 9.11

The pig liver esterase-catalysed asymmetrization of diester **9.18** has been used in the asymmetric total synthesis of aristeromycin and neplanocin A (Fig. 9.11), and that of **9.19** has proved valuable in the synthesis of C-ribosides such as showdomycin and 6-azapseudouridine (Fig. 9.11).[10]

The use of pig liver esterase-catalysed hydrolysis is not confined to cyclic compounds. Some types of acyclic diesters have been shown to be excellent substrates (Fig. 9.12).[11]

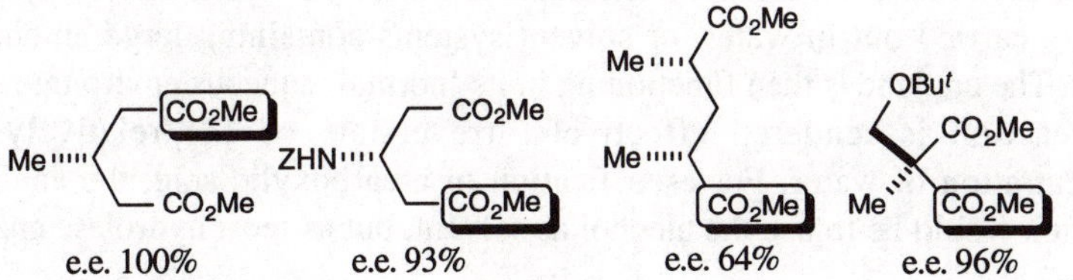

Fig. 9.12

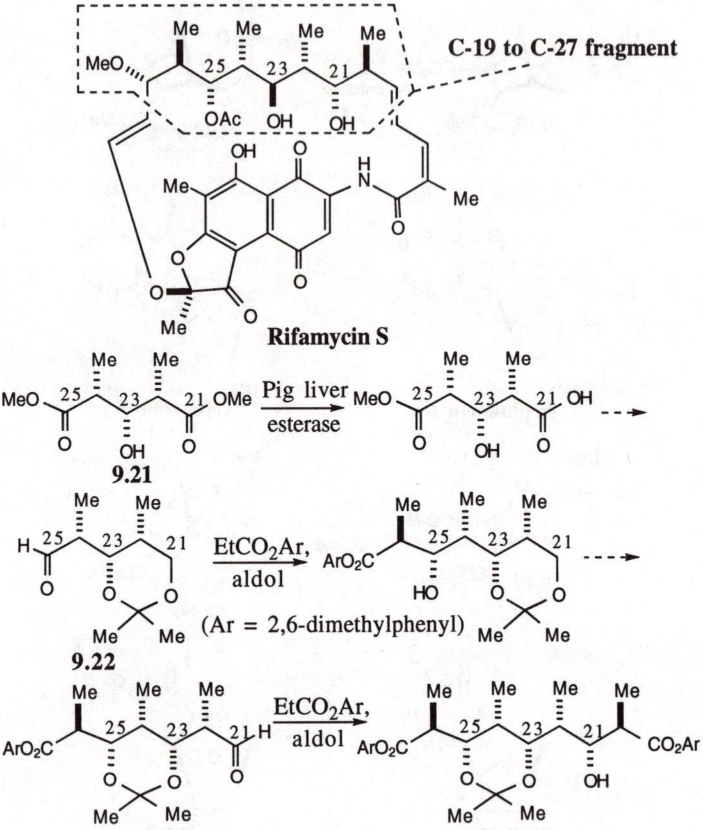

Fig. 9.13

The acetonide **9.20** (Fig. 9.13) which is a potential precursor for the C-19 to C-27 fragment of rifamycin S has been prepared using a route which makes use of the inherent symmetry of this fragment about C-23.[12] The design of this synthesis is such that after asymmetrization of **9.21**, the rest of the chain can be built out from C-21 and C-25 by sequential aldol reactions on aldehydes **9.22** and **9.23** (Fig. 9.13).

As implied by the general scheme shown in Fig. 9.3, reactions which are catalysed by hydrolase enzymes are reversible, and under suitable conditions they can be used in the synthesis of esters, rather than hydrolysis.[13] Hydrolyses are usually carried out in water, or solvent systems containing large amounts of water. The enzyme is then functioning in its 'normal' aqueous environment, and the reaction is rendered effectively irreversible by the relatively high concentration of water. For esterification of a carboxylic acid, the analagous situation would be to use the alcohol as solvent, but as most hydrolase enzymes

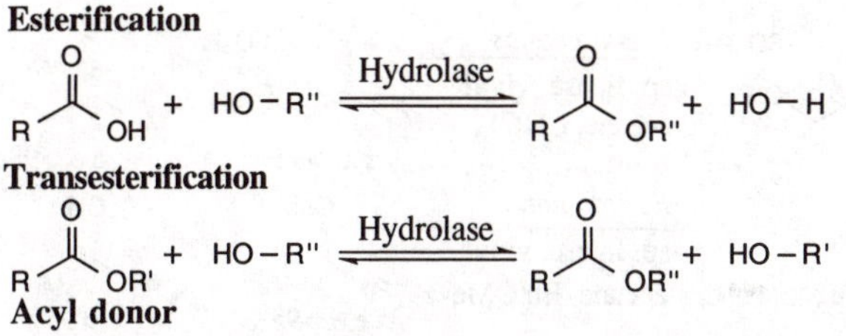

Esterification

Transesterification

Acyl donor

Fig 9.14

are inactivated by such polar hydrophilic organic solvents, this is not usually a useful approach.

Hydrolase enzymes retain catalytic activity in relatively non-polar lipophilic organic solvents such as hydrocarbons, chloroform. dichloromethane, and ethers. This difference between lipophilic and hydrophilic organic solvents is probably due to their effect on the hydration of the enzyme. Enzymes usually contain considerable amounts of associated water which is important in retaining activity. Hydrophilic organic solvents strip away this water which is required for catalysis. In lipophilic organic solvents this water remains associated with the enzyme and catalytic activity is maintained.

The requirement for this type of solvent causes a number of problems. In the esterification of a carboxylic acid, water is produced, which can form a discrete phase which then separates the enzyme from its substrate. Physical or chemical removal of this water is again problematical. Because of these problems, and of other considerations, transesterification is often the method of choice. The advantage is that no water is formed in this process and it is relatively easy to maintain an appropriate amount of water in the system (Fig 9.14).

Transesterification itself is a reversible process, and steps must be taken to ensure that formation of the desired product is effectively irreversible. Several methods have been developed, one of the most effective being the use of irreversible acyl donors.[13]

When vinyl acetate **9.24** is used as the acyl donor, on transesterification the liberated alcohol R'OH (Fig 9.14) is an enol, which rapidly isomerizes to acetaldehyde, effectively making the transesterification step irreversible (Fig 9.15).

This approach of using vinyl acetate, and other related enol esters, has been widely used for the kinetic resolution of racemic primary and secondary alcohols; some examples are illustrated in Fig 9.16.[14]

Vinyl acetate is also useful as an irreversible acyl donor in the hydrolase-catalysed asymmetrization of *meso*-diols. The asymmetrization of diol **9.25**

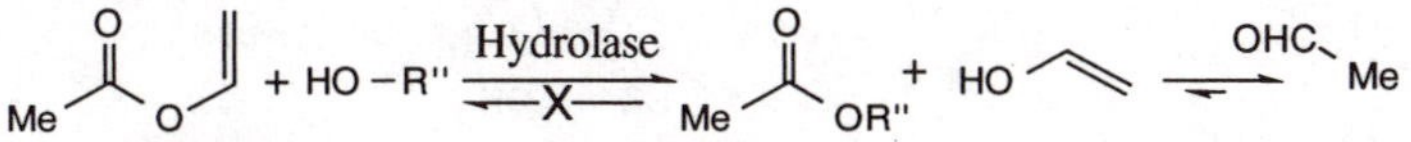

9.24

Fig 9.15

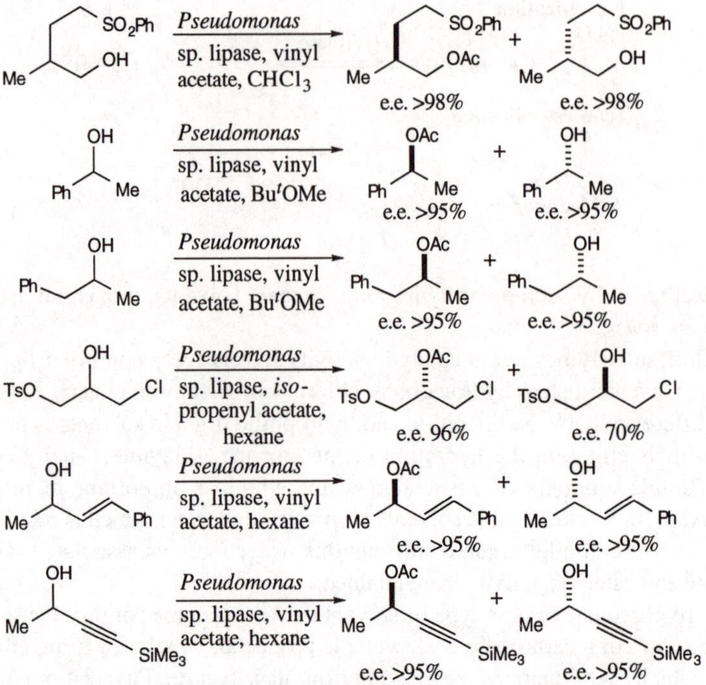

Fig 9.16

(Fig 9.17) in this way gave the monoacetate **9.26** in high yield and high enantiomeric excess. This monoacetate was then converted to lactone **9.27**, a known precursor to aristeromycin, a naturally occurring carbocyclic nucleoside (Fig. 9.17).[15]

Kinetic resolution using hydrolase enzymes, valuable as it is for the

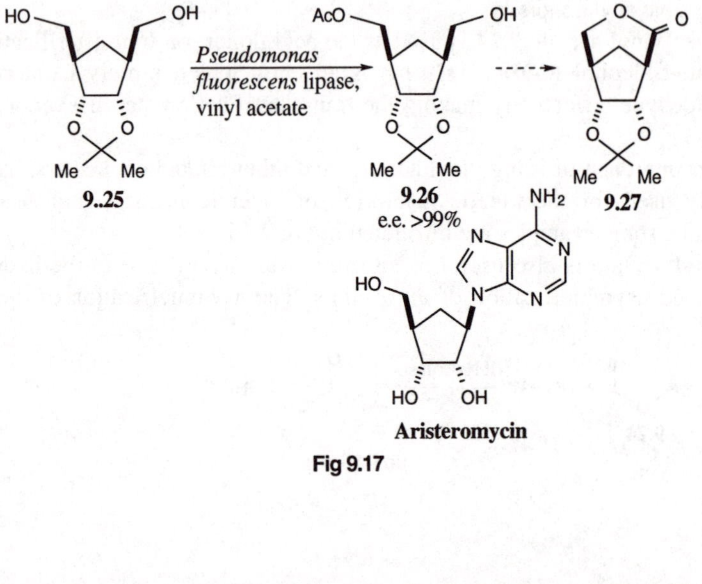

Aristeromycin

Fig 9.17

$$R_{substrate} \underset{\xrightarrow{\hspace{2.5cm}}}{\overset{\text{Racemization}}{\rightleftharpoons}} S_{substrate} \xrightarrow{\text{Enzyme}} S_{product}$$

Fig. 9.18

preparation of enantiomerically enriched compounds, does suffer from the potential limitations alluded to earlier, especially that the yield cannot exceed fifty per cent. In certain cases it is possible to circumvent this by the use of an *in situ* process which causes racemization of the substrate. Under these conditions the enantiomers of the substrate are in equilibrium, and enzyme-catalysed reaction of the faster reacting enantiomer can then effectively convert both enantiomers of the substrate into the desired product, as illustrated schematically in Fig. 9.18 for the hypothetical case in which only the (*S*)-enantiomer of the substrate is hydrolysed by the enzyme.

A real situation is likely to be more complicated than the example illustrated above (Fig. 9.18), as seen by the hydrolysis of the racemate of **9.28** (Fig. 9.19) to produce the (*S*)-**9.29**, the biologically active enantiomer of ketorolac, a potent anti-inflammatory and analgesic agent.[16] For the desired racemization/enzymatic hydrolysis approach to be successful, it is desirable that the rate of racemization is greater than that of 'chemical' hydrolysis ($k_{rac} > k_{chem}$), that the enzymatic hydrolysis proceeds faster than racemization ($k_{enz} > k_{rac}$), and that product racemization should not occur ($k_{rac'} \sim 0$).

These conditions were almost completely fulfilled by the use of the protease from *Streptomyces griseus* at pH 9.7.[16] The racemate (*R*,*S*)-**9.28** was converted into the desired acid in high yield and enantiomeric excess (Fig. 9.20). The (*R*)-enantiomer of the product probably arises from a combination of competing 'chemical' hydrolysis (k_{chem} Fig. 9.19) and product racemization ($k_{rac'}$ Fig. 9.19).

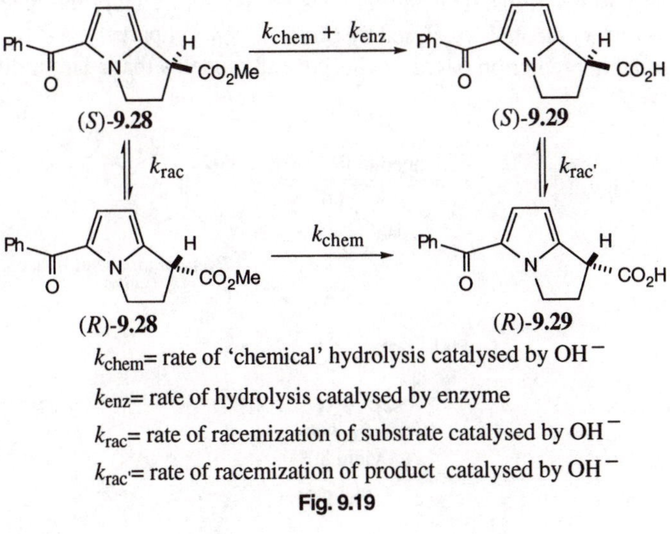

k_{chem} = rate of 'chemical' hydrolysis catalysed by OH^-

k_{enz} = rate of hydrolysis catalysed by enzyme

k_{rac} = rate of racemization of substrate catalysed by OH^-

$k_{rac'}$ = rate of racemization of product catalysed by OH^-

Fig. 9.19

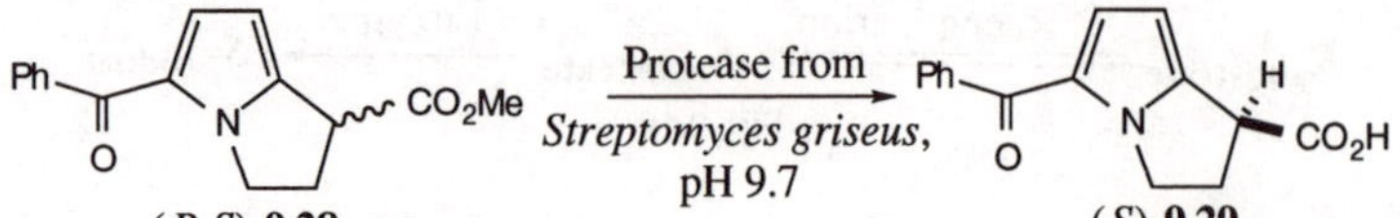

Fig. 9.20

This approach of racemization coupled with an enzymatic kinetic resolution is also possible under esterification conditions. The reversible formation of aromatic cyanohydrins combined with an *in situ* esterification catalysed by *Pseudomonas* sp. M-12-33 lipase is successful in the efficient asymmetric synthesis of cyanohydrin acetates (Fig. 9.21).[17] Acetone cyanohydrin **9.30** functions as a source of HCN (and is not a substrate for the enzyme), the racemization is catalysed by a basic ion-exchange resin, and an irreversible acyl donor is used in the lipase-catalysed transesterification. In this way cyanohydrin acetate **9.31**, the desired enantiomer for the synthesis of pyrethroid insecticides, could be prepared efficiently.

Given the structural complexity of enzymes it is not surprising that prediction of which enantiomeric product might be obtained in an unknown case is difficult. Nevertheless, in some cases significant progress has been made towards developing reliable models to account for existing results and to predict the outcome of a new reaction.

Such a model has been proposed to predict which enantiomer of a secondary alcohol reacts faster in reactions catalysed by cholesterol esterase, *Pseudomonas cepacia* lipase, and *Candida rugosa* lipase.[18] Similar models have been proposed for other lipases.[19] This simple model (Fig 9.22) is based on the relative sizes of the groups attached to the carbinol carbon, and correlates the outcome of more than 130 reactions. This model does not predict reliably the reactions of acyclic secondary alcohols catalysed by *Candida rugosa* lipase.

A reasonable prediction based on this general model is that a larger difference

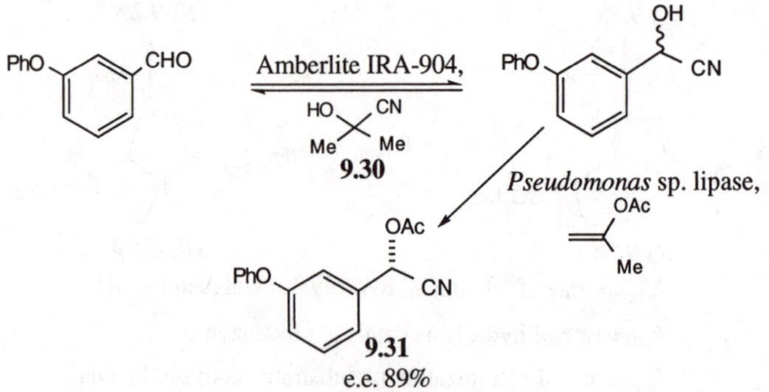

Fig. 9.21

Faster reacting enantiomer in cholesterol esterase catalysed hydrolysis.

Faster reacting enantiomer in *Pseudomonas cepacia* and *Candida rugosa* lipase catalysed esterification.

Fig 9.22

in size between the medium (M) and large (L) substituents should lead to a bigger difference in the relative rates of reaction of the two enantiomers. This idea was useful in the design of an efficient method for the preparation of (*S*)-4-acetoxy-2-cyclohexen-1-one **9.32** (Fig. 9.23).[2]

Asymmetrization of diacetate **9.33** could not be achieved in high enantiomeric excess, and this was attributed to the similarity in size of the two 'substituents' attached to the chiral centre (CH_2CH_2 vs. $CH=CH$). *Trans*-addition of bromine to *meso*-**9.33** gives racemic **9.34**, and the enzyme-catalysed hydrolysis of this diacetate is interesting in that both enantiomers can satisfy the requirements of the model for rapid hydrolysis.

The two acetate groups in **9.34** are non-equivalent, OAc_A and OAc_B, being respectively *cis*- and *trans*- to the adjacent bromine substituent. The above model predicts that for one enantiomer hydrolysis of OAc_A will be rapid, and for the other OAc_B will be cleaved rapidly, which is found to be the case (Fig. 9.23). This results in each enantiomer giving rise to diastereoisomeric products (+)-**9.35** and (−)-**9.36**, both in very high enantiomeric excess. This mixture of diastereoisomers could then be converted into the desired enone **9.32** by reductive elimination of bromine and oxidation to the ketone (Fig. 9.23).

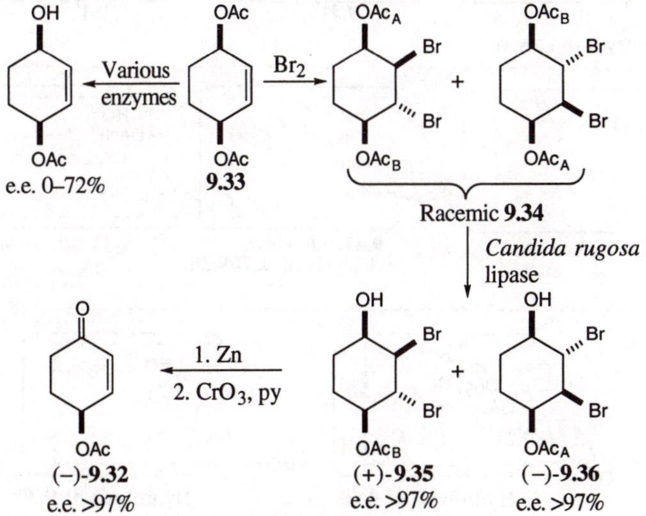

Fig. 9.23

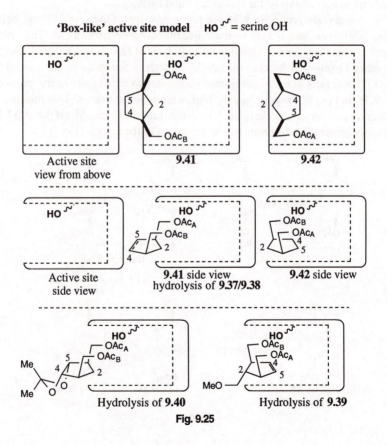

9.37
e.e. 18%

9.38
e.e. 26%

9.39
e.e. >99%

9.40
e.e. >99%

Fig. 9.24

A somewhat more elaborate model has been proposed to account for the results of the asymmetrization of some *meso*-diacetates catalysed by *Rhizopus delemar* lipase (Fig. 9.24).[20] Using this enzyme, unhindered diacetates **9.37** and **9.38** are poor substrates (e.e. <27%), whereas more hindered analogues such as **9.39** and **9.40** give high enantiomeric excesses.

These observations are accommodated by the proposed 'box-like' active site model (Fig. 9.25). In this model the active site serine hydroxyl group is placed at the top of the box as illustrated in Fig. 9.25.

'Box-like' active site model   HO⌁ = serine OH

Active site
view from above

9.41

9.42

Active site
side view

9.41 side view
hydrolysis of **9.37/9.38**

9.42 side view

Hydrolysis of **9.40**

Hydrolysis of **9.39**

Fig. 9.25

This model accounts for the lack of enantioselectivity in the hydrolysis of **9.37** and **9.38**, which have no substituents at C-2, C-4, and C-5, as both modes of binding **9.41** and **9.42** are possible and hydrolysis of both OAc_A and OAc_B can occur. Substitution at C-4 and C-5 should strongly disfavour binding as in **9.42** and allow only hydrolysis of OAc_A. The same argument leads to the conclusion that substitution at C-2 should direct hydrolysis towards OAc_B. The hydrolysis of **9.39** and **9.40** is consistent with this model (Figs 9.24 and 9.25).

References

1. Jones, J. B. (1985), in *Asymmetric Synthesis*, (ed. J. D. Morrison), Vol. 5, pp. 309–344, Academic Press, New York.
2. Gupta, A. K. and Kazlauskas, R. J. (1993), *Tetrahedron Asymmetry*, **4**, 879.
3. Deardorf, D. R., Matthews, A. J., McMeekins, D. S., and Craney, O. L. (1986), *Tetrahedron Lett.*, **27**, 1255.
4. Schneider, M., Engel, M., Honicke, P., Heinemann, G., and Gorisch, H. (1984), *Angew. Chem. Int. Ed. Engl.*, **23**, 67.
5. Griffith, D. A. and Danishefsky, S. J. (1991), *J. Amer. Chem. Soc.*, **113**, 5863.
6. Zhu, L.-M. and Tedford, M. C. (1990), *Tetrahedron*, **46**, 6587.
7. Taken from Ref. 6.
8. Kobayashi, S., Kamiyama, K., and Ohno, M. (1990), *J. Org. Chem.*, **55**, 1169.
9. Sabbioni, G. and Jones, J. B. (1987), *J. Org. Chem.*, **52**, 4565; Mohr, P., Rosslein, L., and Tamm, C. (1987), *Helv. Chim. Acta*, **70**, 142; Laumen, K. and Schneider, M. (1984), *Tetrahedron Lett.*, **25**, 5875; Jones, J. B., Hinks, R. S., and Hultin, H. G. (1985), *Can. J. Chem.*, **63**, 452.
10. Ito, Y., Shibata, T., Arita, M., Sawai, H., and Ohno, M. (1981), *J. Amer. Chem. Soc.*, **103**, 6739; Arita, M., Adachi, K., Ito, Y., Sawai, H., and Ohno, M. (1982), *Nucleic Acids Research, Symposium Series*, **11**, 13; Arita, M., Adachi, K., Ito, Y., Sawai, H., and Ohno, M. (1983), *J. Amer. Chem. Soc.*, **105**, 4049; Ohno, M., Ito, Y., Arita, M., Shibata, T., Adachi, K., and Sawai, H. (1984), *Tetrahedron*, **40**, 145.
11. Horton, D., Machinami, T., and Takagi, Y. (1983), *Carbohydr. Res.*, **121**, 135; Ohno, M., Kobayashi, S., Iimori, T., Wang, Y.-F., and Izawa, T. (1981), *J. Amer. Chem. Soc.*, **103**, 2405; Chen, C.-S., Fujimoto, Y., and Sih, C. J. (1981), *J. Amer. Chem. Soc.*, **103**, 3580; Luyten, M., Muller, S., Herzog, B., and Keese, R. (1987), *Helv. Chim. Acta*, **70**, 1250.
12. Mohr, P., Waespe-Sarceivic, N., Tamm, C., Gawronska, K., and Gawronska, J. K. (1983), *Helv. Chim. Acta*, **66**, 2501; Tschamber, T., Waespe-Sarceivic, N., and Tamm, C. (1986), *Helv. Chim. Acta*, **69**, 621.
13. Faber, K. and Riva, S. (1992), *Synthesis*, 895.
14. Santaniello, E., Ferraboschi, P., and Grisenti, P. (1990), *Tetrahedron Lett.*, **31**, 5657; Laumen, K., Breitgoff, D., and Schneider, M. P. (1988), *J. Chem. Soc., Chem. Commun.*, 1459; Chen, C.-S. and Liu, Y.-C. (1989), *Tetrahedron Lett.*, **30**, 7165; Burgess, K. and Jennings, L.-D. (1990), *J. Amer. Chem. Soc.*, **112**, 7434.
15. Tanaka, M., Yoshioka, M., and Sakai, K. (1992), *J. Chem. Soc., Chem. Commun.*, 1454.
16. Fülling, G. and Sih, C. J. (1987), *J. Amer. Chem. Soc.*, **109**, 2845.

17. Inagaki, M., Hiratake, J., Nishioka, T., and Oda, J. (1991), *J. Amer. Chem. Soc.*, **113**, 9360.
18. Kazlauskas, R. J., Weissfloch, A. N. E., Rappaport, A. T., and Cuccia, L. A. (1991), *J. Org. Chem.*, **56**, 2625.
19. Kim, M.-J. and Cho, H. (1992), *J. Chem. Soc., Chem. Commun.*, 1411.
20. Tanaka, M., Yoshioka, M., and Sakai, K. (1993), *Tetrahedron Asymmetry*, **4**, 981.

Index

DATE DUE			

Procter　　　　266604